MICROORGANISMS AND MAN

MICROORGANISMS AND MAN

Orville Wyss

The University of Texas at Austin

Curtis Eklund

The University of Texas at El Paso

John Wiley & Sons, Inc.

New York | London | Sydney | Toronto

Library of Congress Catalogue Card Number: 70-146674

ISBN 0-471-96900-1

Printed in the United States of America

10 9 8 7 6 5 4 3 2 1

This book is dedicated to the memory of
Dr. O. B. Williams
whose deep concern for good teaching of microbiology
made a lasting impression on both of us.

Preface

This textbook is a study guide for an elementary course in microbiology to be used in conjunction with the traditional lecture, laboratory, and library reading assignments. It is not a compendium in microbiology. This text is a nonexhaustive approach that will stimulate students who are not biology majors to seek additional material and will give them a better picture of the diversity and scope of biological science. We know from experience that this approach is useful in the megauniversity whose large libraries, although they cannot make all reference material available to each student, can supply sufficient diversity that each student can pursue in depth special topics selected from the broad field of microbiology. This same approach appears to work equally well in the institutions with smaller classes but with more restricted library facilities.

Students who complete this course will acquire a sympathetic view of science and of scientists and will learn that the procedures for securing scientific information are within their own capabilities. We hope they will learn that their "social concern" need not be wasted and that the science and technology that make this concern useful are not limited to students who are science majors. This book gives them a basis for understanding the principles of ecology where "life bears on life" and the surviving species are the ones that successfully adapt to an available ecological niche. We hope the book will impel competent and concerned students to pursue further ventures in science and, especially, in microbiology. Students who are concerned about the pollution of our environment will be reassured by the parody, "never underestimate the power of the microbe," which is merely a restatement of Pasteur's admonition, "Gentlemen, the microbe will have the last word."

We thank Mrs. Eleanore Tulley for making many helpful contributions to the manuscript, and Miss Ellen Gaither for secretarial assistance.

Orville Wyss
Curtis Eklund

Contents

ONE

The Nature and Scope of Microbiology

The science of microbiology studies the plant and animal forms not visible to the naked eye. These forms, called microorganisms or microbes, include viruses, rickettsiae, bacteria, yeasts, molds, algae, and protozoa. The term **bacteriology** is often used as a synonym for **microbiology** because the laboratory tools and techniques as well as the theoretical approaches developed for the study of the **bacteria** have been useful in studying other microbes.

In the strict sense bacteriology is a subdivision of microbiology (Fig. 1.1). Other subdivisions include **mycology,** the study of yeasts and molds; **virology,** the study of viruses and rickettsiae; and **immunology,** the study of those chemical modifications in the blood and other tissues which are involved in immunity to disease. Often included in microbiology are **fermentations,** which concern the chemical reactions brought about by microorganisms yielding such useful products as beer, vinegar, and antibiotics. This diverse subject matter is limited and unified by the use of certain basic procedures for securing information which were originally devised by bacteriologists. Cells of animal tissues (e.g., embryonic chick cells and human cancer cells) are cultured routinely in the flasks and test tubes of the microbiologist using methods and equipment identical to those used to grow certain more fastidious bacteria. With this exception and with the exception of certain **protozoa,** the study of microscopic forms of the animal kingdom is usually excluded from microbiology since laboratory facilities and the background training of the scientist in that field are generally that of a specialist in animal biology (**zoology**). Similarly much of the study of algae—and even a part of that field of knowledge dealing with molds—seems to thrive best on an educational

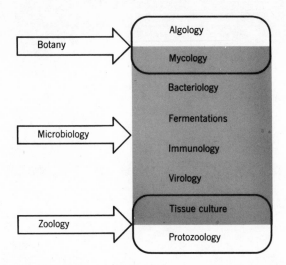

Fig. 1.1. Microbiology includes those disciplines where the methods for getting new information are similar. The tools of the microbiology laboratory have been borrowed by zoologists and botanists for some of their problems, and biological science departments are often organized as indicated.

background in botany. But in recent years, botanists and zoologists dealing with the problems of microorganisms have found it often more convenient to group their studies under a single designation, microbiology, which implies the study of the bacteria and their relatives.

The bacteria are a miscellaneous group of single-celled organisms that never carry on that type of photosynthesis that yields oxygen and whose cellular dimensions are of the order of 1/1000 millimeter (1 micrometer, abbreviated μm). Microbiologists formerly called this unit of measurement the micron (μ). Though the range of the bacteria is between 0.2μm and 2.0μm, the average bacterium is about one micrometer in thickness, which means that if a thousand of them were lined up side by side, they would just reach across a period on a printed page. One trillion (10^{12}) bacteria of average size could be packed into a 1 milliliter (ml) pipette, and, since a teaspoon is about 5 ml, five times that number would fill a teaspoon. Since there are about 2,500,000,000 (2½ billion) people in the world, we find that a teaspoon full of packed bacteria represents 2000 times as many individuals as there are people on the earth. The size relationship of the bacteria as compared with other microscopic forms is shown in Fig. 1.2.

A distinctive feature of bacterial cells is that they lack an organized nucleus with the nuclear membrane characteristic of the cells of higher animals and plants. They are therefore classified as **procaryotes** (pro = primitive; caryon = kernel or nucleus) to indicate a primitive nature and that is a fundamental difference from the **eucaryotic** (eu = true) cells which have true

Microorganisms and Man

nuclei. Nevertheless, bacteria contain organized nucleic acid which serves as the information molecule to program the activities of the cell and to give hereditary determinism just as do the nucleic acids in the chromosomes contained in the complex nuclear structures of eucaryotic organisms. Bacteria are found almost everywhere in nature and they have evolved into such variable forms with such diverse activities that they must be placed centrally in any scheme denoting relationships of living things (Fig. 1.3). In fact, it is quite commonly accepted that the first living organism on earth was similar to certain bacteria that are now in existence.

The **rickettsiae** are smaller procaryotic organisms (about 0.2μm) which live as parasites inside the cells of insect hosts or in those of man or other animals. The **chlamydiae** are even more dependent than the rickettsiae and could have evolved only after the appearance of animals in which they live.

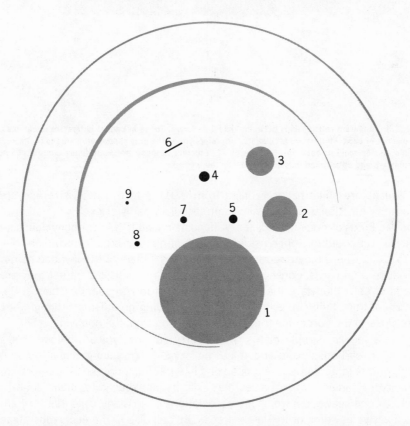

Fig. 1.2. Relative sizes are shown by superimposing microbial forms on a red blood corpuscle: (1) Mold spore. (2) Ordinary bacterium. (3) Small bacterium. (4) Rickettsia. (5) Chlamydia. (6) Tobacco mosaic virus. (7) Bacteriophage. (8) Yellow fever virus. (9) Polio virus.

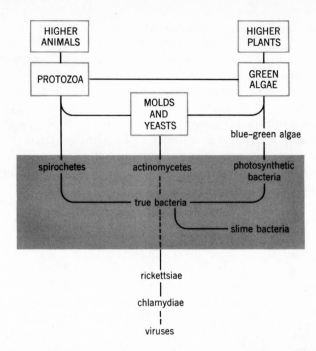

Fig. 1.3. Apparent relationships between biological forms. Those in capital letters are eucaryotes. Those not in capital letters are procaryotes or protophyta. In the gray-shaded area are the Schizomycetes (fission fungi or bacteria). The viruses fit in neither of these groups. In future classifications, rickettsiae and chlamydiae may be placed with the bacteria.

Viruses are still smaller, ranging from 0.01μm to 0.1μm in size and are from the functional point of view, composed primarily of one or the other of the two kinds of nucleic acid, **deoxyribonucleic acid** (DNA) or **ribonucleic acid** (RNA). They multiply only inside the cells of cellular organisms which they penetrate, where they direct the metabolic machinery of the invaded cell to produce more virus. Consequently, viruses are not cells, but rather they are parts of cells that have the ability for a kind of independent existence. Because of this ability, we will refer to them as living organisms although they are close to the borderline that divides the living from the nonliving.

The algae are simple plants which carry out the typical photosynthesis taking in carbon dioxide and releasing oxygen. The **blue-green algae,** one type of this large group, are procaryons like the bacteria, but all the rest are eucaryons ranging from species that exist as single cells to others such as kelps (large seaweeds) which are associations of so many cells that they are as massive as some of the larger trees. The cell size of the **eucaryotic algae** varies but often is in the range of 4–8μm while the blue-green (procaryotic) algae usually have cells of bacterial dimensions.

The **yeasts** and **molds** (**fungi**) do not have chlorophyll and their cells have a diameter of 5-10μm. Filaments may be visible as a mycelium, such as the moldy growth on the surface of vegetables, or as the mushrooms which are masses of filaments shaped into characteristic structures carrying the spores. The fungi that do not grow in filaments, but exist as individual cells, are called yeasts.

Between the bacteria and the filamentous molds are the **actinomycetes** which are moldlike bacteria. Between the protozoa and the bacteria are the protozoalike bacteria with flexible cell walls, the **spirochetes.**

Many biologists would relegate the organisms of interest to the microbiologist to a separate kingdom, the *Protista,* as separate from the kingdoms of *Animalia* and *Planta.* However, under currently used classifications, most microorganisms are classified in the plant kingdom. In the animal kingdom, the phylum *Protozoa* contains the single-celled animals whose microscopic nature and relationship to the bacteria make them suitable objects for study in the bacteriology laboratory. In earlier classification schemes, animals and plants were differentiated by a series of characteristics. Most important, plants generally take in their food in soluble rather than in particulate form; and plants generally have rigid rather than flexible and soft cell walls. According to those criteria, the protozoa were classified with animals while the fungi, algae, and the bacteria were classified with plants (Table 1.1).

In the plant kingdom, the division *Protophyta* contains bacteria, rickettsiae, chlamydiae, viruses, and blue-green algae, while the second division, *Thallophyta,* includes somewhat more complex plants that are not differentiated into tissues such as leaves, roots, and stems; the differentiated plants are classified in other divisions of the plant kingdom not of direct interest to microbiologists. Those thallophytes having green pigment are called algae,

Table 1.1 Classification of Living Forms*

I. Animal Kingdom
    Phylum 1. *Protozoa*
    Phylum 2, 3, etc. Other Animals
II. Plant Kingdom
    Division 1. *Protophyta* — primitive plants without true nuclei
                Class 1.   *Schizophyceae* — the blue-green algae
                Class 2.   *Schizomycetes* — the bacteria, actinomycetes and related forms
                Class 3.   *Microtatobiotes* — the rickettsiae, chlamydiae and viruses
    Division 2. *Thallophyta* — primitive plants with true nuclei
        Subdivision 1.   *Algae* — the green, red, and brown algae and the diatoms
        Subdivision 2.   *Fungi* — the molds and yeasts
    Division 3, 4, etc. Other plants

---

*Those of interest to the microbiologist are in italics.

and those without green pigment are referred to as fungi. The fungi include the yeasts and the molds and are subdivided further as are the algae (see Chapter 10 and 11). The bacteria constitute the *Schizomycetes,* a class of organisms that divide by fission or splitting. Under the *Schizomycetes* a number of further subdivisions include (1) the true bacteria, which are typical of the commonly held concepts of bacteria — probably because members of this group were the first studied seriously by the pioneer scientists, Robert Koch and Louis Pasteur; (2) the actinomycetes, which are moldlike bacteria; (3) the spirochetes, which are the spiral-shaped bacteria with soft, flexible cell walls, a feature which makes them similar to animal organisms such as the protozoa; and (4) the slime bacteria.

In the class *Microtatobiotes* are the transitional forms and the viruses, the smallest of living things. Most microbiologists define "living" as those things which, under proper conditions, have the ability to reproduce and to mutate; of the many characteristics of life, viruses have these two which appear of primary importance even though one might regard them merely as pieces of genetic information which require host cells for translation and action. Clearcut borderlines do not exist between one group of living things and another, yet classification systems are devised because they aid the human mind in comprehending better the vast diversity of living things by grouping together organisms having similar characteristics (Fig. 1.3). This device, though intellectually useful, can be misleading unless it is recognized that between every pair of related groups are transitional forms that might be placed in either.

This brief discussion serves as an introduction to the organisms that are the subject matter of this science. The beginning student trying to get an initial overview of the scope of microbiology will find it useful to consider the human side of the profession, that is, what microbiologists do for a living.

Many microbiologists are employed in jobs that are allied to the practice of medicine; technicians are employed in hospitals, in laboratories, and in doctors' offices. Assisting in the diagnosis of **infectious disease** is among their more demanding activities. In modern medicine many ailments are easily recognizable by clinical symptoms. Situations do occur, however, in which the physician alone is unable to determine the nature of the ailment. In such instances it may be necessary for the microbiologist to aid the physician by carrying out certain tests. In the diagnosis of gonorrhea, pus from the inflamed area of the genitalia is smeared on a glass slide and stained by a special dye. Microscopic examination usually reveals whether the inflammation is a result of the activities of the bacteria causing gonorrhea or due to some other disorder. Similarly, microscopic examination of smears, which are made from lesions of the skin, from the lungs, from intestinal discharges, from the throat, or from the ear canals may produce sufficient information for the physician to make the necessary diagnosis and institute proper treatment. In some situations the organism causing the difficulty cannot be identified by microscopic observation. In such cases, the microbiologist often attempts to grow the organism from the diseased tissue. The physician can usually diagnose whooping cough by observing the behavior of the patient, but occasionally symptoms appear which are not characteristic of the disease, and the physician requires laboratory aid. Since the whooping cough microorganism looks like thousands of other organisms when observed under the microscope, it is necessary to grow the organism which is secured from the patient. The patient is asked to cough against the surface of a jelly made according to a special nutrient formula or, in other cases, swabs from the nasal passages are smeared on the growth surface. The culture is then placed overnight in an incubator in order to mimic body temperature. If the whooping cough organism is present, characteristic masses of cells will develop which the bacteriologist can recognize as this particular organism and, following confirmatory tests, he can advise the physician accordingly (Fig. 1.4).

The blood stream of a healthy individual is ordinarily free from any living microorganisms which would grow in the test tubes of the microbiologist. Therefore, when a patient exhibits symptoms which are not typical of a disease easily identified by the physician, the bacteriologist may be asked to place blood, freshly drawn from the patient, in a test tube containing the proper food material for growing microorganisms. After a waiting period of about 24 hours to permit growth of any bacteria that might be present, a sample of the test tube's contents is observed under the microscope. The presence of bacteria in the blood stream indicates the cause of the patient's disorder. The type of microorganism present may easily lead to identification of the disease and efficacious treatment.

Some organisms, such as the viruses, do not grow in the test tubes of the bacteriologist. In such cases, laboratory animals are inoculated with materials taken from the patient. By inoculating mice, rats, guinea pigs, hamsters, or rabbits, it is often possible to obtain a clear-cut diagnosis for a disease, such as an atypical encephalitis. But, by and large, the cause of the greatest majority of diseases can be determined by the well-trained and practiced physician from his bedside observation. The laboratory work of the bacteriologist is designed to aid in cases where the symptoms are unusual and beyond the experience of the physician, or when he wishes to confirm his judgment by bringing in additional evidence from laboratory work.

The study of **blood serum** from the patient often aids in diagnosis of disease. Relatively soon after the onset of an infectious disease, the patient produces in his blood serum, chemical substances called **antibodies,** which are characteristic of the disease and which can be detected and measured in the microbiological laboratory. For example, a blood test is usually required for a marriage license, for employment as a food handler, upon appli-

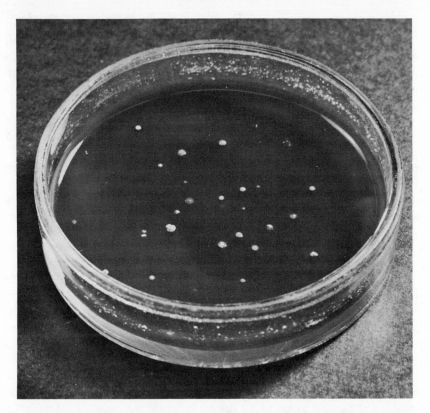

Fig. 1.4. This petri plate has been exposed to a child's cough. The bacteria expelled by the cough impinged on the blood agar and grew during incubation at 37°C to form the distinctive piles or colonies. The pearly, glistening colonies are those of the whooping cough organism, which the skilled microbiologist can distinguish from those of the other harmless bacteria.

cation for life insurance, or upon entrance into the military service of the United States. The blood test involves examination for specific antibodies which result from infection with syphilis. In this way an infection may be detected in a person who is unaware of the illness, and treatment can be instituted for the protection of the person and those he contacts.

The microbiologist's work does not end with assisting in diagnosis; he may be asked to study the nature of the organisms and the probable response of the patient's disorder to one or another drug. The physician then may be advised, for example, that the organism found in the blood stream of the patient is not sensitive to penicillin but that streptomycin worked remarkably well in bringing about destruction of the organism in the test tube, and thus will likely be useful to the patient. Medical microbiologists also follow the progress of the patient during treatment. For example, in a tuberculosis sanatorium it is necessary to determine when patients cease to cough up the organism that causes tuberculosis in order that patients may not be released from treatment too early.

The public health microbiologist who works for state and nationally supported agencies on public health and sanitation problems is closely allied to the medical microbiologist. In the United States the public health is served at several levels. At the federal level are the workers at the National Institutes of Health and regional laboratories who carry out research on unsolved problems of microbiology. All the states have health departments with a centrally located laboratory and many branch laboratories for studying public health problems, encompassing such widespread activities as production of vaccines, the investigation of unusual outbreaks or epidemics of disease, supervision of health tests for food handlers, recommendations for the best practices in the handling of sewage, water, or food, and advisory service to the state legislature on laws for the protection of the health of citizens and livestock. The public health microbiologist also supervises private medical laboratories, gathers statistics on diseases occurring within the state, and publishes these statistics together with bulletins on public health problems. County and city health officers work on local problems and may supervise institutions for indigent patients.

A large number of microbiologists find employment in industry. At the present time one of the major microbiological activities is the manufacture of **antibiotic** substances. These are chemicals which are produced by certain microorganisms and are used in the treatment of diseases caused by other microorganisms. A large fraction of the dollar volume of the drugs sold in drugstores are antibiotics which, prior to 1940, were unknown to the druggist. The aggregate selling price of these approaches one billion dollars per year. The chemical industry extracts antibiotics from fermentation liquor in the huge tanks in which carefully selected strains of microorganisms are grown (Fig. 1.5). The microbiologist supervises the fermentation, maintains and preserves the stock cultures, builds up seed inoculum for the large tanks from his test tube cultures, continually tests his strains, and selects those that are the highest and most efficient producers of the antibiotic materials which he desires (Fig. 1.6). The complexity of this process is illustrated by the flow diagram (Fig. 1.7). He must supervise the extraction of the antibiotics from the broth in which they are produced and test the potency, purity, and safety of the product. This testing guarantees that the individual vials of antibiotics possess the activity stated on the label and that the products are free from any other contaminating microorganism or chemical which would make the use of these substances dangerous.

Another activity of microbiologists in the industrial field is the production of vaccines for the purpose of giving immunity against the diseases for

Fig. 1.5. At Eli Lilly and Company, antibiotics are grown in large fermentation tanks. Because it would require a building several stories high to house the tanks, they are buried in the ground. Photo courtesy Eli Lilly and Company.

which such a procedure has been found workable. The preparation of most vaccines requires the production of large amounts of the specified microorganism. The microorganisms must be either modified or killed, so that they will not produce a serious case of the disease and yet, upon injection into the

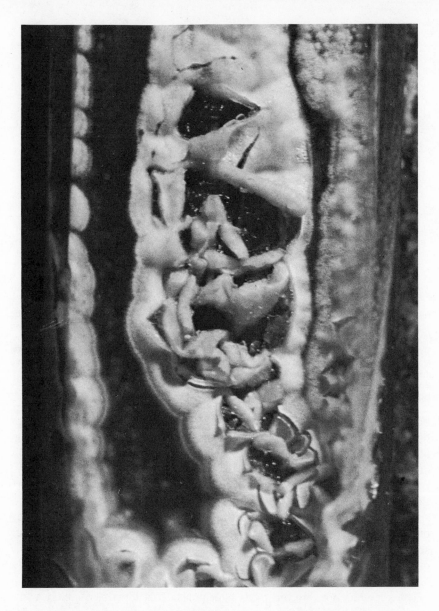

Fig. 1.6. Actinomycete growth—a close-up view. Growing in this test tube is the actinomycete, Streptomyces rimosus, which is used in the production of Terramycin. This antibiotic was discovered in 1949 and is effective in the treatment of over 100 diseases. Photo courtesy Pfizer, Inc.

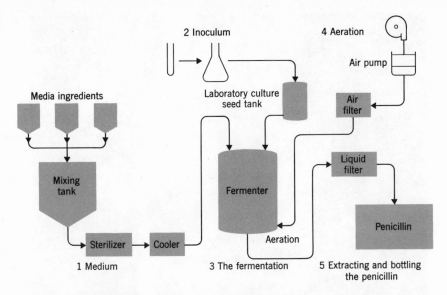

Fig. 1.7. The manufacture of penicillin is shown here schematically. (1) A medium of corn steep liquor, sugar, salts and other ingredients is mixed, sterilized, cooled, and pumped into the fermenter. (2) The mold, Penicillium chrysogenum, is transferred from slant cultures to the "seed" tank. (3) The sterile air is forced through the fermenter during incubation. (4) The introduction of sterile air is to provide the required highly aerobic conditions. (5) After the maximum yield of penicillin is produced, the mold mycelium is removed by filtration and the penicillin is recovered in pure form by a series of manipulations which include precipitation, redissolving, and filtration.

body, will give rise to an **immunity** which will protect the person against infection. In the production of the vaccine against whooping cough, the organisms are grown in broth and then harvested and killed by the addition of chemicals. Each batch must be tested to insure that there are no living organisms present and that vaccinated animals, when subjected to a challenge dose of the living organisms, are protected and will not succumb to the disease. The smallpox vaccine is raised on the embryos in chick or duck eggs (Fig. 1.8) or on the skin of a calf. This vaccine for smallpox is a living virus but has been so selected and treated that it will cause only an isolated lesion at the point of inoculation and at the same time will confer immunity. Polio vaccine is grown on monkey kidney tissue which is cultured in flasks and tubes.

The vaccines are referred to as **biologicals.** This term also includes immune serum which is still occasionally used for the treatment of certain diseases. These serums are nothing more than the liquid part of the blood of humans or animals that have been previously vaccinated and therefore contain the chemical substances providing immediate protection against a developing case of a disease.

The industrial microbiologist may be concerned with the testing of either chemicals or equipment used in the control of microbiological activity in widely diverse environments: treatment of water and sewage, air purification, disinfection of dishwashing rinse water, or of the skin of a patient about to undergo surgery.

Much of the work of microbiologists in industry has nothing to do with the production of substances to be used in medicine. Microbiological products are used widely in the chemical industry. Solvents, such as ethyl, propyl, and butyl alcohol, and organic acids, such as citric, gluconic, lactic, and acetic are produced by growing microorganisms on starches and sugars. At the

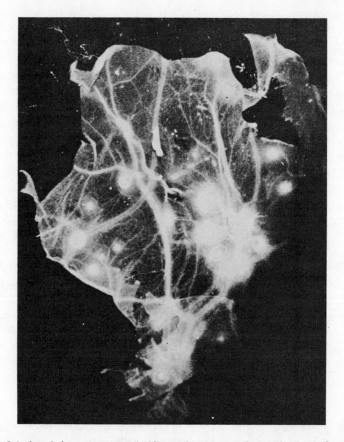

Fig. 1.8. Colonies of virus grow as small white patches on a membrane of a developing chick embryo. Here the membrane has been removed from the shell of the incubating egg and spread out to show the lesions produced when vaccinia virus (used to vaccinate humans against smallpox) is grown in embryonated eggs. Photo from Scientific American, courtesy Chas. Pfizer, Inc.

present time, many of these substances can be produced somewhat more cheaply by chemical procedures using by-products of the petroleum industry as starting materials. However, the limited amount of petroleum on this earth will eventually lead to higher prices, whereas the materials used in microbial fermentation are renewable every year through the growing of crops. Therefore, future development along microbiological lines appears certain, since industrial microbiology may potentially be applied in the synthesis of many other useful products when the price structure renders this profitable.

Some industrial processes employ microorganisms in modifying chemicals to yield some substance more valuable than the starting material. For example, certain sugar factories are prohibited by law from disposing of waste molasses by dumping it into streams or rivers. It is more economical to ferment the molasses to produce solvents — even at a loss — in order to eliminate the waste product. In European countries where petroleum is scarce, even such raw materials as sawdust and waste wood products are converted into sugar by treatment with acid, and the sugar is then fermented by microorganisms yielding useful chemicals such as alcohols and acetone.

A number of the vitamins are produced more cheaply by microorganisms than by chemical synthesis. The vitamin, riboflavin, which is extensively used in chicken feed, is produced to a large extent by a fermentation process. Under this process riboflavin accumulates in the broth in which certain microorganisms are grown. In the manufacture of vitamin C, one step in the process is more easily accomplished by microorganisms than by a standard chemical manipulation in the laboratory. The same is true in the production of the hormone, cortisone. The chemist carrying out the multiple steps in synthesis finds it difficult to introduce a hydroxy group into the proper position on the cortisone molecule, but this step is readily accomplished by the inoculation of the proper microorganism into the material. Then the chemist can proceed with further manipulations which lead to the final active cortisone product. Folic acid, vitamin B_{12}, and some amino acids essential for animal growth are produced largely through fermentation. Specific fermentation processes employing selected high-yielding mutant strains of microorganisms have been designed for these purposes. In such activities the microbiologist seeks to understand the internal mechanisms of the microbe and to manipulate them in such a way as to produce a maximum of those products useful to man.

Some of the earliest manufacturing processes employed microorganisms. Microbes have long been used to release the fibers from the stem of the flax plant, allowing them to be separated and spun into linen thread. Crude microbial processes have also been used to carry out one or more steps in the manufacture of leather. Very fine leathers are even now produced with careful supervision of the activities of bacteria which modify the animal hides in preparation for tanning.

The microbiologist's attention is required on problems of microbial deterioration of any industrial material that remains sufficiently moist to support microbial growth — from fence posts to steel pipes, and from pup tents to the muds used in drilling oil wells.

Bacteriologists working in the dairy industry are faced with a threefold problem. In the first place, several diseases of the cow are transmissible to man by means of milk. Proper control procedures are required to insure that milk is free of these disease-producing organisms as well as possible infectious agents from milk handlers.

Second, the quality of milk is measured in part by the number of organisms in it. The presence of a large number of bacteria indicates careless methods of production and also seriously affects the shelf life of the milk in food stores or in the home. Continued growth of the medically harmless bacteria in milk will soon produce off-flavors, making the milk unacceptable to the consumer. Estimation of the number of bacteria in milk is necessary to determine how much the farmer should be paid, since this is the basis for classifying milk as Grade A or B or C. Finally, in the manufacture of numerous dairy products such as cheese and butter, the activity of microbes is of paramount importance in determining the flavor and texture of the product and must be carefully controlled during the manufacturing process.

Food is a vehicle capable of carrying disease-producing organisms from a food handler to the consumer. Certain microorganisms which grow in food may cause food poisoning or food infection. Whenever a food is produced, a race begins between man and microbe to see which shall consume it. The microbiologist resorts to many artifices to limit the microbe to as little as possible.

Microorganisms are used in the production of a number of food materials such as sauerkraut and soya sauce; yeasts are used in the production of bread and various fermented beverages. Many of these materials were produced long before the existence of microbiology; even today, in some breweries, the brewmaster is regarded as an artist who acquires his knowledge through an apprenticeship. However, most large breweries employ formally trained microbiologists to culture and control the yeasts and to examine the microbiological aspects of the various processes in manufacture. In wine-producing areas, such as California, special laboratories have been equipped to deal with the microbiological problems peculiar to the wine industry. The production of food from microbial cells is now being done commercially. For sustained space flights (longer than one month) it will be necessary to produce food on the way utilizing waste liquids, gases and solids; this has stimulated the development of feasible microbial life support systems for humans.

Microbiologists are employed by water plants to determine whether the water is safe for drinking. In the disposition of the used water supply (i.e., the sewage) of cities, microbiologists concern themselves with the action of the bacteria useful to the disposal process. Industries that produce wastes of such quality that might interfere with the successful operation of the city sewage disposal plant, are required to pretreat these wastes, usually by microbial action, before discharging them into the city sewerage system. Also of concern are the potential dangers of the disposing of sewage in recreation areas, or in areas where shellfish abound, because of the possibility of introducing infectious disease through shellfish consumption.

At agricultural colleges and experiment stations, microbiologists are interested in problems of soil. The soil abounds with microorganisms of many types. Their activities have an effect upon the fertility of the soil, as measured by its ability to produce useful crops. In a few instances, as in the growth of legumes, efficient strains of microorganisms are inoculated into the soil with the seed in order to improve the growth of these plants (Fig. 1.9). We may speculate that in the not too distant future the farmer will no

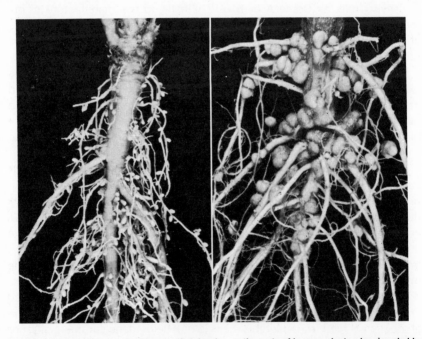

Fig. 1.9. Root nodules are small tumors that develop on the roots of legume plants when invaded by the proper strain of root nodule bacteria. Possibly originating as a plant disease, this has evolved as a symbiosis, where plants growing on nitrogen-poor soil benefit from the ability to fix atmospheric nitrogen in the nodules. Photo courtesy J. C. Burton.

more think of farming with the ordinary natural microorganisms in his soil than he would consider using the wild grasses for pasture and hay, wild plants for grains or fruits, or wild animals as the producers of edible flesh. Microbiologists determine which activities of the microorganisms are desirable in the soil and how these activities can be encouraged. They also study the problem of using chemical substances to destroy microorganisms whose activities may be undesirable in the soil. The production of composts and organic manures is a microbiological problem. The accumulation in the soil of chemicals used as pesticides or herbicides is a problem of agricultural import which has a microbiological solution; in fact, we can expect that many agricultural chemicals not easily consumed by soil microbes will be prohibited for widespread usage. As pollution of the earth with man-made products becomes a serious problem, the microbiologist must be the authority as to what may be tolerated.

Microorganisms are useful in certain assay or measurement procedures to determine the presence of toxic substances or of growth stimulants. For example, it is easy to find a microbe whose growth is limited by the amount of a certain vitamin present in the food material. The amount of growth which the microorganism makes on the food being assayed can be compared with the amount of growth made under the same conditions on a growth medium containing measured amounts of the pure vitamin. Vitamin assay procedures have been standardized so that they are routine for practically all of the water-soluble vitamins. The usefulness of such procedures was illustrated in the discovery of vitamin B_{12}, the antianemia vitamin. The amount of vitamin B_{12} in rich natural sources, such as liver, is so small that the substance cannot be measured there by any known chemical procedure. When the only method of analysis was its administration to humans with anemia, progress in research was exceedingly slow. When B_{12} was purified by extracting it from liver, it was necessary at each step to determine whether the vitamin was in the extract or remained in the residue. Analytical procedures required the participation of a number of anemic patients who would consent to live on a diet low in the vitamin until illness became severe, and then to take either one or the other of the two fractions to determine which contained the active substance. Since many purification steps were required, many patients were needed, and progress in purification was thus almost impossible. However, when it was discovered that certain strains of microorganisms could be obtained that would fail to grow in a medium containing all the food they needed, except antianemia substance, experiments could be done rapidly, and pure crystalline vitamin B_{12} was soon isolated.

Microorganisms are used for the assay of many other chemicals which either stimulate or inhibit their growth or development. These chemicals include not only the organic compounds but also trace amounts of certain minerals. Such tests may be used as an index of soil fertility, since the chemical substances in the soil which are available to the green plant will also be available to the microorganisms. Surveys of the microbial population of the soil have been used as aids in prospecting for petroleum. Where the soil microbes can utilize volatile petroleum products rapidly, it is suspected that seepage of such materials is occurring from an underground pool, and oil or gas may be found in the area.

History records successful use of biological warfare by the Conquistadors, who disseminated the white man's diseases by artfully distributing to the Indians garments from infected persons. Throughout the world, a considerable number of microbiologists are known to be at work on the production of microbes which might render an enemy less effective by producing infections in him, in his animals, and in his plant crops. The ability of the initial infection to multiply is an aggressive feature in biological warfare not present in other weapons. Biological warfare research reverses the usual objectives of microbiology in medicine. Rather than destroying or limiting the spread of microbes, biological warfare seeks to produce large quantities of the infectious agent, to keep it alive and vigorous, and to seek out methods of effective spreading. At the same time, a large program involves working out defensive methods of detecting and dealing with this problem should an enemy attack with biological weapons.

Perhaps the most interesting and, in the long run, most profitable efforts of microbiologists are those concerning the uses of microorganisms in research. The microbe is now used so widely as a research tool that in large universities we can rarely find a department in the natural sciences where one or another of the research workers is not using microorganisms. The similarities among the fundamental processes of living things lead many research workers to employ the microbe as a convenient tool for fruitful basic studies on problems ranging from photosynthesis to cancer. An astonishingly large proportion of the modern biological research programs, which have led to recognition and fame (e.g., Nobel prizes and other national and international awards), have been conducted using microorganisms as research tools (Table 1.2). Finally, many students of microbiology desire to understand the microbe for its own sake, that is, to know what it is and how it functions.

Summary

Microbiology deals with the smaller living creatures and especially with what they are and what they do. It concerns itself with the destructive and constructive activities of microorganisms. The microbiologist is employed by

Table 1.2 Nobel Prize Winners Whose Work Relates to Microbiology

Year	Name	Nationality	Citation
1901	Emil von Behring	German	Serum therapeutics
1902	Sir Ronald Ross	English	Transmission of malaria from mosquito to human
1905	Robert Koch	German	Founded scientific bacteriology and bacterial culture methods.
1907	Charles Louis Alphonse Laveran	French	Protozoans as agents of disease (trypanosomes, etc.)
1908	Paul Ehrlich and Elie Metchnikoff	German Russian	Immunity
1913	Jules Bordet	Belgian	Discoveries in the realm of immunity
1927	Julius Wagner-Jauregg	Austrian	Therapeutic importance of malaria infections in syphilis patients
1928	Charles Nicolle	French	Typhus fever
1930	Karl Landsteiner	American	Human blood groups
1939	Gerhard Domagk	German	Antibacterial effect of prontosil (sulfonamide chemotherapy)
1945	Sir Alexander Fleming Ernst Boris Chain Sir Howard Florey	English British Citz. English	Discovery and development of penicillin
1946	Wendell M. Stanley John H. Northrop	American American	Production of purified enzymes and virus protein
1951	Max Theiler	S. African	Discoveries concerning yellow fever
1952	Selman A. Waksman	American	Discovery of streptomycin
1953	Fritz Albert Lipman	German-American	Discovery of coenzyme A
1954	John F. Enders Thomas H. Weller Frederich C. Robbins	American American American	Discovery that poliomyelitis viruses multiply in tissue cultures
1958	George W. Beadle Joshua Lederberg Edward L. Tatum	American American American	Discovery that genes act by regulating chemical processes

Table 1.2 (*Continued*)

Year	Name	Nationality	Citation
1959	Severo Ochoa Arthur Kornberg	American American	Discoveries of the mechanism in the biological synthesis of ribonucleic and deoxyribonucleic acids
1960	Frank MacFarland Burnet Peter Brian Medawar	Australian British	Discovery of acquired immunological tolerance to tissue transplants
1962	James Watson Francis Crick M. Wilkins	American British British	DNA structure
1965	J. Jacob A. Lwoff J. Monod	French French French	Regulation in cells
1966	C. R. Huggins F. Peyton Rous	American American	Viruses in cancer
1968	M. Nirenberg G. Khorana R. W. Holley	American American American	Genetic code
1969	S. E. Luria M. Delbrück A. Hershey	American American American	Bacteriophages in molecular biology

federal and state public health departments, by the drug industry, by milk and other food manufacturers, by agricultural experiment stations, and by any number of unrelated industrial fields in which microbes are used as research tools. The successful practice of microbiology is based on developing procedures rendering those microorganisms which hurt us less harmful, and making those which help us more useful. The environment in which we live is conditioned far more by microbes than by most efforts of modern human technology.

Bacteria are central to the study of microbiology. From them evolved more complex forms of the kingdom *Protista,* such as the protozoa, the algae, and the higher fungi. Similarly, less complex forms evolved from the early bacterial prototypes. The variability is so great that to develop useful generalizations about microorganisms it is necessary to consider them in subdivisions ranging from the viruses to the higher fungi.

TWO

How It Began

Organisms that we would consider "alive" have existed for almost four billion years, two-thirds of the almost five billion years (approximate age) of the earth. Fossil records of microbes that are indistinguishable in size, appearance, and general make-up from some modern bacteria are found in sedimentary rocks which had solidified over three billion years ago. New methods of eroding away the rock to reveal the details of these microscopic organisms have made it possible to produce convincing electron microscope pictures of organisms that were formerly regarded as too soft to leave fossil records (Fig. 2.1). Use of these procedures on later rocks has produced a continuous record showing the emergence of more complex microbes and extending the record of life on our geological clock (Fig. 2.2). It is apparent that billions of years of evolution and development occurred between the emergence of microorganisms and that of higher forms, and man's recognition of their significance is very recent indeed.

In contrast, the recorded history of microbiology is measured in terms of mere hundreds of years. Just as political history was studied as a chronological record of the rise of statesmen and nations, so scientific history has been studied as the consecutive biographies of scientists and their achievements (Table 2.1). The biographical approach greatly oversimplifies the development of the biological sciences. Many of our "great" discoveries were made and publicized several times before they gained common acceptance. The qualifications of the people we elect to "greatness" are extremely un- . even. Sometimes we honor those who put the keystones in the arches of discoveries. At other times we select those who made an initial contribution upon which the later discovery rested. Occasionally attention is focused

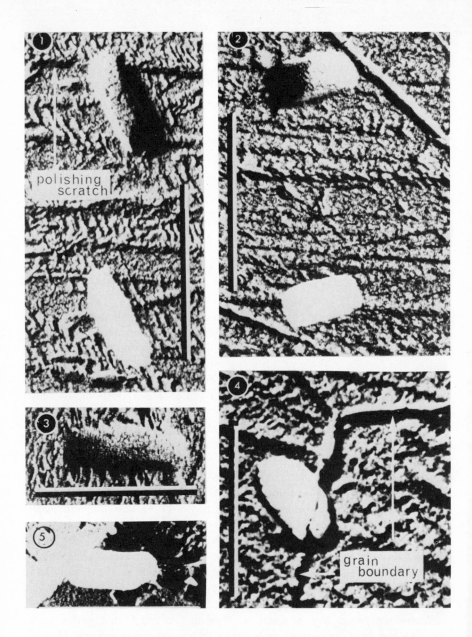

Fig. 2.1. Negative prints of electron micrographs of chert from the Fig Tree Series showing Eobacterium isolatum, preserved both organically and as imprints in the rock surface; line in each figure represents 1 μm. (1) Electron dense, organically preserved rod-shaped cell about 0.6 μm long (white, below) and its imprint in the chert surface (above). During preparation of the sample the bacterium-like fossil was displaced from its original position in the chert matrix. The presence of imprints and

upon an early scientist who made essentially no contribution to the main stream of development, but was important when a renewal of interest caused others to rediscover his contribution in "ancient" scientific writings.

The development of science is regarded by some as similar to a project in map-making. History records only the names of those whose portion of the map contains the great rivers or mountain peaks. Very often it is not until later that we discover that the true wealth of the land lies under the drab, flat areas. It will be observed that Table 2.1 contains few recent citations in spite of the fact that many of the greatest developments in biological sciences in the last quarter century resulted from studies with microorganisms. While there is no doubt that these will eventually be cited as milestones, it is still too early to judge their influence on microbiology.

THE FIRST VIEW OF MICROORGANISMS

Fossil records indicate that disease-producing microorganisms have plagued man from the time of his early appearance on earth. Roman and Arabic writers speculated on the existence of nonvisible forms of life. Some early philosophers even suggested that "germs" of contagion passed from one person to another causing disease. Needless to say, all of these theories remained mere speculation until man devised instruments for the observation of subvisible organisms.

A study of the origin of the microscope leads us through the development of spectacles, the improvement of glass, and the consequent general interest in lenses. The fine art of lens grinding apparently occupied the attention of many artisans in Holland and the low countries around the turn of the sixteenth century. Zacharias Janssen had produced a compound microscope by 1590, but his lenses were too imperfect and his perseverance insufficient to permit him to see the microbes.

---

the fact that they transgress structures of the mineralogical matrix, such as grain boundaries and polishing scratches, indicate that the minute organism is not a contaminant. (2) Organically preserved cell about 0.5 μm long (white, below) displaced from its imprint in the rock surface (above). Note the irregular, granular texture of the cellular imprint. Subparallel, horizontally oriented lineations in the rock surface are polishing scratches; a prominent grain boundary is present to the right of the imprint. (3) Rod-shaped imprint in chert surface. About 0.75 μm long, this is one of the longest imprints of E. isolatum observed. (4) Organic, somewhat flattened bacterium-like cell transgressing chalcedony grain boundary. The continuity of the grain boundary, passing through the fossil organism, demonstrates that E. isolatum is indigenous to the rock and is consistent with a primary origin for the chert. (5) Organically preserved cell of E. isolatum showing the short, broad, rod-shaped morphology of the fossil organism. This well-preserved cell demonstrates the morphological similarity of the ancient organism to modern bacillar bacteria. Photos courtesy E. S. Barghoorn and J. W. Schopf: Science, **152**, No. 3723, 758-763, 1966. Copyright 1966 by the American Association for the Advancement of Science.

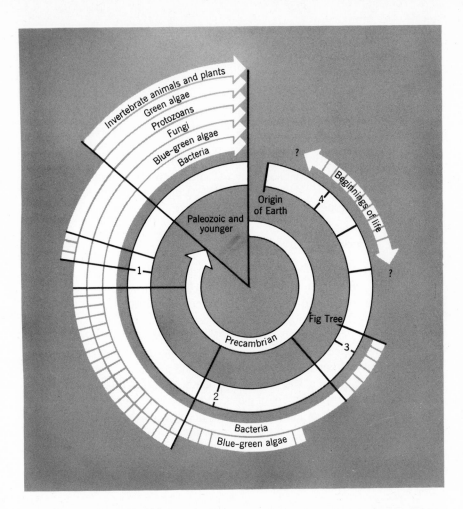

Fig. 2.2. "Geologic clock" in billions of years showing the timing of the "Fig Tree chert bacteria" in Fig. 2.1 and the appearance of other microorganisms. Cross-hatched arrows indicate uncertainties.

Anton van Leeuwenhoek, a tradesman of Delft, Holland, is credited with the first recorded observation of the world of subvisible life. During his leisure hours, this successful Dutch merchant produced simple microscopes consisting of home-ground convex lenses mounted in brass and silver. His lenses were so perfect that by using them, together with keen eyesight and extreme patience, Leeuwenhoek was able to observe the larger microorganisms for the first time. He communicated the descriptions of his microscopical discoveries in a series of letters to the Royal Society of London from about 1676 until his death in 1723. Although these letters served as the first

Microorganisms and Man

record of investigations in microbiology, few of Leeuwenhoek's contemporaries could repeat his studies with the instruments available at the time. During the two centuries which followed, persons interested in microscopy were forced to confine their attention to larger forms of life. Meanwhile, rapidly developing technology led to the production of better and cheaper glass, which, in turn, resulted in the production of excellent lenses and, ultimately, the modern compound microscope.

Table 2.1 Some Reference Points in Microbiological History

Discoveries prior to development of the science

1546 — Fracastorius suggests living agents as cause of disease
1590 — Janssen perfects a compound microscope
1676 — Leeuwenhoek reports first observations of bacteria
1762 — von Plenciz suggests that each disease is caused by a different subvisible organism
1796 — Jenner introduces smallpox vaccination
1810 — Appert wins Napoleon's prize for food canning
1839 — Schönlein suggests the relation of fungus to skin disease in man
1853 — De Bary demonstrates that molds may cause plant disease

The establishment of microbiology and immunology

1857 — Pasteur reports microbial origin of fermentation
1864 — Pasteur resolves the problem of spontaneous generation
1870 — Lister uses antiseptics in surgery
1876 — Koch grows pure cultures of the anthrax organism
1882 — Burrill discovers that plant disease can be caused by bacteria
1884 — Metchnikoff reports the phenomenon of phagocytosis
1885 — Pasteur produces vaccines for rabies and anthrax
1890 — Winogradsky isolated nitrifying bacteria from soil
1892 — Iwanowski shows the viral nature of tobacco mosaic disease
1895 — Ehrlich introduces the antigen-antibody theory of immunity
1898 — Löffler and Frosch prove the viral nature of hoof and mouth disease
1906 — Ehrlich produces the chemotherapeutic agent "606"
1909 — The discovery of rickettsiae
1911 — Rous associates viruses with some tumors
1915 — Twort and d'Herelle (1917) discover viruses which attack bacteria

Some modern milestones

1926 — Kluyver points out that most of the fundamental chemistry of all living systems can be inferred from studies with microbes
1928 — Fleming recognizes potential medical usefulness of penicillin
1931 — Goodpasture uses embryonated eggs for virus culture
1935 — Domagk presents the first sulfonamide
1944 — Electron microscopes are made available
1944 — Avery shows that genetic control in microbes resides in the DNA
1949 — Enders develops tissue-culture methods for viruses

The first observation of microorganisms revived speculation concerning the origin of living things. Although the concept of the voluntary generation of higher plants and animals from nonliving materials had been largely abandoned by the time of Leeuwenhoek, it was a far more difficult task to prove that microbes were not produced spontaneously. Some of the first true experimentation with these newly discovered microbes was directed toward an explanation of their mysterious appearance in organic infusions.

Early in the eighteenth century an Italian priest named Spallanzani boiled beef broth in containers and then sealed them against air. Since no growth appeared in his airtight containers, the critics of the period claimed that the absence of spontaneous generation in this case merely resulted from the exclusion of air which was vital for the process. Nearly 100 years later, a retired French confectioner named Appert made practical use of Spallanzani's experiment by devising a method for preserving food. He boiled airtight containers of food and, on opening the containers later, he found the food well preserved.

Toward the end of the eighteenth century it became apparent that oxygen was essential for the life of higher animals. At this time it seemed possible that the airtight seal of Spallanzani and Appert prevented growth because it excluded oxygen rather than microbial contamination. A series of ingenious studies were then initiated which strongly suggested that growth of microbes would not occur in properly heated infusions which, though exposed to air, had been treated to remove or destroy all microbial life. In 1837, Schwann passed air into boiled infusions through red-hot tubes, and thirteen years later Schroeder and von Dusch performed more decisive experiments by filtering air through cotton into the boiled flasks.

In 1864, the brilliant French chemist, Louis Pasteur, reported his investigations which conclusively demonstrated the faulty logic of the investigators who defended spontaneous generation. He simply boiled infusions in flasks with long, narrow, gooseneck openings (Fig. 2.3). Untreated air passed freely in and out, but microbes settled in the gooseneck and no microbial growth developed in the broth. Certainly Pasteur's true genius lay at least in part in his ability to devise and carry through the convincing crucial experiment.

By 1851, previous to the above studies, Pasteur had established a name for himself by other experimentation in the field of microbial decomposition. He had observed that, unlike other living things, some microorganisms preferred to live in the absence of oxygen. He called these organisms **anaerobes,** and he named the anaerobic decomposition of sugars **fermentation.** His work on fermentations led him to an investigation of the "sick" wines of France. The inferior flavors of these beverages rendered them unsuitable for competition with the better wines produced in the cooler regions of Germany. He showed that "sickness" was due to the growth of flavor-destroying

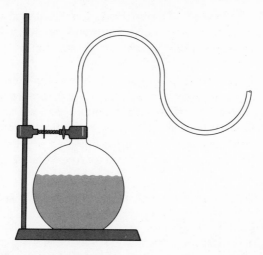

Fig. 2.3. The famous swan-neck flask used by Pasteur during his studies on spontaneous generation. The construction of the neck permitted free access to air to the flask contents but prevented entry of microorganisms present in the air.

bacteria. He devised a heating process which is now called **pasteurization** by which the harmful bacteria were destroyed, but the flavor of the wine preserved. He thus guided the French vintner to domination of the wine industry of the world.

THE GROWTH OF MEDICAL MICROBIOLOGY

Although disease epidemics have been witnessed during all periods of recorded history, it was not until the middle of the eighteenth century that man began to comprehend the true nature of their cause. The idea of communication of disease by contact was expressed many times by early thinkers, but, since these suppositions could not be founded upon direct observation of the disease producers, they did not greatly influence the train of thought of the times. For example, in 1546 Fracastorius theorized that disease was the result of "infection" passing from one individual to another in the form of "seeds" or "germs." In 1762, von Plenciz further insisted that each disease was due to a different kind of invisible, living thing.

In 1839, the German physician Johann Schönlein discovered the parasitic fungus that caused the skin infection known as favus. Thus, for the first time, the relationship between a microorganism and a human disease was brought to light. In 1853, Anton De Bary demonstrated that certain molds caused the smuts and rusts of a number of food plants.

For the master scientist, Pasteur, it was only a transitional step from "sick" wines to "sick" silkworms. After five years of exhaustive study, he was able to prove that pebrine, a disease of silkworms which threatened to destroy the silk industry, was due to a microscopic protozoan. He devised methods for its prevention, rescued the industry, and saved France several million dollars yearly.

Another great figure, contemporary to Pasteur, was the German physician, Robert Koch (Fig. 2.4). His great successes stemmed from the techniques he devised. He developed a liquefiable, solid culture medium and worked out methods for isolating pure strains of disease-producing microbes free from contamination by ordinary organisms of the dust and air. His investigations of single kinds of organisms in test tubes were free from the confusion that

Fig. 2.4. Dr. Robert Koch (1843–1910) in search of the rinderpest microbe. The elaborate laboratory equipment which was so useful to Koch in developing methods for solving many problems in microbiology will not aid him here; rinderpest is caused by a virus which is invisible under the light microscope and which fails to grow outside of animals cells. Photo courtesy The Bettmann Archive.

results from studying mixtures. He was the first to prove unequivocally the cause and effect relationship of bacteria to disease. This feat was accomplished in 1876 with the disease anthrax.

Koch's techniques were so useful that students from all parts of the world flocked to his laboratory to learn methods. The decade and a half following 1876 is commonly called the Golden Age of Bacteriology. In the words of Koch, "New discoveries come like apples falling from a tree." During this period, most of the bacteria which cause disease were isolated, described, and studied in the test tubes of the bacteriologist or in his laboratory animals. Koch himself isolated the tubercle bacillus, the causative agent of tuberculosis, and published a classic monograph on this work, which served as a model for his students and later investigators.

Although certain fungi had already been reported as plant-disease producers, it was not until 1882 that the American scientist Burrill discovered that fire blight in pears was caused by bacteria. This finding stimulated the development of plant pathology (to which other Americans, including Erwin F. Smith, made significant contributions) and stimulated the development of the science of microbiology in the United States.

In 1892, Iwanowski reported the first experiments which led to knowledge of the viruses. In an attempt to isolate the causative agent for the tobacco mosaic disease, he found that Koch's methods failed to reveal an organism that could be viewed under the microscope. He passed the juice from an infected area of a plant through a filter with a pore-size small enough to hold back bacteria. Since the filtered juice still retained its disease-producing capacity, Iwanowski concluded that he was working with organisms too small to be observed by the microscopes of that time. He named the newly discovered microbe the **filtrable virus.**

In 1898, Löffler and Frosch demonstrated that hoof and mouth disease of cattle was caused by a virus-like agent. During the decade following 1900, the virus nature of a variety of common diseases was demonstrated. Twort, in 1915, and d'Herelle, in 1917, created a furor when they announced the discovery of viruses that could attack bacteria. These investigators found that broth cultures of certain bacteria could be dissolved by the addition of a bacteria-free filtrate from sewage. The name **bacteriophage** was proposed by d'Herelle because he considered the filtrable agent to be a living microbe which "ate" bacteria.

THE DEVELOPMENT OF IMMUNOLOGY

One of the greatest contributions to preventive medicine was made in 1796 when Edward Jenner transferred material from a human cowpox pustule to the arm of a healthy boy. A few weeks later the boy was inoculated with infectious matter from a smallpox patient; he failed to develop the disease.

By 1800, several thousand persons had been similarly inoculated against smallpox with a vaccine grown on the belly of calves, and the science of **immunology** was born (Fig. 2.5).

By 1880, Pasteur had isolated a microorganism which he considered the cause of a disease known as fowl cholera. By accident, he discovered that chickens would not die of the disease when they were inoculated with the organisms that had been standing in his laboratory test tubes for several weeks. Furthermore, when the same chickens were reinoculated with fresh vigorous cultures, they were resistant to the disease. It was evident to Pasteur that he had in his hands a procedure for extending to other diseases a vaccination process similar to the one which had been outlined by Jenner

Fig. 2.5. Tiny virus. These four ovals are air-dried vaccinia virus particles magnified many times by an electron microscope at the Virus Laboratory of the University of California (Berkeley). Vaccinia, or cowpox, is the material used in vaccinations to immunize against smallpox. Photo courtesy Pfizer, Inc.

years before. Pasteur then subjected other disease-producing organisms to altered environmental conditions such as aging, drying, growth at abnormal temperatures, and treatment with chemicals. By 1885, he had developed methods for the production of successful vaccines for anthrax in cattle and sheep, and for rabies in humans and dogs.

The work with vaccines led to extensive research and speculation on the whole problem of immunity. In 1884, Elie Metchnikoff demonstrated the role of white blood cells as destroyers of disease-producing bacteria. He named these special blood cells **phagocytes** and advanced the theory that they were the agents which produced immunity. A few years later, Paul Ehrlich presented a theoretical explanation of immunity based on specific chemical changes in certain soluble substances in blood serum. Our present knowledge indicates that both mechanisms play roles in the immune response, but the arguments which arose between the followers of the two theories did much to put immunology on a scientific basis.

THE DEVELOPMENT OF CHEMOTHERAPY

By 1870, the English physician Joseph Lister had shown that wound infections by microorganisms could often be prevented by using dressings saturated with a solution of carbolic acid. Together with Koch, he devised ingenious tests for determining the necessary strengths of chemical substances used to destroy microorganisms outside the body.

Meanwhile, Ehrlich turned his attention to the possibility of using chemical substances to destroy microorganisms within the body. However, the search for materials that could be introduced into an animal to destroy specific microbes without poisoning the animal tissues proved to be a particularly difficult one. His study led him to the investigation of chemicals that were particularly successful in the treatment of syphilis. In 1906, after 606 trials, he discovered an arsenic compound that would destroy the syphilis microbe and yet not cause particular harm to the patient. His failure to produce compounds effective against other microbial disease producers did not discourage other investigators from continuing the search.

Over 25 years later, Gerhard Domagk discovered that a red azo-dye compound named **prontosil** was active in curing a number of bacterial infections. Later experiments showed that the antibacterial properties of the dye resided in the colorless **sulfanilamide** portion of the compound. The successful **sulfonamides** that followed demonstrated clearly that it was possible to obtain chemical agents against a wide variety of microbial infections, and that these chemicals could be introduced into the body without producing serious toxic symptoms.

In 1929, Alexander Fleming reported that the mold *Penicillium notatum* produced a substance which prevented the growth of certain bacteria on laboratory culture media. He named the antibacterial substance **penicillin** (Fig. 2.6) and suggested that it could be useful in medicine. His discovery went unnoticed until technology had developed sufficiently to perceive the true value of the antibiotic. Just as the development of the microscope awaited the technology of glass making, so the production of penicillin awaited other developments. It awaited the development of a chemical technology to isolate it from natural material; it awaited the stimulus following the successful development of sulfonamide chemotherapy; it awaited a knowledge of procedures in genetics for the development of mutant strains of *Penicillium,* yields of which were high enough for economic production; and it awaited a political climate in sympathy with subsidizing the construction of facilities for such a speculative venture.

As soon as the success of penicillin was assured, great numbers of scien-

Fig. 2.6. Penicillin mold. This symmetrical colony of green mold is Penicillium chrysogenum, **a mutant form of which now produces almost all of the world's commercial penicillin.** Photo courtesy Pfizer, Inc.

tists began searching among the microorganisms for strains and cultures that would produce antibacterial chemicals. In the decade beginning in 1940, a great number of such cultures were discovered. Some have proved to be so useful in medicine that many of the problems of bacterial infections are now successfully controlled.

RECENT ADVANCES IN VIRUS RESEARCH

In spite of the success of the smallpox vaccination program, the lack of methods for culturing other viruses thwarted progress in this field. In 1931, Goodpasture and his associates discovered that the developing chick embryo was an excellent medium for growth of a number of viruses. The use of the embryonated hen's egg was soon standard procedure for the detection, isolation, and study of many common viruses.

Meanwhile, workers in the field of botany and zoology had perfected methods for growing bits of tissue in test tubes. The procedures were patterned after the culture of bacteria and nutritional studies of microorganisms. In 1949, J. F. Enders and his co-workers developed methods for growing viruses on such test-tube cultures of living tissues. This new technique permitted cultivation of those viruses which failed to develop in chick embryos, selection of tissues for optimum virus growth and large-scale propagation of viruses. There are now useful vaccines for most virus diseases including poliomyelitis, measles, mumps and yellow fever.

With the development of the electron microscope (Fig. 2.7), man was able to view the virus for the first time. In addition, he was permitted to study, in much greater detail, the **cytology** of the larger microbial forms. However, in spite of this tremendous advance in the field of microscopy, the interpretation of structural details is still difficult.

Summary

The history of microbiology, probably as much as any other science, emphasizes the dependence of each scientist upon his predecessors. No great discovery is made without previous discovery, and, in this science especially, technical developments almost always precede waves of advance. The development of the microscope stimulated a wave of scientific attention to microscopic observation. The early workers, Koch and Pasteur, provoked waves of discovery as a result of their work on methods and theory. The rise of biochemistry, the creation of the first synthetic antibacterial agent, the development of antibiotics, the use of the chick embryo, and the procedures for tissue culture are all interrelated in the complex of tech-

Fig. 2.7. **(a) A modern electron microscope.** Photo courtesy Forgflo Corp. **(b) Details can be seen in bacteria examined with it. An ultrathin section of the bacterium,** Azotobacter vinelandii **showing a longitudinal section of a dividing cell. Below it is a transverse section of a cell. Different cellular structures may be observed.** Photo courtesy L. Pope.

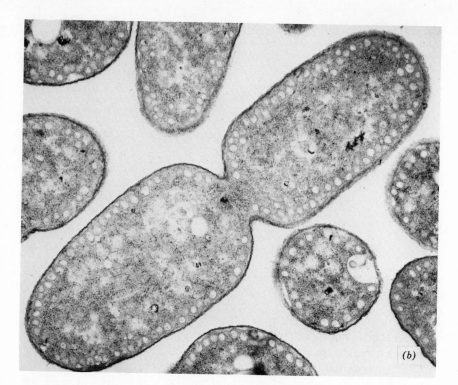

Fig. 2.7. (b)

niques which make up modern science. Even now the fundamental research of today's scientists, which often seems unlikely to benefit mankind, will furnish the basis for the practical applications (useful discoveries) of the future. Conversely, men whose contributions receive great current attention may, a few generations from now, appear as obscure as the inventor of the spermaceti candle.

THREE

Techniques and Methodology

Microbiology is a working and experimental science. In the laboratory, the microbiologist utilizes various techniques, materials, and technical pieces of equipment in experimental and routine investigations.

Although much of the microbiologist's work is performed without a microscope, the growth and development of microbiology depended upon the development of effective, undistorted magnification of very small objects. Mastery of the use of the microscope is basic for thorough understanding of this science.

FUNDAMENTALS OF OPTICS

When slanting rays of light pass from a material through which they travel easily into a material through which they travel with greater difficulty, the light rays are bent or refracted toward the perpendicular. This principle, the **law of refraction,** operates when light passes from air into glass. The rays are bent towards a line perpendicular to the surface of the glass. The light traveling easily through air may be compared to an automobile driving freely over the surface of a dry pavement. If the right front wheel suddenly slips into the muddy shoulder of the road, the car will swerve in the direction of the mud, through which it travels with greater difficulty. The more oblique the angle of entrance into the mud, the sharper the angle of the swerve becomes. On the other hand, if the car is driven directly across the road, both front wheels enter the mud at the same time and the line of travel is not bent at all (Fig. 3.1).

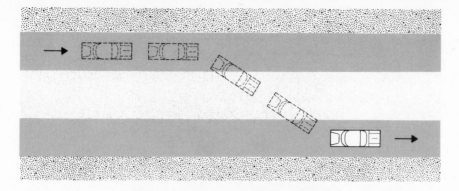

Fig. 3.1. When an unguided automobile enters at an angle on a material which offers greater resistance to its forward progress (such as the unpaved center of this dual highway), it swerves in the direction of the material offering greater resistance. When it leaves the center it swerves again. This is analogous to the bending of light rays encountering an air-glass interface.

A portion of glass with a curved surface bends most of the rays which pass through the air-glass boundary. If this lens is properly ground, the rays of light illuminating the specimen and passing through the lens can be focused to form an image. The rays of light reflected from the two ends of a small object will be farther apart when they come into focus as an image than they were as reflected from the object (Fig. 3.2). The **point of focus** depends on the radius of curvature of the lens. The image of the magnified object appears blurred if it is viewed from any point along the path of reflection other than the **focal point.** The modern microscope depends on a series of lenses to bring a considerable area into focus at one time without distortions. The lenses of compound microscopes are so arranged that one will further magnify the image formed by another.

The more powerful lenses have the greater curvature, which is obtained by making the lenses progressively smaller. The size is finally limited as very small lenses do not pass enough light for the human eye to detect the object. However, powerful lights and condensers underneath the microscope stage can direct more light up through the lens. The most powerful lens of the light microscope is the **oil-immersion lens;** a drop of oil placed between the lens and the object eliminates the loss of slanting rays which would otherwise be refracted. The oil has a refractive index the same or close to that of glass.

Even if sufficient light were available, the nature of light itself imposes limits to magnification. The wavelengths of light visible to the human eye are between 0.4 and 0.7 μm. Objects smaller than half the wavelength of light do not ordinarily block enough of the light to produce a discernible image. The smallest objects that can be resolved by light microscopes are, consequently, two-tenths micrometers in diameter.

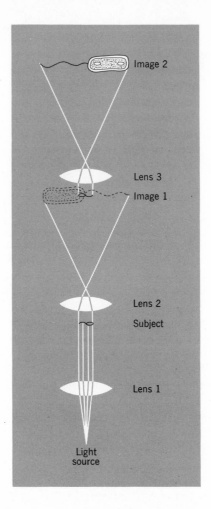

Fig. 3.2. In a complex light microscope the light rays, made parallel by lens 1, pass through the subject, and are magnified by lens 2 to give an image. This image is magnified once more by lens 3.

The modern compound microscope usually has an ocular or eyepiece lens, which magnifies ten times, and three objective lenses attached to a rotary turret at the opposite end of the microscope barrel. The most commonly used objective lenses are: the **low-power objective,** which is 16 millimeters in focal distance and magnifies ten times; the **high-power objective,** which is 4 millimeters in focal distance and magnifies around 45 times; and the **oil-immersion objective,** which is 1.8 millimeters in focal distance and magnifies 97 to 100 times. The total magnification for a given setting is computed by multiplying the magnification of the ocular by that of the objective lens being used. Thus, the total magnification would be 100× with low power, about 450× with high power, and about 1000× with oil immersion. Each of these optical systems is composed of several lenses precisely ground to correct distortions and aberrations.

The other parts of the microscope serve to support the specimen and to allow for the adjustment of the optical system. The **stage** holds the specimen in position relative to the lenses. The **coarse** and **fine adjustment knobs** regulate the distance between the lens and the stage so that the specimen may be brought into sharp focus. The **condenser** beneath the stage concentrates the light rays on the specimen. The **iris diaphragm** shuts out those light rays which pass through the outer edge of the objective lens, thus limiting the view to the center of the lens which gives the truest image with most microscopes.

OTHER TYPES OF MICROSCOPY

Small objects may be better observed under a microscope using **dark-field illumination.** Particles of dust, which are completely invisible to the naked eye, become visible in a beam of light in a dark room. The small particles are not seen but rather the flashes of light reflected from them. The same principle is involved with dark-field illumination. This type of illumination requires a special condenser with a black dot in the center to block out all the rays of light which would pass directly into the objective lens. The resulting dark field allows observation of the flashes of light reflected by small objects on the slide as they are hit by slanting rays of light (Fig. 3.3). A clean slide would appear dark. Although the resolution is no greater than that of the light-field microscope, the dark-field technique makes it possible to observe and photograph objects as small as bacterial flagella, which are not seen with the ordinary light microscope. Such pictures are, of course, not of the flagella themselves but the reflected light.

Special lenses and a condenser added to the ordinary light microscope convert it to a **phase-contrast microscope.** The specialized condenser contains concentric rings which illuminate the specimen by cylinders not vibrating in the same phase. The objective lenses also contain concentric rings in exact alignment with those in the condenser. Minor differences in the density of bacterial cell structures are insufficient to render them visible under the ordinary light microscope. However, these differences are exaggerated under the phase-contrast microscope and appear as distinct structures (Fig. 4.5). The internal structure of bacterial cells can be studied in killed, stained preparations when examined under the light microscope.

Fluorescent microscopy is of special value in diagnostic microbiology. **Fluorescence** is the emission of visible light by a substance, **fluorochrome,** which

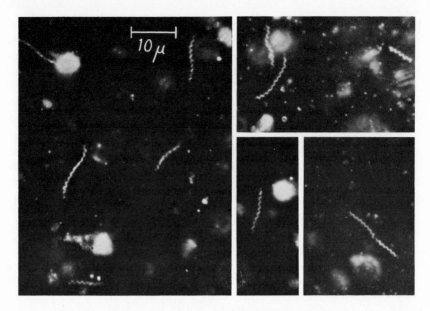

Fig. 3.3. Four photomicrographs of dark-field preparations of Treponema pallidum, **the causative organism of syphilis.** Photo by Theodor Rosebury, courtesy A.S.M. LS-327.

has been excited by high-energy photons (light quanta) of shorter wavelength. Fluorochromes, including fluorescent dyes, possess certain electrons which can be excited by absorbed photons to orbitals of high energy levels which are unstable. As these electrons return to their original orbitals of lower energy level, their energy is emitted as photons of energy quanta lower than that of the incident photons, hence, of longer wavelength. With ultraviolet as the light source, the emitted light is usually in the visible green or yellow range. Blue-violet light is usually used for excitation since regular glass optics may be used with this light source. After passing through the microscope condenser lens, the light strikes the specimen, exciting its fluorochrome to emit a wave band of visible light. A barrier filter screens out the near ultraviolet light and serves to protect the eyes from injury. Thus, the eye sees only the wave band of emitted visible light. Experience has shown that fluorescent objects are best viewed with a dark-field microscope. Recently, differential fluorochrome staining of acid-fast bacilli in sputum concentrates has become popular again due to the availability of better equipment which was not the case when this method enjoyed brief popularity several decades ago.

The method of magnification of the **electron microscope** does not differ fundamentally from that of the ordinary light microscope (Fig. 3.4). A beam of electrons rather than a beam of light is passed through the object to be viewed. Three or more ring-shaped electromagnets serve as lenses to bend the electron beams. Like light rays, the electrons are absorbed or refracted to varying degrees by areas of different electron density in the specimen. The emergent electrons can be focused on a photographic plate or upon a fluorescent screen, thus producing a visible image. Since electrons are readily stopped by all forms of matter, studies are carried out in a vacuum. Living materials may not be used since the specimens must be dried before they are placed in the vacuum. Microbial materials are often mounted on thin films of cellophane or collodion. Ultrathin slices or sections of bacterial cells facilitate electron microscope studies by revealing the nature of intra-cellular bodies and structures (Fig. 3.5). Special **microtomes** have been de-vised to cut sections thinner than one-tenth the diameter of a bacterial cell. The cells are first mounted in plastic and the sections are cut with a dia-mond or glass knife. Magnifications of 10,000 to 80,000 times are possible, and photomicrographs obtained at lower magnifications can be enlarged up to ten times. The electron microscope, in making possible the first observa-tions of the smaller microorganisms, contributed greatly to the under-standing of the external structures of the microorganisms. Ordinary dyes are

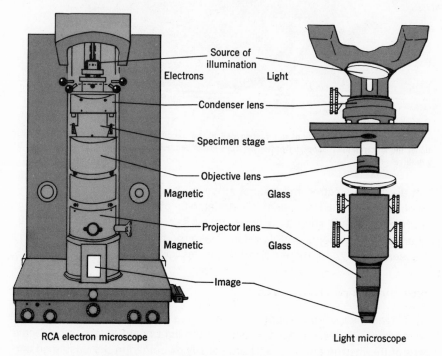

Source of illumination

Electrons   Light

Condenser lens

Specimen stage

Objective lens

Magnetic   Glass

Projector lens

Magnetic   Glass

Image

RCA electron microscope   Light microscope

Fig. 3.4. **The analogous parts of the electron microscope and the inverted light microscope used in the author's laboratory.** Redrawn from diagram, courtesy RCA.

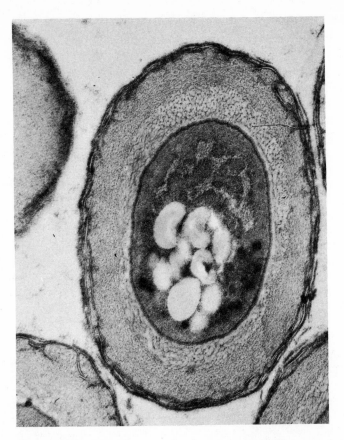

Fig. 3.5. The internal details of this resting cell of the bacterium Azotobacter are demonstrable under the electron microscope only after the cells are imbedded in plastic and ultrathin slices (0.05 μm thick) are cut.

not used with electron microscopy since the observations can be seen only in black and white. However, the material to be observed is often treated with a chemical agent such as a heavy metal salt that has affinity for certain chemical components and therefore make them more electron dense. This frequently brings out more refined details for observation.

PREPARATION OF LIVING CELLS

In a light microscopic examination, bacteria are placed on a glass slide which is positioned on the stage so that the light beam enters the objective lens of the instrument. The cells are usually examined in the unstained condition. Living organisms from a broth culture or those suspended in a drop of water are placed on a slide. A cover-slip is placed on the drop to prevent excessive evaporation and then is examined. This is often referred to as a **wet mount.** Or the suspension of organisms may be placed on a cover-slip which is then inverted and placed on a hanging-drop slide which is hollowed to accommodate the drop. It is then observed with the high objective. The **hanging-drop method** is particularly useful for detecting bacterial motility by observing definite movement of the cells within the microscopic field even though the bacterial flagella are below the resolving power of the light microscope. Nonmotile cells will frequently demonstrate a vibrating motion or **Brownian movement** due to the bombardment of the cells by molecules of the suspending fluid.

Liquid preparations of living mold cultures are of little value since their identification depends on the presence of specific morphological characteristics. Various preparations may be made that permit microscopic observation of growing molds. In general, they entail the inoculation of a suitable solid growth medium on a slide with the desired mold. A warm cover-slip is then placed over the inoculated medium and the slide preparation is placed in a sterile covered dish containing a small amount of sterile water to prevent drying out of the culture. The growing undisturbed mold culture can be examined at different time intervals by the high-power objective of the light microscope.

PREPARATION OF STAINED MATERIAL

The microbiologist's study of bacterial cells more frequently employs stained preparations for microscopic observation. Staining procedures, of course, kill the bacteria and often cause distortion of the cells. In general, it permits a better viewing of their cellular morphological characteristics and often reveals more than the mere structural arrangement of the cells.

In the preparation of a **bacterial smear,** a light suspension of the organism is prepared in water on a glass slide and then smeared to obtain a thin film. After it is allowed to air dry, it is subjected to gentle heat which causes the cells to adhere or "fix" to the slide so that they will not be washed off during staining procedures. The smear is then covered with the desired staining solutions.

Basic (alkaline) dyes are usually used to stain bacterial cells. Methylene blue, crystal violet, safranine, and basic fuchsin are such basic dyes. Acid dyes, such as acid fuchsin, are rarely used in microbiology. In preparation of a **simple stain,** an alcohol or water solution of a single dye is flooded over the fixed film and permitted to remain there for several minutes. The slide is then rinsed with water, blotted dry, and observed under the microscope. Since bacteria contain a higher proportion of RNA, DNA, and protein than cells of higher plants and animals, the entire bacterial cell takes up the dye, yielding more evenly stained cells than do higher cellular forms. Simple stains are primarily used to indicate the general shape, grouping, and size of bacterial cells, i.e., **cellular morphology.** Rarely do they reveal any internal structural details. Nuclear material, for example, takes up the stain so that it cannot be differentiated from deeply stained cytoplasm. Certain chemicals, called **mordants,** are frequently used to demonstrate specific structures since they cause dyes to become fixed to them. For example, tannic acid, used as a mordant, reveals the bacterial cell wall in conjunction with a simple stain. Bacterial flagella can be demonstrated by adding a mordant which precipitates around each flagellum. This procedure, in reality, causes an artifact sufficiently thick that it is visible under the light microscope when stained.

Colloidal materials such as Nigrosin or Congo red do not penetrate bacterial cells. Called **negative stains,** they dye the background on the slides, leaving the cells colorless in the midst of a dense precipitate. To demonstrate the presence of a capsule and its size, a second dye of a different color may be applied. Clear halos surrounding the stained cells are indicative of capsules, which are not stained by ordinary basic dyes. The possession of a capsule by an organism is routinely demonstrated by suspending cells into a drop of India ink. A cover-slip is placed on the mixture and pressed slightly to make a thin preparation. Since India ink molecules do not permeate the cells, the capsule would appear as a clear halo against a black background surrounding a slightly more dense cell. Some bacterial cytologists believe that negative stains give a more accurate picture of the size and structure of bacterial cells because the method does not require a fixing procedure which could distort the cells, nor does it necessarily involve initial penetration of the cells by the dye.

Some staining procedures distinguish between different types of bacteria. These **differential** or **diagnostic** stains are specialized to demonstrate certain features that serve definite differential significances. For example, the presence of lipid storage granules, endospores, or capsules may distinguish one bacterium from others that are similar or identical in all characteristics except the absence or presence of these structures. Such information is used in part in the identification of pure cultures, that is, cultures containing only one type of organism.

Lipid storage granules may be stained with certain lipid-specific dyes such as Sudan black B. Highly-resistant spores require drastic treatment for staining (Fig. 3.6). The smear, covered with the primary stain malachite green, must be steamed for a period of time to force the dye into these structures. Or a suitable chemical incorporated with the dye is used to cause impregnation of the resistant structures. Once the endospores have taken up

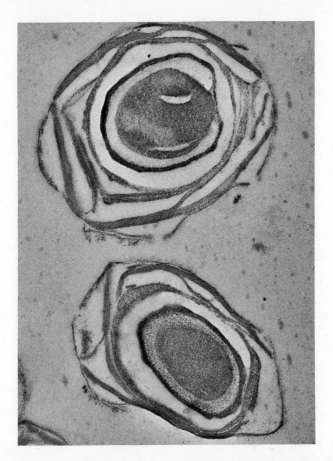

Fig. 3.6. The heavy coats protecting the spore from penetration by chemicals (such as dyes or disinfectants) are shown in this electron micrograph of an ultrathin section of a bacterial endospore.

the stain, they are equally difficult to destain, as they are unaffected by the gentle washing with water or even acid or alcohol which otherwise removes stain from the vegetative cells. After destaining the vegetative cells with one of these types of washings, a second dye or counterstain of a different color, for example, safranine, is applied. Since only the vegetative cells are stained, this permits easier differentiation.

In some bacteria, differential staining with a solution of alkaline methylene blue reveals metachromatic granules, storage granules of inorganic polyphosphate. Their presence serves as presumptive identification of a number of bacteria.

The tubercle bacilli differ from most other pathogenic microorganisms in that they contain large amounts of wax within the cell wall and are, therefore, almost as difficult to stain as the endospores. They require heat or a suitable chemical to serve as an intensifier during the staining procedure to force the primary stain into the cells. Once the cell is stained, however, washing with a dilute mineral acid-alcohol solution will not remove it. Since most other cells are destained by this procedure, the tubercle bacilli can be distinguished from the multitude of other bacterial cells present in sputum, in the throat, or in the tissues. This **acid-fast stain** is routinely used in the medical bacteriology laboratory in diagnosing tuberculosis and in following the progress of the patient in treatment.

A widely used differential staining procedure in bacteriology, the **gram stain,** is employed to divide the bacteria into two major groups; gram positive and gram negative. Four principle steps involved in the basic method are given in Table 3.1.

Many theories have been advanced to explain these characteristics. It has been suggested that the electrical charges on the surface of the bacterial cell are responsible for the affinity for crystal violet. The gram-positive forms are more negatively charged and hence are not as readily decolorized by alcohol. According to another theory, the gram-positive characteristic is directly related to the permeability of the cell membrane, or more likely, the cell wall. Certainly, the cell walls of gram-negative bacteria are thinner and have a unique chemical composition in that their cell walls contain about 20 percent fat. This is 10 times that found in the cell walls of gram-positive bacteria. Although this fatty layer resists penetration by water solutions of certain dyes, alcohol, a lipid solvent used in this procedure, can penetrate it readily and remove the crystal violet-iodine complex, thus decolorizing the cells.

Careful techniques and culturing conditions are required for the successful use of this staining method since false results will occur if the standard procedure is not followed. For example, only young cultures should be used since older gram-positive cells have a tendency to destain. It is advisable to use a sugarless culture medium to grow the cells since the culture may be able to convert sugar to acids. Acidic cultural conditions can cause gram-positive cells to lose their correct staining characteristic.

CULTURE MEDIA

Only a relatively small part of the everyday activity of the modern microbiologist involves examining organisms under the microscope. Many laboratory efforts involve the growing or culturing of microorganisms which require food materials for maintenance, growth, and reproduction. Food materials used for this purpose are called **culture media.** If bacteria are removed from

Table 3.1 The Gram Stain

Steps	Gram-Positive Bacteria	Gram-Negative Bacteria
1. Crystal violet (Buffered at slightly alkaline pH)	Cells stain violet.	Cells stain violet.
2. Iodine solution	Iodine reacts with the crystal violet and fixes the dye in the cell.	Iodine reacts with the crystal violet and fixes the dye in the cell.
3. Alcohol	The alcohol does not remove the crystal violet.	The lipid in cell wall is dissolved in the alcohol, permitting the removal of the crystal violet.
4. Safranine	The cells are already stained violet by the crystal violet; hence, the red safranine is not apparent.	The decolorized cells are stained red by the safranine.

their natural habitat and are to be cultivated in the laboratory, the necessary nutrients must be supplied for them. The culture medium must contain a utilizable source of the elements or food materials of which the cell is made. Not only are culture media used for the isolation and maintenance of pure cultures, but for studying various biochemical and physiological characteristics used for identification purposes.

Culture media must contain the suitable carbon and nitrogen sources, minerals, and, in some cases, "growth factors." A few bacteria can use carbon dioxide but the majority require a more complex organic compound (sugars, alcohols) as their carbon source. Some bacteria can utilize gaseous nitrogen; others can make use of ammonia as their nitrogen source. Most bacteria, however, require a more complex nitrogen source such as amino acids. The majority of bacteria obtain mineral elements from simple salts of sulfur, phosphorous, calcium, potassium, iron, magnesium, sodium, zinc, copper, and so forth.

Certain organisms are unable to make low molecular weight materials (amino acids, nucleotides, vitamins, etc.) that are used as building blocks in the manufacturing of more complex molecules, such as proteins and nucleic acids. In such cases, these simpler materials (**growth factors**) must be added in the preformed state to the culture medium before growth is possible. Bacteria can obtain hydrogen and oxygen from water.

The culture media used by Pasteur were prepared from meat broths. There are much easier methods used today. There are two types of culture media, **complex** and **chemically-defined.** Complex media supply an excess of all nutrients required for growth of the organism and involve mixtures of chemicals, usually natural materials or modifications of natural materials whose chemical make-up is not known. Several nutritionally-rich ingredients are often used to compound these media. **Peptones,** which are rich in amino acids, are frequently ingredients in complex media. They are obtained through the partial digestion of proteinaceous animal or plant material by protein-digesting enzymes. **Infusions** are aqueous extracts of ground meat prepared by infusing meat in water, then steaming to remove heat-coaguable proteins. Infusions have a high content of vitamins, minerals, nitrogen bases, and amino acids. Peptones and infusions are frequently used together in complex media.

Nutrients from beef, liver, or yeast are extracted by boiling and then concentrated to a paste or dried. Such **extracts** are used as a substitute for fresh infusions although not as satisfactory in nutritional content.

To cultivate many disease-causing bacteria, a small amount of **blood** is added to the medium for the purpose of supplying many growth factors and aids in detoxifying various chemical materials.

Chemically-defined media are of known composition, both qualitatively and quantitatively, which is not the case with complex media. Inorganic and/or organic constituents are used. These media are not used to any great extent in growing disease-producing organisms but are principally used in studying nutritional requirements and other metabolic activities of organisms.

In addition to supplying the organism adequate nutrients, other environmental factors must be supplied for growth such as presence of moisture, optimum temperature, and oxygen requirements. The medium cannot be too acid or alkaline. The **pH scale** measures the acidity or alkalinity of a medium by checking on a logarithmic scale the concentration of hydrogen ions (H_2O $\longleftrightarrow H^+ + OH^-$). This scale ranges from 0 to 14. Most bacteria grow best around the neutral pH 7.0 (the number of hydrogen ions equals the hydroxyl ions). Values below 7.0 (more H^+ than OH^-) indicate acidity, and those above 7.0 (more OH^- than H^+) indicate alkalinity. Electrometric instruments or pH indicators which change colors at different hydrogen-ion concentrations may be used to measure pH. If a culture medium is too acid or alkaline, adjustments can be made by the appropriate addition of a suitable amount of acid or alkali.

Liquid culture media, termed **broths,** were used by the earliest bacteriologists. When bacteria grow in broth, the once clear medium becomes cloudy or turbid. The degree of turbidity is a measure of the number of organisms present (Fig. 3.7). Some bacteria form thin films called **pellicles** on the surface of the broth; some form a sediment at the bottom of the culture tube; others are evenly dispersed throughout the broth.

For many purposes, a solid culture medium is needed instead of a liquid medium. **Agar** is the most widely used solidifying agent. It is an impure complex carbohydrate gum obtained from certain marine algae and will melt at 98 C and will remain liquid until cooled to around 40 C. The concentrations added to a liquid medium depends on its use: agar plates, 1.5%; semisolid media, 0.5%.

SPECIAL MEDIA

While standard bacteriological culture media are used principally for cultivation of microorganisms, certain organisms and special experiments demand

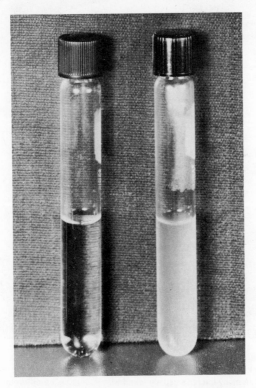

Fig. 3.7.   Left, a control tube of sterile broth. Right, a turbid bacterial culture.

the use of other types of culture media. Tomato juice and other vegetable juices provide an excellent liquid medium for the cultivation of organisms characteristically found in plants. **Differential media** are employed to differentiate between types of bacteria. For example, on an agar medium containing the chemical, bismuth sulfite, the typhoid fever-causing bacterium will grow as a black colony of cells and can be distinguished from a mixture of other organisms present not displaying this coloration (Fig. 3.8). **Selective media** may contain substances which will allow the growth of one type of organism that is desired for isolation, yet will inhibit unwanted organisms. The selective device may involve the addition of a dye, such as crystal violet, that inhibits most gram-positive bacteria but permits the growth of gram-negative bacteria. It may involve the addition of an acid to lower the pH to a

Fig. 3.8.   A streaked plate of Salmonella typhi on bismuth sulfite agar. The isolated colonies in the upper left can be picked as pure cultures. This is a selective medium because the typhoid bacterium grows better on it than many bacteria usually found with it; it is also a differential medium since the typhoid colonies on it show a distinctive shiny surface and a halo.

value unfavorable for the growth of skin bacteria but not for the ringworm mold.

Various chemicals may stimulate growth by lowering an unfavorable oxidation-reduction potential in the culture media. Others bind toxic heavy metal ions (Mg^{++}, Cu^{++}). Thioglycollate, capable of reducing oxygen, is added to media which are used to grow organisms that do not grow in the presence of oxygen.

STERILIZATION

Since bacteria and other microorganisms are found everywhere in nature where life is expected, the bacteriologist who prepares a culture medium will find that immediately hundreds of microorganisms from the air, glassware, and chemicals will begin to grow in this material. Any bacteriological experiment which he might attempt would be hopelessly confused by the presence of this mixture of unwanted organisms, called **contaminants.** The battle against contamination constitutes much of the effort in the microbiology laboratory. After a culture medium is made, it is placed in containers that will prevent the further entrance of other microorganisms from the air, dust, or other contaminating environment. The classical method of keeping contaminants out of glass and tubes is by plugging them with cotton. With moderately compressed cotton plugs, the passage of air and other gases in and out of the container is not impeded, but microorganisms are effectively prevented from either entering or leaving. Modern-day plastics are now frequently used as enclosures for flasks and tubes.

Next, the contaminants already present in the medium must be destroyed. The procedure used to free the culture medium and equipment from all living reproductive cells is referred to as **sterilization.** A number of different methods are routinely employed in the bacteriological laboratory for this purpose. For empty glassware, such as test tubes, pipettes, and flasks, **dry heat,** such as that produced by an oven is ordinarily applied. When 160 C heat is applied for two hours, the organisms found in the culture equipment are completely destroyed; the material is **sterile.** Enclosed tubes and flasks, petri plates and pipettes that are wrapped in paper or contained in another suitable container may be heat-sterilized in the hot air oven since this heating will not cause charring of the paper or cotton.

Heat-stable culture media cannot be sterilized in the oven at 160 C because they would evaporate. Such materials are sterilized utilizing **moist heat.** They are placed for 20 minutes in a large enclosed container, similar to a large pressure cooker and referred to as an **autoclave,** live steam at fifteen pounds of pressure per square inch is applied, and contaminants are destroyed (Fig. 3.9). The pressure has no direct effect upon the sterilization

Fig. 3.9. The autoclave is used for routine sterilization in the microbiology laboratory. Photo courtesy American Sterilizer Co.

Microorganisms and Man

process, but at fifteen pounds of pressure, the temperature of the steam can reach 121 C. Since this produces sterility quite as effectively as the dry-heat oven treatment, it is evident that moist heat is more efficient for destroying microorganisms. Not only does a lower temperature suffice, but also a shorter time is required. When steam contacts cool objects, it liberates large amounts of latent heat, which promptly raises the temperature of the articles to that of the steam. Therefore, moisture, together with the rapidly elevated temperature, results in the effective sterilization in the autoclave. Moist heat is destructive through its ability to coagulate protein of the contaminating microbes; dry heat is also destructive through its ability to oxidize organic components of the living cells.

Equipment and materials that are not damaged by direct heat may be sterilized by a flame, which destroys microorganisms by **incineration.** This procedure is used to sterilize the inoculating needle, which is a piece of nichrome or platinum wire attached to a handle and used to transfer microorganisms from one test tube or petri plate to another. Some materials are sterilized by the use of chemicals. For example, the desk tops on which microbiologists perform their procedures are ordinarily washed down with a disinfecting chemical, such as phenol or cresol. Glass rods that are used to spread microorganisms over solid medium surfaces are first dipped in alcohol, which is then burned off before use. An efficient cold sterilizing procedure is to hold materials in a closed container of ethylene oxide gas for several hours.

Solutions of chemicals that are heat sensitive can be rendered sterile by **filtration.** A culture medium containing heat-labile substances may be sterilized by passing it through a membrane filter having pores of such size that all bacteria are filtered out. Heat-sensitive liquids such as blood serum, vitamins, and amino acids may be sterilized in this manner.

GROWTH ON LIVING SYSTEMS

Some microorganisms will not grow on nonliving artificial culture media and are, therefore, termed **obligate parasites.** The viruses that cause disease in plants are cultured in the greenhouse on growing plants protected from contamination by plastic or glass hoods. As the host plants age and die, the greenhouse cultures of viruses are transferred to healthy young plants, and the stock culture is maintained. Similarly, microorganisms are grown in experimental laboratory animals such as monkeys, guinea pigs, rabbits, mice, and rats.

The most convenient laboratory animal is the developing **chick embryo** since it is enclosed in its own covering, the egg shell, which keeps out contaminating organisms. When a fertile egg has been incubated for a few days, one can observe the development of the baby chick. This embryo serves admirably for the growth of many microorganisms that are not easily grown in any other type of laboratory media.

Tissue culture involves the growth of animal cells in test tubes, plates, and flasks by application of the procedures worked out earlier for bacteria (Fig. 3.10). Although very complex nutrients must be compounded for these media, it is now no great problem to grow skin tissue, tissues from certain types of cancers, and kidney or heart tissues in sterilized containers with the added proper nutritional environment. Microorganisms requiring living cells for growth may be grown and studied under these conditions without the complications of using a whole living animal.

Many kinds of culture media are required, although microorganisms are the most flexible of all the living forms in their ability to thrive under diverse conditions. Man and other higher animals live under moderate temperatures and pressures, in contact with relatively inert gases and chemicals. Microorganisms are not so restricted. Many species, it is true, favor the conditions under which higher plants and animals thrive. Other species, however, may prefer temperatures much higher or lower than those which we regard as being compatible with life. For example, some will thrive in strong acid or under high pressure. The microbiologist utilizes knowledge of these factors

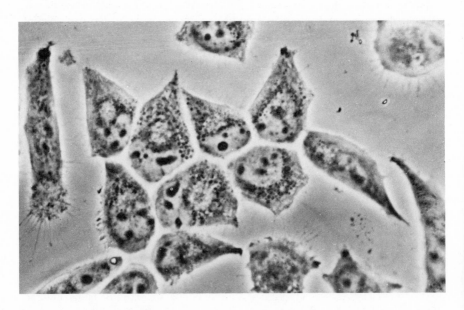

Fig. 3.10. A monolayer of mouse fibroblastic cells, strain L929, growing in a complex medium and cultured in a petri plate. Photo courtesy Drug-Plastic Research and Toxicology, The University of Texas at Austin.

in preparing culture media and in regulating the growth and death of micro-organisms.

GROWTH OF BACTERIA

When microorganisms are inoculated into a sterile medium and placed in suitable environmental conditions, growth occurs. The term **growth** as applied to bacteria almost always connotes increase in numbers, that is, multiplication. The individual organism grows both before and after cell division occurs, but the growth of the culture involves the production of more cells and is most often measured by determining the number of cells produced. However, techniques are available for measuring cell mass, and the growth of multicellular microorganisms, such as molds, is often followed in that manner. By measuring the inoculum and then by determining the amount or number of microbes at different time intervals, the growth rate may be estimated or determined. Procedures are available for measuring either the total amount of protoplasm produced or for measuring the amount of viable growth, that is, living protoplasm. The methods are summarized in Table 3.2.

Bacterial cells may be counted directly under the microscope in much the same manner that blood cells are counted in the hospital laboratory. A **counting chamber,** such as the Petroff-Hauser bacteria counter, is a thick glass slide with a small depression carefully machined to 0.02 mm deep (Fig. 3.11). On the bottom of the chamber are etched lines which divide it into squares. When the liquid containing the bacterial cells is placed in the depression and covered with a cover-slip, the slide may be examined under the microscope, and the bacteria in each of a number of the squares may be counted. Since there is a definite volume of liquid above each square, the number of bacteria per milliliter can be computed from the average count per square. The **direct microscopic count** cannot be used with turbid materials because particles of solids, such as soil, make counting impossible. Thus, this technique is restricted to determining the number of organisms in a broth culture or clear liquids; however, because living and dead cells appear the same, it does not distinguish between them. In certain instances stains applied to a culture may indicate whether cells are living or dead. For example, dead yeast cells stain deeply when suspended in a solution of methylene blue while living cells are colorless. By such staining, the quality of a yeast preparation may be determined.

Sometimes a rough estimate of the total number of bacteria is all that is required. By mixing a drop of the bacterial suspension with an equal volume of human blood, smearing it on a glass slide, and using proper staining, a proportional count can be made. A comparison of the number of bacteria and the number of red blood corpuscles is made under the microscope. If

Table 3.2. Summary of Methods for Determining Bacterial Growth

Total Numbers (Multiplication)	Total Mass (Growth)
Total Quantity of Growth	
1. Counting chamber	1. Turbidity
2. Stained slides	2. Total cell volume
3. Proportional count	3. Weight
4. Electronic particle counter	4. Colony diameter
Viable Quantity of Growth	
1. Plate count	1. Oxygen uptake
2. Dilution count	2. Nutrients (sugars) consumed
3. Membrane filters	3. Metabolic products (acid, CO_2, or NH_3)

bacteria and cells are present in equal numbers, the original material contains about five billion bacteria per milliliter, since that is approximately the number of red corpuscles in normal human blood. Similarly, a proportional count of viruses can be made by taking electron microscope pictures of a mixture of the viruses with a finely divided colloid in which the number of particles is known.

Bacteria are regarded as living only if they can multiply; that is, produce an infection in an animal, a plant, or a tissue culture; make a tube of broth become turbid; or produce a colony of cells on a solid plating medium.

The number of living cells is most often determined by a cultural method such as the **plate count** or **colony count.** A measured amount of the material containing the microorganisms to be counted is mixed with melted, cooled nutrient agar. After the microorganisms are thoroughly mixed with the nutrient agar, the mixture is poured into a culture dish or **petri plate.** Upon solidification, each cell or clump of cells is trapped in a definite place within the gel. Since nutrient agar contains many of the materials necessary for microbial growth, the organisms, whose nutritional and environmental requirements are satisfied, will begin to multiply when the plate is incubated at the proper temperature. After 24 to 48 hours, wherever an organism had been immobilized in the gel, there will be a pile of cells or **colony,** which can be readily seen by the naked eye. A count of the number of colonies reveals how many organisms there were in the original material. However, if too many organisms are planted on a plate, there will be neither space nor food material for each to form a visible discernable colony. In that case, the whole plate is covered with tiny microcolonies, which run together in a solid mass of confluent growth. For good plate counts, it is desirable to have less than 300 colonies per plate. Materials containing more than 300 organisms per millimeter must, therefore, be diluted in sterile, suitable diluent so that a countable number can be plated. This procedure is called a **dilution plate**

count. It is used to determine the number of viable cells in a culture or in other bacteria-containing materials. Less than 30 colonies per plate is undesirable since with such small numbers the chance variations become large.

Although each colony is often regarded as the progeny of a single cell, actually organisms stuck together in a clump will give rise to only one colony. Plate counts on milk usually show about one-quarter as many organisms as are counted by the direct microscopic method due to clumps and dead cells. Only a relatively small fraction of the bacteria of soil may appear on the plate count since it is impossible to have the nutritional needs and the environmental conditions for cultivation favorable to all organisms that are present. Therefore, the total plate count procedure yields relative, not absolute counts of the organisms present.

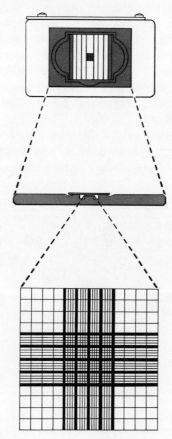

Fig. 3.11. A Petroff-Hauser counting chamber for making direct microscopic counts. The top representation is the front view, showing area which has etched rulings; in the center is a side view, and at the bottom is a diagram of the ruled area.

Counts of living bacterial cells can be obtained by other means such as the **dilution method** (Fig. 3.12). Tubes of a suitable nutrient broth are planted with a series of dilutions of the material containing the bacteria. After incubation, the highest dilution showing growth gives a basis for estimating the number of organisms originally present. For example, if growth occurs in a tube inoculated with a milliliter of sewage that was diluted 1:1000, but not in a tube inoculated with the dilution 1:10,000, it is evident that at least 1000 organisms but not as many as 10,000 organisms were present per milliliter of the original sewage. By using several tubes at each dilution, it is possible to obtain a reasonably accurate estimate of the bacterial population. This method can sometimes be used for determining the numbers of a particular kind of organism when it occurs together with a larger number of other bacteria. In water bacteriology, a count of organisms capable of producing gas from the sugar, lactose, can be made in a sample containing many other bacteria. A number of dilutions of water are inoculated into broth containing lactose. After incubation only those tubes producing gas are recorded. For example, growth may occur in the 1:1,000,000 dilution, but if only the 1:10 dilutions show gas, the conclusion is that, although there are many organisms present, only about 10 organisms per milliliter of the gas producers are to be found in the sample.

When culturing bacteria from the air, fairly pure water, and various other liquids, membrane filters can be used to trap bacteria, molds, and other microorganisms on the filter surface. The membranes, which are thin sheets

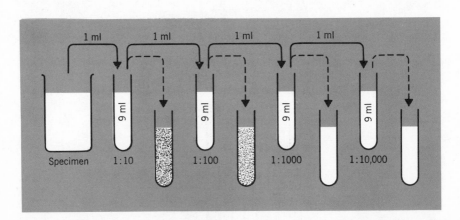

Fig. 3.12. Estimation of bacterial population by serial dilutions. A 1-ml portion is transferred to the succeeding dilution tubes and another 1-ml portion is inoculated into a tube of broth. In this example, the broth tubes inoculated with the 1:10 and the 1:100 dilutions became turbid, while those inoculated with the 1:1000 and the 1:10,000 dilutions remained sterile. Therefore, the specimen has more than 100 bacteria per ml but less than 1000. Greater precision can be obtained by using multiple tubes.

Microorganisms and Man

made of cellulose acetate or similar materials, have uniform pores with diameters sufficiently small to sieve these microorganisms from air or liquids. Measured quantities of gas or liquid are usually drawn by suction through the filters. The membrane is then placed on the surface of an absorbent pad soaked with a suitable nutrient, differential, or selective medium. The medium soaks through the membrane and nourishes the cells, which develop into colonies so that they may be counted. The membrane filter also has numerous applications in applied microbiology and in research.

Plate or **plaque** counts of viruses can be made by mixing or by spreading properly diluted virus preparations over a layer of living, susceptible animal, plant, or bacterial cells. During incubation, each virus particle will infect a cell, multiply, lyse the cell, and release new infectious virus particles that subsequently infect and lyse neighboring cells. At these areas, a circular clear zone of lysed cells occurs. This is called a **plaque** and represents one original infectious virus unit or plaque-forming unit. The plaques may be counted and multiplied by the dilution factor to obtain the number of infectious units in the original, undiluted suspension of viruses.

There are a number of means by which the total cell mass can be measured. These are discussed below.

Turbidity. Growth in broth cultures is determined by the turbidity resulting from the presence of bacterial cells. With most bacterial cells, the broth becomes slightly clouded when it contains about 10 million cells per milliliter. In a nutritious broth, an average bacterial culture attains a maximum population of the order of several hundred million to one billion cells per milliliter. When a single cell grows into a fully developed culture, more than 20 cell divisions occur before visible turbidity results. Turbidity is a useful measure of the suitability of a culture medium. For accurate measurements, photoelectric devices are used to determine the extent of the turbidity by measuring the interference offered to the passage of a beam of light.

Volume. Centrifuge tubes are made with a graduated capillary in the bottom so that when 10 ml of the microbial cultural fluid is sedimented by centrifugation, the volume of the cell mass is evident. This test is valuable in the yeast industry where interest centers on the amount of marketable cell material.

Weight. Centrifuged cells can be dried and weighed. Weight is used as a measure of mold growth, for most molds grow as a pellicle on the surface of a liquid medium, and the pellicle can be fished out, rolled into a ball, and subsequently weighed.

Colony diameter. When a microorganism is inoculated on an agar plate, its rate of growth can be followed by the increase in colony size or by the rate of progression of the edge of the colony across the agar surface. This procedure is useful especially in measuring mold growth. Depending on the mold and cultural condition, the growth may vary from several millimeters per hour to barely measurable progress per day.

The activity of the protoplasm does not always parallel its mass. With older cultures much dead or relatively nonreactive material may be present. For example, molds generally contain more supportive and structural cells and less metabolically active cells than bacteria. The most common measurements involve the uptake of oxygen, the utilization of food, and the accumulation of metabolic products. These are discussed below.

Oxygen uptake. Special instruments employing flasks attached to manometers are used extensively for determining oxygen uptake by microorganisms. The Warburg respirometer is an example. The growth of the organisms (i.e., the increase in metabolically active cells) can be followed by measuring the rate of increase of oxygen utilization.

Disappearance of food. In fermentation industries, the microbiologist is often concerned with the conversion of sugar into some useful product. Under such circumstances, the disappearance of the sugar measures the progress of the microbial population development; when the sugar is all gone, the fermentation is complete.

Accumulation of metabolic products. Measurement of the metabolic products of bacteria is often simpler and of more immediate interest than numbers or amounts of the microbes. For assaying vitamins, the amount of acid produced in a given time is a measure of the amount of active organisms, and this in turn, is directly proportional to the amount of essential vitamin originally present in the culture medium. Acid is easy to measure by titration with alkali using an indicator dye to determine the end point. The gas, carbon dioxide, is produced by many organisms, and since the amount produced is dependent upon the amount of active cell material, measurement of carbon dioxide serves as a convenient estimate of active cell numbers or active cell mass. Under special conditions, other metabolic products may be measured to determine the rate of development for a microbial population.

ISOLATION OF PURE CULTURES

Well-isolated bacterial colonies on the surface of an agar medium serve as a starting point for the isolation of a pure culture. The entire cell population in the colony is presumed to have descended from a single cell. Since it is easier to pick surface colonies, plates for the isolation of pure cultures are most commonly prepared by streaking the surface of an agar medium with an inoculating loop needle in such a manner to obtain a "physical" dilution of the cells so as to obtain isolated colonies (Fig. 3.13). To insure the absence of latent contaminants, however, a small amount of growth is usually restreaked on another plate of agar medium at least one additional time before picking it for the isolation of a pure culture. A small amount of the colony is transferred aseptically to a sterile tube of a suitable nutrient medium and serves as the inoculum for the new culture.

Summary

Much of the microbiologist's work involves microscopy and the culturing of microorganisms on various types of culture media for study and observa-

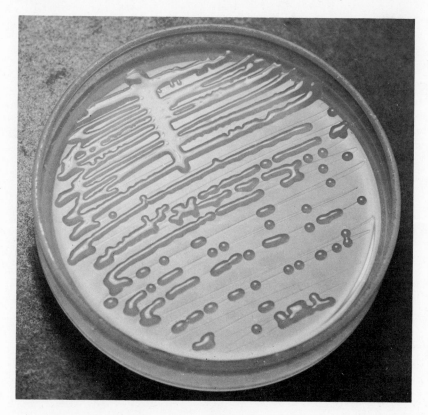

Fig. 3.13.   **A streak plate showing isolated colonies of** Azotobacter vinelandii.

tion. However, he uses methods and equipment which are commonplace in other scientific disciplines. Like the biologist he grows microorganisms in animals; draws blood; studies effects on biological behavior; and cultures microorganisms in germ-free animals. Like the chemist he uses analyses for the identification of chemical compounds; column and paper chromatography; and the radioactive labeling of chemical compounds. Like the biophysicist he utilizes ultracentrifugation for the separation and purification of cellular structures and even macromolecules.

FOUR

Bacterial Anatomy

Although they were once believed to be a "bag of enzymes," it is now known that the eubacteria or true bacteria possess a highly organized cell very different in nature than those of the higher protists and plants and animals. Bacteria are **unicellular** even though two or more cells are frequently connected. The important unicellular characteristic is the lack of active association among these cells. Each cell is capable of independent life, and each carries out its life processes, growth, multiplication, and death without depending on the others which may be attached to it in chains or clumps. A single cell detached from a multicellular mold and placed in proper conditions for growth will yield a new mold, but, in many of the molds, the cell groups show a definite specialization of function. Some serve for attachment to the food material while others secure moisture and oxygen from the air or bear the fruiting bodies. Such specialization is not characteristic of bacteria.

The most common shape of the bacterial cell is that of a cylinder or rod, and organisms of that shape are called **bacilli** (singular = **bacillus**). In some bacillus forms, the length of the cell is only slightly greater than its diameter; in others, the cell may be up to ten times as long as it is thick. The ends of the cell may be square-cut or rounded. Some bacilli tend to remain stuck together after division and thus produce long chains of cells, while others break apart rather easily and appear as individual cells. Some cells tend to bulge in the middle, while others may bulge on one end, appearing definitely club shaped. Most bacilli are slightly curved rather than being perfectly straight.

A spherically-shaped cell is referred to as a **coccus** (plural = **cocci**). Certain cocci appear as perfect little balls, while others have shapes like beans, arrowheads, or even cones. When the length and the width of the cell are about equal, and there is no evidence of a definite cylindrical shape, the cells are "true" cocci. The arrangement or grouping of the cocci is distinctive. For example, the bacteria causing the most common type of pneumonia frequently are observed as two attached cocci. Such a pair is referred to as **diplococcus.** When cocci divide in one plane and all progeny continue to divide in a parallel plane, chains are formed; this grouping is referred to as **streptococcus.** Sometimes the organisms divide in three planes and adhere together in packets making a cube of eight cells; this is termed a **sarcina.** If cocci divide in different planes so that the spherical cells adhere together like a bunch of grapes, the arrangement is called **staphylococcus.** When cells separate after division and each is found singly, the cell is called a **micrococcus.**

A third cell shape that exists with the true bacteria is the helical or spiral-shaped organism. Such cells are referred to as **spirilla** (singular = **spirillum**) (Fig. 4.1). If there is only one curve and it appears comma shaped, the term **vibrio** is used. The spirillum differs from the spirochete in that spirilla have

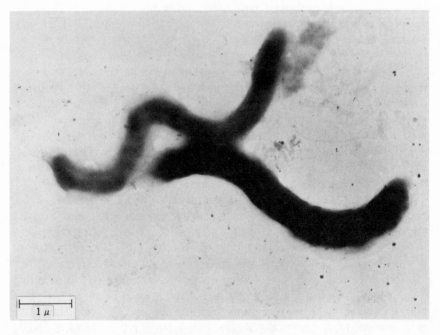

$\vdash\!\!\!-\!\!\!-\!\!\!\dashv$ $1\,\mu$

Fig. 4.1. Electron micrograph of Spirillum rubrum. **Two cells are shown. One is shaped like the letter "S"; the other is shaped like the figure "3."** Photo courtesy K. Polevitzky and R. Picard; A.S.M. LS-117.

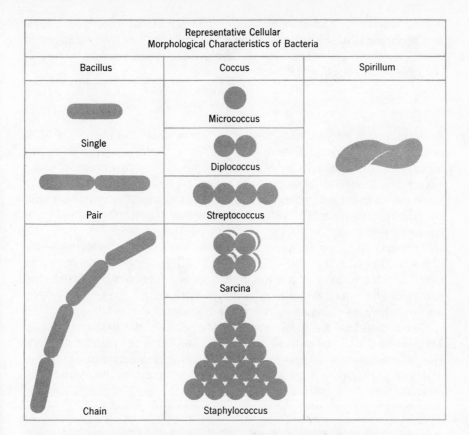

Representative Cellular Morphological Characteristics of Bacteria		
Bacillus	Coccus	Spirillum
Single	Micrococcus	
	Diplococcus	
Pair	Streptococcus	
	Sarcina	
Chain	Staphylococcus	

Fig. 4.2.   Representative cellular morphological characteristics of bacteria.

rigid cell walls and move by flagella; the spirochetes are flexible and move by axial filaments. The **cellular morphologic characteristics,** that is, size, shape, and grouping are important aids in the identification of bacteria (Fig. 4.2).

The general cellular structure of bacteria is to some degree like that of higher plants and animals. The **protoplast** consists of organized protoplasm enclosed by a cell membrane. A cell wall surrounds the protoplast. Bacteria, along with the blue-green algae and rickettsiae, are characterized by the more primitive, **procaryotic** cell type. The bacterial body is small in comparison to other cells but its structure is well organized, as revealed by the electron microscope. Various structures that may be found in the true bacteria will be discussed with respect to their physical and chemical natures, and their function(s). Some of them are nonessential, since not all bacteria possess them and their removal does not affect the cell's viability.

The bacterial anatomists group the various structures into three categories: **appendages** (flagella, pili); **cell envelope** (capsule, cell wall, cytoplasmic membrane); and cytoplasm with the **cytoplasmic inclusions** (nuclear material, ribosomes, etc.) (Fig. 4.3).

FLAGELLA

Essentially all of the spirilla and about half of the bacilli are capable of independent motion. Any particle may be carried by water currents or be imparted vibratory motion as a result of being bombarded by molecules of the suspending fluid. But these bacteria are able to move independently in a liquid environment due to **flagella,** organelles of motion. The motile bacteria swim actively by means of a flagellum or several flagella. Only a few cocci possess flagella.

The flagella are long, thread-like structures, often several times the length of the cell to which they are attached (Fig. 4.4). From their origin within the cytoplasm, the flagella pass through the cell envelope and project into the surrounding environment. Although they may be longer than the cell, they are ordinarily less than one-twentieth the diameter of the cell (less than 0.05 μm) and, therefore, are below the resolving power of the light microscope. Their presence may be detected microscopically if the cells are treated by a special procedure which precipitates material around each flagellum. The true size and shape of the flagellum is demonstrated with the electron microscope.

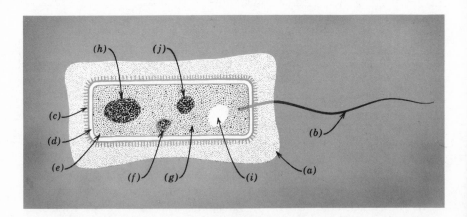

Fig. 4.3. This artist's concept of the bacterial cell reveals the parts to be observed but shows the nucleoid as a well-circumscribed body rather than as diffuse lobes shown by electron microscopy. (a) Capsule. (b) Flagellum. (c) Pili. (d) Cell wall. (e) Cell membrane. (f) Mesosome. (g) Ribosomes. (h) Nucleoid. (i) Endospore. (j) Storage granule.

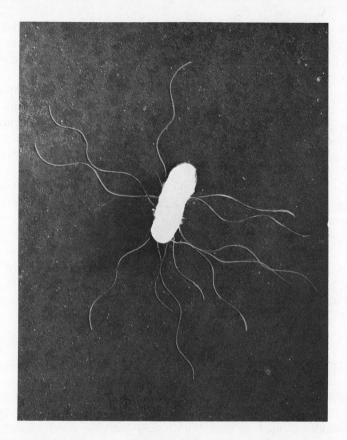

Fig. 4.4. An electron micrograph of a bacterial cell showing peritrichous flagella. Since this cell has not been cut into ultrathin sections, no internal detail is evident.

Motility of an organism may also be observed in a liquid medium under the microscope. Whereas nonmotile forms vibrate back and forth in almost the same area (**Brownian movement**) or move about in currents of water, the motile cells have definite motion in relation to each other and can be observed moving about as if proceeding to some destination. A semisolid medium is frequently used as a cultural means for determining motility since motile cells will move away from the line of inoculation.

Chemically, the flagellum is composed of protein, sometimes called **flagellin,** which is similar to protein found in muscle. With high magnification, a bundle of three or more parallel or intertwined longitudinal fibers make up the flagellum. They may be removed by vigorous shaking with no effect on the cell's viability. Under proper growth conditions, the flagella on such artificially deflagellated cells grow back.

The flagella seem to act by contraction and rotation, moving about in the rear of the bacterium like a propeller. Not only do they drive the bacterium forward, but they also act as rudders, governing the direction of movement.

The number and arrangement of flagella on a cell is of taxonomic significance. An organism with a single flagellum is referred to as **monotrichous,** while a cell with a tuft of flagella on one end is called **lophotrichous.** If the flagella are scattered over the entire surface of the cell, it is a **peritrichous** arrangement.

Many bacteriologists believe that the most primitive bacteria were probably the spiral-shaped water bacteria, which were monotrichous. Evolving from these were rod-shaped bacteria with a single flagellum, followed by spiral- and rod-shaped bacteria with tufts of flagella. These organelles permitted swimming even as the organisms moved to less watery environments. The motile bacteria found in such places as the intestinal tract of man usually are peritrichous. Such an adaptation permits them to move around in the films of water found on dryer particles. Other bacteria which have moved out of the water to a terrestrial existence, such as organisms found on the surface of plants, have abandoned motility entirely and exhibit no flagella.

PILI

Some bacterial cells possess many small hairlike structures on their surfaces that are not related to motility (Fig. 4.5). A cell may have about 150 of them, giving the appearance of a fringe. They were called fimbriae, but now the word **pili** (hairs) is used more commonly. A single **pilus** is an extremely fine appendage, much smaller than a flagellum. Pili differ in length and diameter, and a cell may possess more than one type. These unique appendages, which originate in the cell envelope and are composed of protein, are found only on some gram-negative species.

Their function(s) is not truly understood. Some bacteriologists believe that they function in adhesion to surfaces and are involved in the absorption of nutrients. One type, the **F-pilus,** is known to be the conjugation tube used in the transfer of genetic material like an **episome,** an extra piece of genetic material that a cell may possess. The F-pilus is also the site of attachment for certain bacterial viruses.

THE CAPSULE

Many bacteria and most blue-green algae possess a layer of material that lies external to the cell wall and is referred to as the **capsule.** It can differ in thickness, density, and adherence to the cell. If it is very thin and difficult to detect microscopically, it is a **microcapsule.** When the layer is thick enough

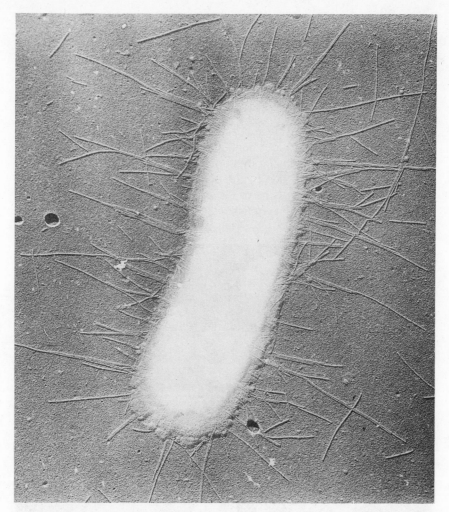

Fig. 4.5. Electron micrograph of Escherichia coli **demonstrating pili.** Photo courtesy C. Brinton: Biochemistry of Bacterial Growth, J. Mandelstam and K. McQuillen (eds). Blackwell Scientific Publications, Great Britain, 1968.

to be detected, it is called a capsule. True capsules show some structure when observed by sensitive techniques under the electron microscope. Certain bacteria produce very large quantities of **slime** which is chemically similar to capsular material but which sloughs off the cell. Capsular material and slime are produced in greater amounts when the bacteria are grown in the presence of abundance of food materials.

Generally speaking, the capsule is made of a gummy complex carbohydrate material or polysaccharide of which there are many different types synthesized by bacteria. In certain gram-positive bacilli, the capsule material is a proteinaceous material. In a few cases, the capsule may be a mixture of the two types of chemical compounds. Colonies of encapsulated bacteria are more mucoid and moist in comparison to colonies of nonencapsulated cells.

Since the capsule does not stain with the more common dyes used in the laboratory, its detection is now easily demonstrated by making an India ink preparation. The clear capsule appears as a halo surrounding the denser cell as seen against the black background afforded by the ink particles (Fig. 4.6).

The possession of a capsule by certain disease-producing bacteria protects the cell from destruction by the phagocytic white blood cells and certain other defense mechanisms of the animal body. In soil organisms, the capsule seems to protect the organism against drying. The capsular material

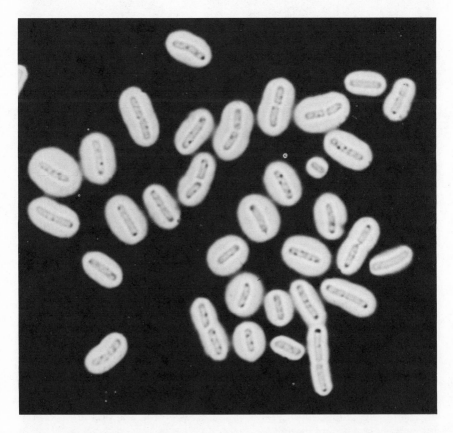

Fig. 4.6. India ink preparation of Bacillus megaterium **viewed with a phase-contrast microscope. The clear capsule appears as a halo surrounding the denser cell.** Photo courtesy C. F. Robinow.

in the soil is of considerable importance in breaking up the soil into small aggregates and thus contributes to what is called the tilth of the soil.

From capsules of a certain bacterium, a blood thickener or a blood plasma substitute has been prepared to be administered to shock victims who are losing their blood liquids in the tissue. If whole blood or blood plasma is not available, solutions of the capsular polysaccharide (dextrans) are more useful than physiological saline or glucose transfusions since they, like natural blood, are more readily retained by the capillaries. In certain industries the development of a large number of slime-producing bacteria is a nuisance and causes economic loss. In the sugar industry, they may multiply rapidly and clog pipes, and in paper mills, encapsulated bacteria stick to the rollers and cause uneven deposition of the pulp.

CELL WALL

The bacterial **cell wall,** though only about 0.01 μm, is extremely strong in proportion to its size. It is a well-defined structure and in some cases may constitute as much as 35% of the dry weight of a cell. This rigid wall, which gives the cell its shape, can be observed in the thin-sectioned cell but does not show up well under the light microscope unless given special chemical treatment. When dried or placed in strong salt solution, the internal contents of the cell may collapse away from the cell wall, but the bacterium will retain its original appearance because the cell wall is not readily distorted. The wall enables the bacterium to retain its shape and integrity even when exposed to great variations in the liquid in which it lives.

Although the cell wall is an essential structure for survival by most bacteria, definite physical and chemical differences are known to exist between gram-positive and gram-negative cell walls. The gram-positive cell wall is a homogeneous layer and is somewhat thicker than the gram-negative bacterial cell wall, which is composed of several layers.

A unique sugar-containing protein, referred to as **peptidoglycan** or **murein,** is found only in the cell walls of bacteria, rickettsiae, and blue-green algae and is responsible for the rigid nature of the structure. The backbone of this macromolecule is composed of the amino sugars, glucosamine and muramic acid, occurring in alternating molecules, and linked together in a chain by specific chemical bonds. Branching off the muramic acid molecules are numerous side chains, each composed of specific amino acids. The number of amino acids in a side chain is limited, usually three or four, and the types of amino acids contained within depend on the bacterial species. An unusual amino acid, **diaminopimelic acid,** is frequently one of these amino acids but so far has been found only in procaryotic cell wall material. Further crosslinkage may be formed by chemical bonding between these side chains resulting in stabilizing the murein as well as helping to establish a two-to-three dimensional meshwork.

Cell walls of gram-positive bacteria possess larger quantities of murein than of the gram-negative bacteria. In addition to murein, the gram-positive cell wall may contain polysaccharide and teichoic acids (chains of sugars connected by phosphoric acid). Little or no lipid is found. In contrast, the gram-negative cell wall is more complex chemically. Lipid is plentiful and is often complexed in the form of lipopolysaccharides, present in large quantities in the outer layer, and some lipoprotein. Murein makes up the inner layer. Other types of organic compounds are also frequently present but in smaller amounts.

An enzyme called **lysozyme,** found in tears, egg white, and other natural sources, will break the chemical bonds between the amino sugars of murein and cause a general breaking of the rigid cell wall material making the cell osmotically sensitive. If not in a solution with osmotic properties approximating that of the cytoplasm, the cell will burst. In a 10% sugar solution, the cell will not burst, but when the cell wall is ruptured by lysozyme, will assume a spherical shape. Spherical lysozyme-treated gram-positive cells are called **gymnoplasts** (protoplasts) while osmotically-sensitive gram-negative cells are referred to as **spheroplasts** since the rounded-up cells still possess much cell wall material. Penicillin is an effective antibiotic used to treat many bacterial infections or diseases. Its effectiveness resides in the fact that its presence prevents proper formation of murein and, therefore, growing cells are made osmotically-sensitive and thus vulnerable to lysis in the body's tissue.

There exist in nature an unusual group of bacteria called *Mycoplasma* that are unable to synthesize cell wall materials and are, therefore, naturally occurring protoplasts. For cultivation in the laboratory, 20% serum is used in the medium to render a suitable osmotic pressure necessary for their survival. These unusual organisms are further discussed in a later chapter.

Other properties or characteristics of cell walls will be discussed in later chapters. For example, the cell wall is the attachment site for most bacterial viruses prior to infection.

THE CELL MEMBRANE

Against the internal side of the cell wall is the **cell** or **cytoplasmic membrane** (Fig. 4.7). It is delicate and thin (about 0.005 μm). Often the membrane constitutes about 10% of the dry weight of the cell. The major chemical composition is phospholipid and protein. As in higher organisms, the membrane appears to be a triple-layered profile; this has been explained as a phospholipid layer sandwiched between two protein layers. Recent studies indicate

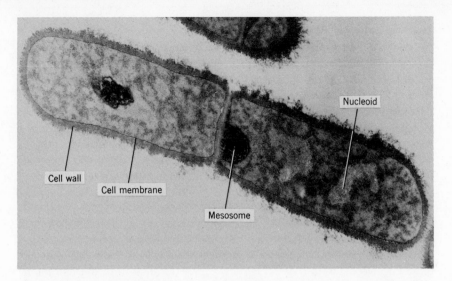

Fig. 4.7. Ultrathin section of Bacillus megaterium in which the cell wall and membrane are particularly defined. Photo courtesy L. Pope.

the presence of nucleic acids, especially in the areas of the membrane where rapid growth takes place. The concentration of nucleic acids in the growth areas supports the hypothesis that they serve as templates (forms or patterns) for making proteins.

As a living membrane, it is differentially permeable, allowing some molecules to pass but not others. Many smaller molecules are able to pass through when their concentration is lower within the sample. This simple diffusion is sometimes called **passive transport** since no expenditure of energy is required. Microbes must, by their nature, live immersed in their nutrient material; the cell must have the ability to control the concentration of substances which enter the cell, whether these be toxic substances or those needed for structural materials or energy production.

Active transport is the movement of molecules into the cell by means of the energy-requiring enzymes or **permeases** which are localized in the cell membrane. These unite with substances outside the cell, carry these through the membrane, and discharge them into the cytoplasm. Such transport is necessary for the entrance of molecules of larger size and for molecules whose concentration outside the cell is lower than that within the cell. The membrane is also involved actively or passively with outward transport of waste materials and other substances excreted by the cell. Since mitochondria are absent in bacteria the respiratory enzymes are associated with the bacterial membranes.

Inside the cytoplasmic membrane is the **cytoplasm,** a protoplasmic material quite similar to that found in animal and plant cells. It consists largely of a water suspension of enzyme proteins and ribonucleic acid. Also it contains a metabolic pool of ions and other nutrients to be used in making larger, complex compounds to be used in various structures and energy production. Various wastes reside here until expelled. Within the cytoplasm, there are various structures that function in various activities of maintenance, growth, and division of the cell. These cytoplasmic inclusions are discussed next. Not all are essential structures, and therefore are not found in all bacteria.

THE MESOSOMES

Large, irregular, convoluted invaginations of the cell membrane extend into the cytoplasm of many bacteria; especially they have been observed in gram-positive bacteria (Fig. 4.7). They appear to arise from the cell membrane, and are presumed to have the same chemical nature. Various functions have been associated with these structures. For example, their presence would increase the total membrane surface of the cell. It is also believed that they contain the enzymes required for cell wall synthesis, especially the septum, since the mesosome is frequently situated near its formation. In addition, enzymes involved in energy production have been detected in these membranous structures. Strong evidence suggests that they function in aiding the segregation of replicated nuclear material prior to cell division.

THE RIBOSOMES

Much of the ribonucleic acid is organized, in association with protein, into granular structures, the ribosomes, which are distributed evenly throughout the cytoplasm. The bacterial ribosomes are smaller in size than those found in higher organisms, yet these essential structures are able to synthesize the necessary proteins of the cell.

STORAGE INCLUSIONS

Certain bacteria are able to manufacture certain **storage inclusions** or **granules** to be used as nutrients in the time of need. For example, *Corynebacterium diphtheriae,* the bacterium that causes the disease diphtheria, is able to form granules of **volutin** which is formed from the linkage of many molecules of phosphate. These bodies stain pink with methylene blue, and their detection is an aid to this organism's identification. Other storage inclusions produced are **lipids** which may be quite ordinary fats or special polymers found mainly in bacteria such as **poly-beta-hydroxybutyric acid, glycogen, granulose,** and even elemental **sulfur** (Fig. 4.8).

THE CHROMATOPHORES

Due to the absence of chloroplasts in the bacterial cells capable of using light as an energy source, membranous structures have been found to be

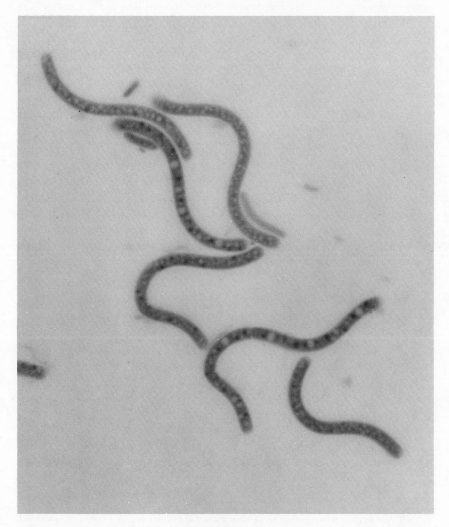

Fig. 4.8. Spirillum volutans, **a long spiral-shaped bacterium, when stained with methylene blue displays numerous storage granules.** Photo courtesy R. G. E. Murray.

present in photosynthetic bacteria. Within these, the pigments and enzymes required for light utilization are localized.

THE NUCLEOID

Due to the fact that bacteria do not possess the characteristics of the nucleus in higher cells, the term **nucleoid** was coined to specifically designate the bacterial nucleus. A nuclear membrane is absent; the genetic material is thus in constant physical contact with the cytoplasm. The genetic material is not organized into the characteristic chromosomes. Yet, it is composed of the same important type of chemical substance, deoxyribonucleic acid (DNA), that makes up the heredity-governing units (the genes) in the cells of higher animals and plants and which directs the activities of the cell. The DNA exists as a single thread-like molecule about 1000 times as long as the cell itself. In many cases, electron micrographs reveal no free ends of the DNA fibril, indicating that it exists as a circular molecule packed in an irregular manner in the cytoplasm and attached to the cell membrane (Fig. 4.9). After division of the nuclear material, growth of the cell membrane moves the daughter nucleoids apart. Often, there are two or more nucleoids per cell since cell division generally lags behind nucleoid division. Since there is only one thread for each nucleoid, the complex events such as spindle fiber formation that occurs in mitosis are unnecessary. The DNA molecule replicates and by simple cell growth the genetic material is separated in such a manner that the descendent progeny inherit equal shares of the material controlling the heritable characteristics of the parent cell.

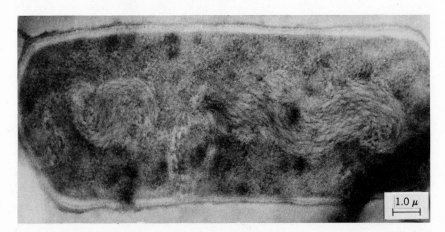

Fig. 4.9.   An electron micrograph of an ultrathin section of Bacillus subtilis **showing two nucleoids. The fibrillar nature of the DNA is evident.** Photo courtesy G. W. Fuhs: Bacteriol. Reviews, **29,** 277, 1965.

Fig. 4.10. An electron micrograph of an ultrathin section of a Clostridium cochlearium **sporangium.** **The terminal spore's diameter being greater than the cell causes the cell to bulge.** Photo courtesy L. Pope, D. P. Yolton, and L. J. Rode: J. Bacteriol., **96,** 1859, 1968.

ENDOSPORES

Some of the rod-shaped bacteria, especially those of the genera *Bacillus* and *Clostridium,* form spherical or oval bodies called **spores.** Since they are formed within the cytoplasm, they are referred to as **endospores.** A vegetative cell that forms a spore is a **sporangium.** Cells that are genetically-endowed to produce spores usually do so in older cultures and when readily utilizable food is no longer available. Spore production or sporulation involves, in general, a condensation of nuclear material and cytoplasm surrounded by the formation of a double-membrane system. This forespore undergoes further maturation by the synthesis of several outer layers between the membranes so as to surround the spore sap or central body (Fig. 4.10). The mature spore is released from the sporangium to exist as a free spore.

Not only is the spore different in its anatomy, but it also differs in other respects from the vegetative cell. Although alive, spores are inert as they

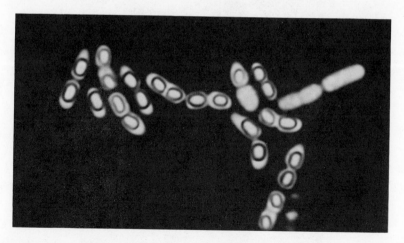

Fig. 4.11. **Nigrosin film of** Bacillus megaterium **showing the refractile endospores.** Photo courtesy C. F. Robinow.

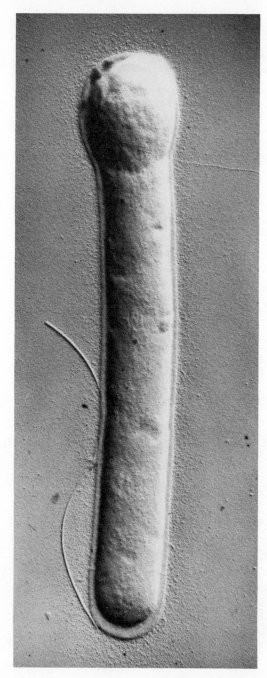

Fig. 4.12. **Carbon replica of** Clostridium cochlearium **showing the presence of a bulging endospore.**
Photo courtesy L. Pope.

Microorganisms and Man

exhibit little or no metabolism. The spore is dehydrated, highly refractile (Fig. 4.11), and contains a unique chemical, **dipicolinic acid,** as well as a high calcium content. Due to their physical organization, dehydrated state, and perhaps unusual chemical composition, spores, in contrast to their mother cells, are highly resistant to heat, dyes, dryness, irradiation, and various chemical agents. It has been reported that spores may remain dormant for 70 or more years.

Spores may develop in various sites within the cytoplasm, but their position is constant for a given species. In some species the spores are located in the center of the cell, while in others they may be near the terminal end of the cell. Most spores are no larger than the diameter of the cell but, in a few species, they may be large enough to cause a bulge in the cell wall, thus appearing club-shaped (Fig. 4.12).

Under suitable conditions, a spore may develop or **germinate** into a vegetative cell. This process seems to occur when the spore's outer layers are rendered more permeable by various means. For example, exposure to heat or to various chemicals is known to induce spore germination. When this occurs, the spore becomes hydrated, loses its refractility, its unique chemicals, and its resistance to the various adverse conditions. An outgrowth phase then takes place that leads to vegetative cell formation and eventual cell division.

The spore permits the organism to survive through periods and in places where the conditions for growth are temporarily undesirable. As a spore, the organism may be spread throughout the world because spores are readily carried by air currents and are not killed by the drying process. Since one spore is formed per cell, and only one cell develops from a spore, sporulation cannot be considered as a means of reproduction with increase in numbers. The extreme resistance of spores presents some serious problems with food spoilage especially in the food canning industry. In any situation where elimination of all microbes is sought, the treatment employed must be sufficiently drastic to destroy the most resistant microbial form, that is the bacterial spore.

Summary

Modern procedures, including the electron microscope, have shown that bacterial cells are more complex than once believed. Although their procaryotic cytology is considerably different from eucaryotic organisms, bacteria possess analogous structures for carrying out the necessary functions of growth, maintenance, and reproduction.

FIVE

Functioning of the Bacterial Cell

Microorganisms are organized living units which grow, consume food, reproduce, and finally die. These processes are influenced by the external environment of the cell and by the microbiologist's method of study such as the development of cultures in broth and the formation of colonies on agar. Microbial physiology studies the processes and functions of microbes. Although this chapter deals with bacteria, the principles discussed are relevant to all microbiology and even to all of biology.

Just as the cell is the basic structural unit both of bacteria and of plants and animals, so certain basic physiological processes take place in all organisms, giving them those properties we describe under the term "living." Much of our knowledge of the cellular biochemistry and physiology of animals and plants is derived from studies on bacteria by scientists who have had little interest in the bacteria themselves. The outstanding characteristic of microorganisms is their tremendous biochemical activity, which is almost solely responsible for their importance in industrial processes, in the soil, and in disease.

At the cellular and biochemical level, any fundamental property found in higher animals and plants will have its counterpart somewhere among the microorganisms, often with several modifications. For example, the many steps in the process by which lactic acid is made from sugar in the cells of muscle, liver, and brain are well known; among the microorganisms, some bacteria carry out this process in precisely the same manner. Other microorganisms, however, produce the same lactic acid from the same sugar in a completely different manner. This diversity of processes among the microbes is explained by evolutionary theory. Numerous procedures may be adequate for life at the microbial level, but the processes that survived in the evolutionary development of higher animals and plants are more limited. Higher animals and plants have great uniformity of cellular environment. The microbial cell is naked and unprotected in a highly variable environment; the development of complex structures in plants results in the plant cells being bathed in a sap of fairly uniform composition; the animal cell is exposed to an environment that is even more strictly regulated in pH, osmotic pressure, chemical composition, and, in the case of the warm-blooded animals, also in temperature.

Microbes are referred to as "simple" forms of life because they lack such complicated plant and animal structures as tissues and organs. However, the microbial cell, itself, is certainly no less complex than any individual cell of a multicellular organism. The single microbial cell contains all of the elements necessary for independent existence.

Physiologically, the microorganism consists of structural parts such as walls and membranes, **reactor molecules** called **enzymes,** and a **nuclear director apparatus,** which is an arrangement of genes. The genes determine the characteristics of an organism; each microorganism has thousands of genes acquired originally by some remote ancestor. These control heredity and thus determine whether the organism will be staphylococcus, a tubercle bacillus, a mold, or a rickettsia. Prior to cell division the genes in an individual microbe are reproduced exactly, and then the cell splits by a complex and precise procedure which insures that each daughter gets exactly the same complement of genes as the parent (Fig. 5.1).

More pertinent to this chapter, the genes supervise the activities of the cell. Generally, each gene directs the cell to produce a reactor molecule, an enzyme. The enzymes catalyze chemical reactions involved in such life processes as nutrition, growth, energy utilization, multiplication, and even death. In its simplest form a living system must have three fundamental parts: (1) a code, (2) a translating mechanism to transfer information, and (3) a reactor which acts in response to the transferred information. This model of a living cell finds a simple analogy in the phonograph, where the code is on the record or tape, and the pick-up arm translates and carries the information to the sound-producing reactor device. In addition, living cells have a sophisticated set of brakes which slow down or speed up and, in fact, starts and stops the actions of the three components. These brakes are set or released by the external environment which gives the cell sensitivity and responsiveness to its surroundings—the quality that is called "alive." The code is composed of DNA. The translating device in living cells is composed of several types of ribonucleic acid, acting with structures called ribosomes to make the enzyme proteins coded by the DNA. Until further refinements have been

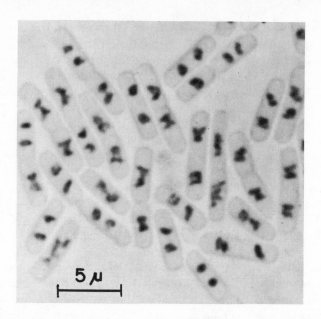

Fig. 5.1. The log-phase cells were stained with a nuclear stain which reveals where the DNA molecules are concentrated. The nuclear bodies divide prior to cell division. Photo by C. F. Robinow, courtesy Academic Press.

made in electron microscopy, knowledge of the existence of genes is inferred from a study of the reactor mechanisms, the enzymes produced under their direction.

Enzymes

NATURE OF ENZYMES

Enzymes are organic **catalysts** produced by the living cell. In the cell sap they form colloidal suspensions which are quite sensitive (labile) to deleterious influences. A substantial part of the protein of cells consists of enzymes. Some enzymes are pure proteins; others are proteins attached to a nonprotein component known as a **prosthetic group** or associated more loosely with a nonprotein **coenzyme.** The protein part of these associations is the **apoenzyme.** A large number of enzymes have been purified and many have been isolated in crystalline form and their characteristics studied by the methods of the protein chemists. Like other proteins, they have high molecular weights and are denatured by heat and certain chemical agents.

The building blocks of the proteins are called **amino acids** of which there are 20 different molecular structures, each containing both an **amino group**

(—NH₂) and a **carboxyl group** (—COOH). The amino acids, glutamic acid and aspartic acid, have two carboxyl groups in each molecule while the other amino acids have only one. Lysine has two amino groups and one carboxyl group. Two of the amino acids, methionine and cysteine, contain a sulfur atom in each molecule. The amino acids are hooked together in chains by a **peptide linkage** which unites the amino group of one to the carboxyl group of another; chains that attain a length of 200 to 300 amino acids have the characteristics of proteins. The type and nature of protein is defined not only by the length (number of amino acids) but by the sequence; a change of one amino acid in the chain may make a completely different protein. This results because the chain is arranged in a winding helix whose coils are held in place by loose chemical bonds; the coils, in turn, are folded into globular packages held together by bonds between the sulfurs of the sulfur-containing amino acids. It is the shape of these packages that determines enzyme activity; the primary and secondary folding may be different if the sequence of amino acids is changed even slightly, so the code which determines the amino acid sequence automatically controls the final shape and hence the activity of the product (Fig. 5.2).

Some enzyme protein molecules are suspended in the cell fluids and have activity in their free state; others are aggregated into structures consisting of several protein subunits which together have coenzyme attachment sites. In bacterial cells the respiratory enzymes are attached to the cell membrane

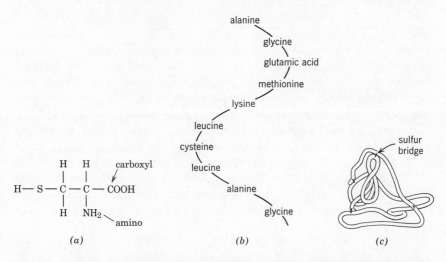

Fig. 5.2. (a) Formula for the amino acid, cysteine, showing an amino group and a carboxyl group. (b) A segment of the peptide chain (amino acids connected by peptide bonds) twisted into a helix. (c) The helix folds into a globular shape bringing together the active groups of amino acids that are quite far apart on the helical chain. The shape of the globule determines what substrates will fit in the active groups and permits an infinite variety.

while in the eucaryotic cells they are organized in the membranes of organ-elles which are called **mitochondria.** In all cells the protein-building enzymes (proteins that make proteins) are organized into organelles called **ribosomes.**

ENZYMAL ACTION

Enzymes working as catalysts permit the organism to complete its reactions rapidly and at moderate temperatures, approximately neutral pH, and low concentration of substrate. The chemist can carry out the same reactions in the test tube in a reasonable time only by resorting to high temperatures, strong acids or bases, and concentrated solutions. Though the enzymal mechanism of action is not completely understood, it proceeds in at least two distinct steps. First, the enzyme reacts with the substrate to produce a complex consisting of a firmly bound combination of the two:

enzyme + substrate ⟶ enzyme-substrate complex.

The existence of such a complex during enzyme action has been demon-strated for a variety of enzymes. Actually the enzyme associates so closely with the substrate that the forces in the internal structure of the latter are modified and the substrate will undergo chemical reaction which it would not do in the unattached form. The enzyme-substrate complex may break down to give a new compound, which is the product of the reaction, and the enzyme is recovered unchanged, thus:

enzyme-substrate complex ⟶ enzyme + product.

In this way, one enzyme molecule can catalyze the conversion of thousands of substrate molecules into product molecules without becoming depleted. The reactions are reversible; under suitable conditions the enzyme may react with the product and, after forming the complex, may yield the original substrate. The active part of the enzyme protein which reacts with the sub-strate may be sulfhydryl (SH), amino (NH_2), hydroxyl (OH), or carboxyl (COOH) groups on the protein. The folding of the molecule places these components from the different amino acids at exactly the right distance to "fit" the substrate. Also, the prosthetic group attached to the protein has ac-tive components capable of combining with the substrate. Many of the coen-zymes are composed either wholly or in part of substances commonly known as vitamins.

FACTORS AFFECTING ENZYME ACTION

The activity of enzymes is measured by the rate of disappearance of the substrate or the accumulation of the end product. The activity of enzymes is determined by a variety of factors:

1. Enzymes are specific for certain substrates; for example, an enzyme which will act on cellulose will not act on protein. Enzymes that hook together two sugars will not hook together two amino acids. This specificity is not absolute, for in some instances closely similar substances are acted upon by a single enzyme.

2. Enzymes exhibit a definite maximum, minimum, and optimum temperature. In fact, the relationship of microorganisms to temperature depends on the effect of temperature on the correlated, functioning enzymes of the cells. Two factors are involved; one is the increase in chemical reaction rates with increased temperatures; the other is the destructive action of heat on proteins. Thus enzyme reaction rates increase as the temperature rises, until that range is approached where the enzyme is partially destroyed by heat. Above that point the activity of the enzyme drops sharply (Fig. 5.3).

3. Activity of enzymes is influenced by the pH of the solution. Again, a maximum, optimum, and minimum are observed, although the pH inside a cell need not always be that of the medium. An organism's enzymes are protected to some degree by the semipermeable cytoplasmic membrane. The concentration of the active form of an enzyme may be altered by pH changes. In addition, the substrate may also be capable of ionization and, as the pH changes, yield more or less that configuration subject to attack by an enzyme. Metal cofactors or activators are also affected by pH. At high pH values magnesium ions form an insoluble, and therefore inactive, hydroxide. The combined actions of these effects define optimum, maximum and minimum pH values for enzyme action.

4. The activity of an enzyme is proportional to its concentration.

5. The activity of an enzyme is affected by the substrate concentration. At very low substrate concentrations the reaction may not proceed. As the substrate concentration is increased, so is the rate, because the molecules of enzyme encounter substrate molecules more frequently and react with them. With concentration increase, sufficient substrate molecules are present to contact every free enzyme molecule. At such concentration of substrate, all of the enzyme is, at every moment, actively engaged. Further increase in substrate concentration will have no effect on the rate of the reaction.

6. The amount of activity is affected by the time permitted for reaction. There is a definite relationship between temperature and time, especially at high temperatures, where tremendous enzymatic action may occur in very short times but, because of destruction of the enzyme, the average activity over longer time intervals may be very low.

7. Activity of enzymes is influenced by the presence of dissolved salts. A certain salt concentration is essential for maintaining the colloidal state, but salts of the toxic heavy metals (mercury, silver, lead) may form inactive complexes with enzyme proteins, thus decreasing enzyme activity.

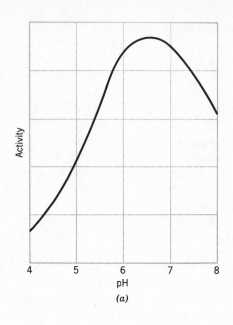

(a)

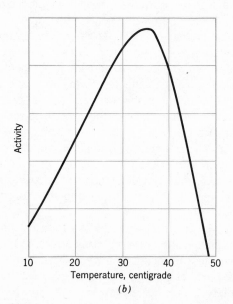

(b)

Fig. 5.3. Every enzyme shows minimum, optimum, and maximum points when activity is measured at varying (a) pH levels and (b) temperatures. These studies were made with a capsule-digesting enzyme from a soil bacterium.

Almost all kinds of enzymes encountered in biology are produced by microorganisms, and every organism contains hundreds of enzymes. Of course, every microorganism does not have all these enzymes, and its reactions depend on which enzymes are present. Physiological reactions which aid us in classifying bacteria are usually qualitative tests for the presence of specific enzymes. Anaerobic organisms are deficient in the enzymes found in aerobes which permit the latter to utilize oxygen. Since such characteristics are inherited, the ultimate control of enzyme formation resides in the genetic apparatus, and anaerobes are deficient in the genes controlling the formation of the enzymes necessary for life in oxygen.

A certain minimum number of enzymes must be present at all times to maintain life, regardless of the composition of the medium. Enzymes that are always present are termed **constitutive enzymes; inducible enzymes** (also called adaptive enzymes) appear only when microorganisms are grown on a medium containing the substrate for that enzyme. For example, some microorganisms never produce the enzyme which ferments the sugar, lactose; others produce the enzyme only when that sugar is present in the medium. If lactose is absent from the medium during growth but later is added to the grown culture, no fermentation takes place for about an hour, during which time lactose "induces" the enzyme formation. Obviously the gene controlling the formation of the lactose-fermenting enzyme is present all the time, but the enzyme is formed only when "induced" by a substrate that interferes with a repressor gene. In spite of the minute size of the bacterial cell, it may be capable of using a wide variety of substrates. To do this, it maintains repressors (brakes) for those enzymes which are not needed at all times. When the enzyme is needed the brakes are released.

Most enzymes are produced by the organism and retained in the cell. They are called **intracellular** or endoenzymes to distinguish them from those elaborated into the medium, the **extracellular** or exoenzymes. The extracellular enzymes digest large molecules in the medium into smaller fragments which can be transported into the cell: gelatinase, cellulase, and amylase are examples. Starch will not pass through the cell membrane; the only organisms which can use starch for food are those producing the extracellular enzyme, amylase, which breaks the large starch molecule down to the relatively small, diffusible molecules of the sugar, maltose. When bacterial cells are filtered from broth in which they have grown, the exoenzymes remain in the broth filtrate.

Although the algae, like other green plants, are important synthesizers of organic materials (more than two-thirds of the newly synthesized organic matter on this planet is of algal origin), the primary role of molds, yeasts, actinomycetes, and bacteria is one of decomposition and decay. In spite of increasing contributions from man's burning of fossil fuels, nine-tenths of the carbon dioxide released to the atmosphere comes from microorganisms. The cycle of the elements proceeds because sequences of microbes are endowed with extracellular and intracellular enzymes to reduce the large molecules to the stable inorganic forms. A vast array of enzymes are essential to all life so that they occur in all microorganisms. On the other hand, there are enzymes that are found only in a very few species. For example, very few microorganisms can convert ammonia to nitrate, attack the large molecules in petroleum, destroy the heartwood of an oak tree, or reduce sulfate to hydrogen sulfide. Such specialization of function by virtue of unique enzymes opens ecological niches for occupancy by those microorganisms so endowed.

In the naming of enzymes, the suffix "-ase" is attached either to the name of the substrate being attacked or to the name of the reaction taking place. The latter method is generally used for groups of enzymes, for example, hydrolases are the group of enzymes that catalyze the splitting of a molecule by hydrolysis; thus the name implies the addition of water. Specific enzymes such as urease split urea into carbon dioxide and ammonia by the addition of water; lipase similarly splits lipid or fats. Since fats belong to that class of chemical compounds known as esters, lipase is also called an esterase. Some enzymes have retained their classical names such as rennin, trypsin, and ptyalin.

Physiologically, enzymes may be grouped according to five major activities: (1) splitting, (2) adding, (3) isomerizing, (4) transferring, and (5) oxidizing. The splitting enzymes include the extracellular ones, such as amylase and gelatinase, as well as intracellular enzymes which split carbohydrate, fat, and protein molecules into fragments containing only one, two or three carbon atoms together with attached hydrogen, oxygen, etc. atoms. Such fragments are added to other fragments through the catalytic action of adding enzymes to build large molecules needed for constructing new cell material or cell products, a process referred to as **biosynthesis.**

Some molecules are isomerized (parts of the molecule are shifted around, but the size or composition is not changed) in order to render them more susceptible to further enzymatic activity. The transferring enzymes may shift molecular parts from one molecule to another; an amino (NH_2) group on an amino acid present in excess may be transferred to a precursor to complete another amino acid which is in short supply.

Oxidizing enzymes carry out oxidations and reductions in the cell. Oxidation is accomplished by three methods.

1. *Removing electrons.* Ferrous iron, found in the prosthetic group of many oxidizing enzymes, is oxidized to the ferric form:

$$Fe^{2+} \leftrightarrows Fe^{3+} \text{ and 1 electron}$$

2. *Removing hydrogen atoms.* This process is used in the oxidation of many organic molecules. Actually, in this process an electron is also removed, but the hydrogen ion combines with it to make a hydrogen atom. In aerobes, pairs of electrons unite with oxygen and a pair of hydrogen ions to form water. In anaerobes the electrons with their accompanying hydrogens are finally attached to some molecule other than oxygen.

3. *Adding oxygen directly to the molecule being oxidized.* This is not the normal procedure of the oxidizing enzymes and occurs in few biological systems, for example, in the oxidation of hydrocarbons. Usually extra oxygen in an enzymatic product is introduced by the addition of water and followed by removal of the two hydrogens by oxidizing enzymes.

When oxidizing enzymes produce the reverse reaction (electrons or hydrogen atoms are put on a molecule rather than removed from it), the process is called reduction.

The microbial cell is not merely a sack containing endoenzymes distributed as a random mixture in the cell sap; rather, the cell has a definite geography of enzyme distribution. The enzymes involved in respiration are associated with the cell membrane; the ones involved in protein synthesis are associated with ribosomes. Other associations also exist; for example, the several enzymes involved in the successive chemical steps in making a vitamin may be bound in a complex in the cytoplasm.

IMPORTANCE OF ENZYMES IN MICROBIOLOGY

Correlated enzyme activity offers the best explanation yet available for life processes. Enzyme action accounts for the destructive activities of microbes, the procedures by which they produce disease, as well as their formation of useful products.

After studying the morphology of the bacteria (which is itself a result of enzyme activity), the taxonomist studies **physiological characteristics,** reflections of the complement of enzymes. Of course, he cannot study them all; usually about a dozen is sufficient. He classifies the microorganism by checking those enzymes which in simple tests demonstrate the disappearance of substrate or the formation of products. Since each enzyme is coded by DNA, the study of the enzyme content is an indirect way of studying the chemistry of the genetic system, a logical procedure in identifying living organisms.

Bacterial and fungal enzymes are produced in industry for use in such diverse processes as removing protein spots from laundry, tanning of fine leathers, clarification of fruit juice, and conversion of starch to sugar. Penicillinase, an enzyme found in many drug-resistant bacteria, is used to destroy penicillin in blood cultures from penicillin-treated patients so that the infectious agent will grow in the culture medium.

DNA

One of the great generalizations of biology is the universality of the genetic code. The genetic code constitutes a sort of chemical "language of life," according to which instructions are "written" in the molecular structure of the hereditary material for determining the form and function of all living things — a language essentially identical for all life, from microbe to man.

The discovery and deciphering of the genetic code, which won a Nobel Prize in 1968 (see Table 1.2), is generally regarded as the last of the five major milestones in the science of genetics up to now. First was the discovery, in 1866, by Gregor Mendel, that hereditary traits are transmitted to "packets," later to be called "genes." The second milestone was the discovery, in 1866, by Gregor Mendel, that hereditary traits are transmitted as protein molecules, called enzymes. That discovery, which won a Nobel Prize in 1958, was of crucial significance because enzymes mediate most of the chemical reactions in a cell, thereby endowing it with properties known collectively as "life." The third milestone was laid down in 1944 by a group of scientists, headed by Oswald T. Avery, who demonstrated that the hereditary material — the stuff that genes are made of — was a substance called deoxyribonucleic acid, or DNA. In 1953, James D. Watson and Francis H. C. Crick produced the fourth milestone, which was awarded a Nobel Prize nine years later. This was the description of DNA's structure — essentially, a string of subunits of four kinds, called nucleotides, linked together in the form of a double helix, or a twisted ladder shape.

According to the genetic code, sequences of those four DNA subunits, taken three at a time, correspond to protein subunits, the amino acids. In this way, a gene's chain of DNA subunits can specify the sequence of amino acids in a protein and, as a consequence, determine that protein's unique structure and its function as an enzyme. A defect in a gene, called a **mutation,** can cause defective or absent enzymes, resulting often in death or a modified life process.

As can be seen in Fig. 5.4, the twisted ladder (double helix) is made of two chains of alternate phosphate and sugar molecules. The sugar is a five-carbon compound, deoxyribose, from which the name of deoxyribonucleic acid is derived. These are connected by the four chemical pairs, guanine-cytosine and adenine-thymine, and their reverse, cytosine-guanine and thymine-adenine arranged like rungs on a ladder holding strands of the sugar phosphate in coils about .002 μm apart. There are about 10 cross connections in each full turn of the coil which covers a distance of about .0034 μm; the entire DNA of a bacterium is a strand containing about 3,500,000 nucleotide pairs. This would, if stretched out, extend about 1000 μm, which is

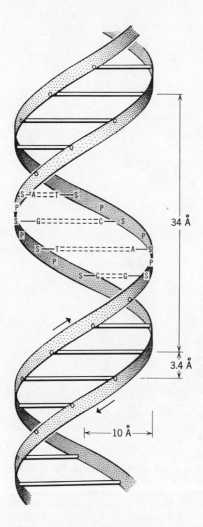

34 Å

3.4 Å

10 Å

Fig. 5.4. The Watson-Crick model of DNA. The molecule is composed of two chains of alternate sugar (S) and phosphate (P) held apart by nitrogenous bases A (adenine), T (thymine), G (guanine) and C (cytosine). The double-chained structure is coiled in a helix. The width of the molecule, the distance between adjacent nucleotides, and the length of one complete coil is rigidly defined and identical in the DNA of all living things.

about 500 times as long as the cell. When carefully separated from the bacterium, the DNA is an endless thread which means it is in the form of a loop. The information on the DNA thread is coded by different arrangements of the four nucleotides; the diagram (Fig. 5.5) shows all the four pairs in a row,

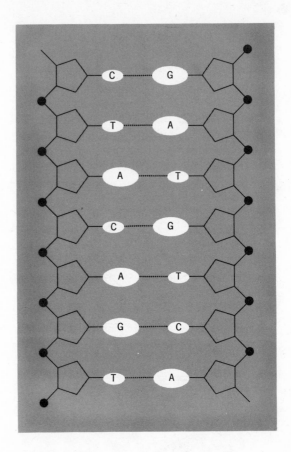

Fig. 5.5. Magnified portion of a DNA molecule uncoiled. The molecule has a ladderlike structure, with the two uprights composed of alternating sugar and phosphate groups and the cross rungs composed of paired nitrogenous bases. Note that each cross rung has one purine base (large oval) and one pyrimidine base (small oval). When the purine is guanine (G), then the pyrimidine with which it is paired is always cytosine (C); when the purine is adenine (A), then the pyrimidine is thymine (T). The order in which these bases occur as one goes up the ladder codes the information.

but in actual situations they occur in random order. The sequence of the order is the code.

The DNA has two functions. First, it is replicated exactly prior to cell division so that each daughter cell gets precisely that information held in the parental DNA. This must be a very rapid process; the bacterial cell may divide every 30 minutes and consequently the turns of the helix must be reproduced at the rate of 140 revolutions per second. Second, the information on the various parts of the DNA molecule must be reproduced by copying it in the form of a complementary strand of RNA thus transcribing the information for the synthesis of enzyme proteins.

RNA

Ribonucleic acid (RNA) is a transcription (generally single stranded) from part of one strand of the DNA molecule. It has the same backbone of alternating sugar and phosphate but the sugar is ribose instead of deoxyribose. Of the four attached bases, three are identical but thymine is replaced with uracil. Three types of RNA are involved in the cell, **ribosomal** RNA, **messenger** RNA, and **transfer** RNA. More than half of the RNA of the cell is ribosomal in that it is bound in small organelles, called ribosomes, which serve as the assembling sites for proteins. The ribosomes of bacteria and blue-green algae (the procaryotic types) are slightly smaller than those of eucaryotic cells; in an ultracentrifuge the former have a sedimentation constant (measured in S units called **Svedbergs** after the inventor of an early ultracentrifuge) of 70S as compared with a constant of 80S for the ribosomes of the latter.

The messenger RNA is a strand of nucleotides whose sequence is determined by that section of the DNA from which it was copied. Moving into the cytoplasm it combines with a group of ribosomes; this combination is called a **polysome,** and it is on this polysome that proteins are assembled. The amino acid sequence in the new protein is determined by the sequence of the nucleotide bases in the messenger. For example, three uracils in a row determine one particular amino acid, an adenine followed by two uracils determines another amino acid, and so forth. This is a **triplet code** composed of the four bases (adenine, cytosine, guanine and uracil) taken 3 at a time. If you make a chart you would observe that there are 64 possible arrangements of 4 things taken 3 at a time. Since there are only 20 amino acids this leaves certain combinations for punctuation of the code, that is, they indicate the starting and stopping points for the protein chain; there is also some **redundancy,** that is, some amino acids are coded by more than one triplet (Fig. 5.6).

The actual amino acid is brought to the polysome by still another type of RNA, the transfer RNA; these shorter chains are of 20 different types and each type unites with a specific amino acid in the cytoplasmic fluid and brings it to the proper place on the messenger (Fig. 5.7). In this manner chains of amino acids are arranged in the proper order to produce the protein coded for in the DNA of the cell; an enzyme fastens the chains by bonding the amino group of one to the carboxyl group of the next. As pointed out earlier, the code seems to be universal. The details may differ, but the same general principles apply in all living things.

THE GENETIC CODE

1st ↓ 2nd →	U	C	A	G	3rd ↓
U	PHE	SER	TYR	CYS	U
	PHE	SER	TYR	CYS	C
	LEU	SER	Ochre	?	A
	LEU	SER	Amber	TRP	G
C	LEU	PRO	HIS	ARG	U
	LEU	PRO	HIS	ARG	C
	LEU	PRO	GLUN	ARG	A
	LEU	PRO	GLUN	ARG	G
A	ILEU	THR	ASPN	SER	U
	ILEU	THR	ASPN	SER	C
	ILEU	THR	LYS	ARG	A
	MET	THR	LYS	ARG	G
G	VAL	ALA	ASP	GLY	U
	VAL	ALA	ASP	GLY	C
	VAL	ALA	GLU	GLY	A
	VAL	ALA	GLU	GLY	G

Fig. 5.6. This chart shows the "best allocations" of the 64 codons at this time. Some of these allocations are less certain than others. The two codons marked ochre and amber are believed to signal the termination of the polypeptide chain. The evidence used to produce this table comes mainly from the bacterium, E. coli. It is likely that the genetic code in other organisms is either very similar or identical to that shown here.

The Functioning Cell Membrane

Semipermeable membranes in living cells contain enzyme-like proteins which utilize energy to move substances in and out of the cell. This is called active transport to distinguish it from simple diffusion through the membrane; the latter process is dependent on the concentration of the transported substance on the two sides of the membrane. Active transport, however, can be carried out against a concentration gradient. One explanation is that the membrane contains specific proteins called permeases which unite with substances at the membrane surface much like an enzyme-substrate combination. However, instead of producing a chemical change to yield a

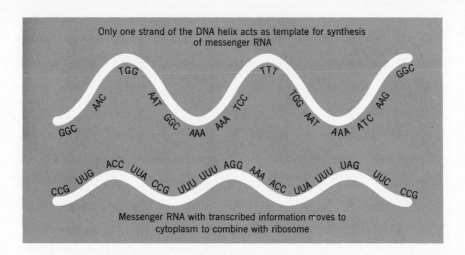

Only one strand of the DNA helix acts as template for synthesis of messenger RNA

Messenger RNA with transcribed information moves to cytoplasm to combine with ribosome

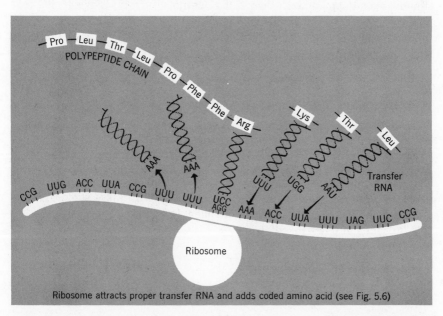

POLYPEPTIDE CHAIN

Transfer RNA

Ribosome

Ribosome attracts proper transfer RNA and adds coded amino acid (see Fig. 5.6)

Fig. 5.7. Protein synthesis. Messenger RNA is synthesized on one of the polynucleotide chains of the DNA of the gene. Actually the bases are equally spaced along the DNA and RNA chains but they are grouped in triplets here for illustrative purposes. Note the transcription. G on DNA is read as C on RNA, A is read as U, T as A, and C as G. These bases are complementary to each other. When transfer RNA in the cytoplasm pick up the amino acids for which they are specific and bring them to the ribosome as it moves along the messenger RNA, transfer RNA bonds to the messenger RNA at a point where a triplet of bases are complementary. The amino acids are then linked by peptide bonds. Synthesis of the polypeptide chain thus proceeds one amino acid at a time in an orderly sequence as the ribosome moves along the messenger RNA. As each transfer RNA donates its amino acid to the growing polypeptide chain, it uncouples from the messenger RNA and moves away into the cytoplasm, where it can be used again. Long polypeptide chains are proteins.

product, the "enzyme-substrate" moves to the other side of the membrane where the substrate is released. Membranes also function mechanically by invaginating to enclose a portion of the liquid outside of the cell. The invagination then pinches off to become a vacuole inside the cell. This is called **pinocytosis** (drinking), and it was observed by the early microscopists working with protozoa; it is now believed to be common even in bacteria and other cells where the membrane is enclosed in a rigid cell wall.

At the present time the older "sandwich" structure of membranes (a continuous layer of fatty material, two molecules thick, with a protein coating on each side) is being challenged by evidence that the membrane may be made of two rows of globular proteins (see Fig. 5.2) whose crevices are packed with lipid. When the final result is determined, it is likely that membrane structure will be another universal, that is, all membranes are fundamentally the same. The internal membranes of procaryotic cells seem to be limited to mesosomes and other extensions of the cytoplasmic membranes. Recent speculation on the origin of the internal structures of eucaryotic cells suggests that the mitochondria originated from invading bacteria and the chloroplasts originated from invading blue-green algae. The membranes of both of these structures are similar to those of the possible progenitors.

Summary

Cell physiology involves the functioning within a membrane of both isolated and organized enzyme systems whose protein structure is coded in the DNA, transcribed to the messenger RNA, and translated in the ribosomes. The chemistry of the cell is a modified version of the chemistry of a non-living system. Although responsive to the environment, it is somewhat protected from it. The reactions obey chemical laws, but are made complex by the presence of powerful catalysts.

SIX

Microbial Growth and Metabolism

Microbial physiology is studied at two levels; at the individual level we determine how the cell grows or dies, and at the population level we consider the growth and death of the culture. As will be seen from subsequent discussion, the tendency to regard the culture as an individual may lead to error. The individual cell may be counted by several methods, but most other measurements of its activity must be made on masses of individuals. In the last chapter we considered the dynamic parts of a cell model. We now consider reproduction and growth of the bacteria.

Reproduction of Bacteria

Microbes would be of little consequence if it were not for their prodigious rate of reproduction when placed in a suitable environment. Bacteria reproduce by **transverse binary fission,** splitting crossways, never longitudinally, and dividing into two approximately equal parts. This process of splitting is preceded by replication of the thread of DNA which must be reproduced before the cell divides.

During replication of both of the strands of the double helix, two loops are formed. Each is attached to the cell membrane at or near a mesosome. Growth of the membrane and the mesosome move the two loops apart and a cross wall is laid down. Additions are made to the cell wall at the place of division until the cells split apart to give two daughter cells (Fig. 6.1). When the culture is growing rapidly, new loops of DNA are begun on each of the

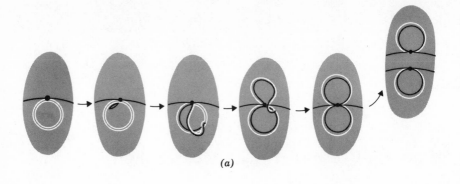

(a)

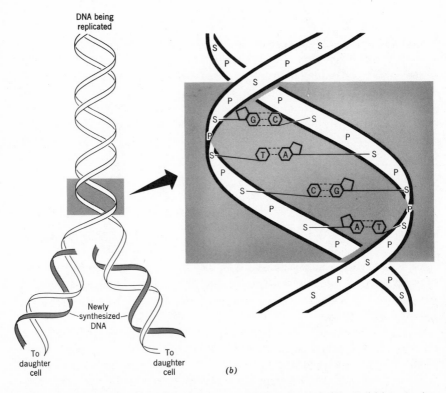

DNA being
replicated

Newly
synthesized
DNA

To
daughter
cell

To
daughter
cell

(b)

Fig. 6.1. (a) Formation and separation of new loops of DNA during bacterial cell division, showing attachment of DNA to the growing cell membrane. (b) Magnification of a segment of the DNA strand showing replication of each strand of the double helix.

forming loops before the original DNA thread has completed its first division. In this manner, the time for the completed reproduction of the entire thread of DNA may actually be longer than the time for cell division, since the time for the replication of each loop may overlap two cell divisions.

In eucaryotic cells similar reproduction of the DNA must occur, but the DNA is incorporated into several structures called **chromosomes** which are enclosed in a nuclear membrane. The nucleus goes through a complicated procedure called **mitosis** which involves replication of the chromosomes. An elaborate mechanism for moving the two sets of chromosomes to opposite sides of the cell is implemented in eucaryotic cells by formation of spindle fibers attached to opposite poles of the cell. Each set undergoes reorganization into a new nuclear body, after which the cell splits in half; each daughter cell then contains a nucleus with a complement of heredity-controlling components identical to that of the mother cell. Such a process is unnecessary in bacterial cells since the single nuclear threads which substitute for eucaryotic nuclei are moved apart to the daughter cells by the simple process of membrane growth.

At some stages during bacterial growth, the nuclear division may proceed more rapidly than cell division; each bacterial cell may then contain two or even four nuclear bodies. But generally, after the nuclei divide, a cell plate grows between them, starting from the cell wall and growing toward the center. This cell plate splits to form the end walls of two new cells, though the cells may be connected by a thread of cytoplasm through the center of the walls after the other processes of division are complete. This thread, called the **plasmodesm,** has been observed in bacteria such as streptococci and chain-forming bacilli.

After the cell has divided, each daughter cell grows until it, in turn, is able to divide its nuclear material and undergo a cell division. Under ideal conditions bacteria have been known to divide once every nine minutes, but the typical species of bacteria commonly dealt with in the laboratory usually have a generation time of about 30 minutes. The **generation time** is the time it takes for the bacteria to undergo separation from a sister cell and fission into two daughter cells. Certain slow-growing bacteria, even under the optimum conditions for growth, divide no more often than once every three to four hours as do many yeasts, algae and protozoa. Cells in cultures of human tissues may divide only once or twice a day.

If a single bacterium in a test tube containing one milliliter of broth began to grow at its optimum rate, it would take a maximum of 30 divisions to produce a population of a billion cells; if it were undergoing fission every 30 minutes, this population would be achieved in 15 hours. After that time the organism would stop growing because of the exhaustion of food material or because of the toxic nature of the waste materials produced. If we follow the population resulting from binary fission where division occurs of one organism into two and the two into four, the four into eight, and so forth, we observe that an infinite series of 1, 2, 4, 8, 16, etc. is produced in the succeeding generations. A little arithmetic will show us that in 10 generations one cell will become 1024 cells. Therefore, if instead of one cell 1000 are placed in the test tube, the process will take 10 generations less (or five hours less in bacteria with a 30 minute generation time) to achieve the maximum population of one billion per tube. If 500 million organisms are placed in the tube, the population will arrive at the figure of one billion in the next cell division, 30 minutes later. This type of reproduction is called asexual because it does not involve the combination of hereditary materials from two individuals.

In certain strains of some species of bacteria, a mating of organisms has been demonstrated resulting in inheritance of some of the characteristics of each of the parents (see Chapter 8). This sexual process is probably not very widespread in nature, but it demonstrates once more the important generalization that most fundamental processes in higher animals and plants have their counterpart among the bacteria.

When growth starts from a single cell and the number of cells doubles regularly the increase in viable count occurs in steps. Such cell divisions are said to be **synchronized.** Soon, however, the cells get out of step, and a graph, made of the number of living cells plotted against time, ceases to show steps but proceeds upward in a reasonably smooth curve. Certain chemical studies on bacteria cannot be accomplished on one or even a few cells, but require a large population growing synchronously. By holding the cells at a temperature slightly below that required for cell division, the individual cells continue to grow slowly but each stops at some point prior to division. When such cells are placed at a more desirable temperature, a sudden doubling of the population results as all the cells divide at about the same time. After a period equal to the generation time, a second doubling of the population occurs. Synchronous growth will occur for several such doubling steps, but the population of bacteria exhibits slight generation-time variation with individual cells and soon gets out of phase. Another method of obtaining a temporary synchrony of some bacteria is to use a filter that holds back only the larger cells; the smaller cells have just divided and will on further incubation show a step-wise growth.

The Bacterial Growth Curve

Under normal conditions when microbes are inoculated into a new environment, the developing bacterial culture follows a typical growth pattern. The cultural history is represented by a graph, which shows the logarithms of the cell numbers on the vertical axis and the time on the horizontal axis (Fig.

6.2). The typical growth curve shows first a period of delayed reproduction, which is called the **lag phase.**

The lag phase may be very short if a large number of actively growing bacteria are inoculated into a medium that is ideal for their growth. In the cultivation of microbes in industry every effort is made to create conditions which shorten the time of the unproductive lag phase. When only a few cells or cells that are old or injured are inoculated, the lag phase is longer and may be prolonged still more if the growth conditions are unfavorable.

In the dairy and food industry, quality control depends on limiting the initial count of microbes and holding the products at low temperatures to prolong the lag phase of those organisms that cannot be excluded. Transfer to a different type of medium lengthens the lag phase because adjustments must be made within the organisms, and often only a few individuals in the inoculum can grow well in the new environment. When an unwashed milk container is again filled with milk, the lag phase may be very short, not only because of the large inoculum, but also because the population is being introduced into an environment with which it has had previous experience.

The **log phase** is that part of the culture history represented as a straight line slanting upward, when the logarithm of number of organisms is plotted against the time; during this time the organisms are increasing in number **exponentially.** The rate of growth during this time is represented by the steepness of the slope; a steep slope indicates a short generation time determined by the suitability of the medium, the conditions for growth, and the type of organism. The explosive nature of exponential growth yields hun-

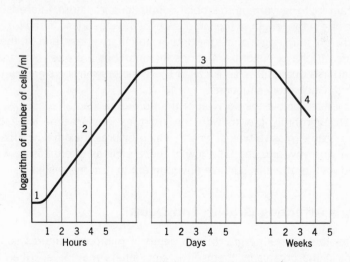

Fig. 6.2. The growth curve of a bacterial population. 1, lag phase; 2, log phase; 3, maximum stationary phase; 4, death phase.

dreds of millions of new organisms within a few hours when proper conditions are available. Characteristics such as motility and the gram reaction are studied using cells within this phase of growth.

When the organisms become numerous their growth is limited by the exhaustion of food, by their toxic excretion products, and, in natural situations, by the presence of their living enemies—protozoa that eat the bacteria and viruses that infect them. When the multiplications and the deaths exactly balance, the culture enters the **maximum stationary phase.** The height of the maximum stationary phase is a measure of the total crop of cells. This number can be increased to some extent by supplying additional food material or by neutralizing toxic excretions. Eventually the culture must arrive at this resting state after which the death and dissolution phase follows. The maximum stationary phase is so short in a few species that transfers to new medium must be carried out daily if the stock is to be maintained. With most other species, this stationary phase may persist for weeks or months. It can be lengthened by placing the cultures in a refrigerator.

The last phase, the **death phase,** often follows an exponential course, and the graph shows a straight line sloping downward. In some dying cultures, abnormal distorted cells appear, which are called **involution** forms; in others, the cells digest themselves by a process called **autolysis.** In nature they rarely die off to the last cell before a change in the environment permits a new surge of growth. To the microbes every crumb of soil, every hair follicle on the skin, or every particle of waste in the intestine may represent a microenvironment for such cycles of growth and death.

The classical growth curve is not restricted to laboratory cultures but is a significant factor governing microbial populations in nature. Interest usually centers on (1) the length of the lag phase, (2) the steepness of the log phase, (3) the height and length of the maximum stationary phase, and (4) the slope of the death phase. Natural situations for microbes are often less than optimal and in a quantitative sense all of these factors may be significantly different from laboratory situations.

Two factors explain the survival of the numerous bacterial species in nature: (1) logarithmic growth with the short generation time exhibited by bacteria is an explosive type of growth which insures maximal occupancy of any available environment; (2) a few of these millions of organisms will survive and wait for the next suitable environment to become available.

Factors Affecting Growth and Death

Of the factors regulating growth and death of microbial cultures some are inherent in the organisms themselves, and others may be ascribed to the specific physical and chemical environment.

Even under optimal conditions the rate of growth of different microorganisms varies considerably. Some organisms are inherently slow growers, while others grow more rapidly. On the other hand, environmental factors which accelerate growth may also accelerate the death processes. Under natural conditions any environment is occupied primarily by the organisms best suited for survival. Because of the ubiquitous occurrence of microbes, new groups are ready to move in as soon as the environment becomes unfavorable for the current occupant. In pure cultures, however, where contaminants are excluded, the effects of environmental factors have been worked out in considerable detail. If culture conditions are ideal for a number of types of organisms, the smaller will grow faster than the larger; for example, under conditions equally suitable for cultures of molds, yeasts, and bacteria, the bacteria will outgrow the yeasts, and the yeasts will outgrow the molds. Organisms preferring high temperatures grow more rapidly than those preferring low temperatures. Organisms that build all their cell constituents from simple carbon and nitrogen compounds grow more slowly than those that require a medium containing more complex organic materials and use partially preformed building blocks.

FOOD

Microbial cells like those of higher animals and plants are constructed from distinctive chemical elements. Those mineral elements found in minor amounts include iron, magnesium, manganese, calcium, potassium, copper, cobalt, zinc, and molybdenum. A few other trace-mineral elements are secured from the medium in the form of soluble salts; however, they are not essential parts of the living cell. The major elements making up the cell substances are carbon, hydrogen, oxygen, nitrogen, sulfur, and phosphorus. Phosphorus and usually sulfur are also utilized as the soluble phosphate or sulfate salts. Hydrogen and oxygen are obtained readily from water. Microorganisms are so versatile that few forms of carbon and nitrogen naturally occurring on this earth cannot be utilized by some species. It has been said that a chemist can not create an organic chemical that some microbe cannot take apart.

Organisms need food for building blocks and for energy. Exhaustion of the food supply is an ultimate factor limiting microbial activity, since the amount of food regulates microbial growth. In this situation the law of the minimum operates; if a bacterium needs a certain vitamin for growth, it will stop growing when the supply of the vitamin is exhausted, even though the broth may still contain large amounts of sugar and peptones (Fig. 6.3). The amount and kind of food available determine the types of organisms that will thrive

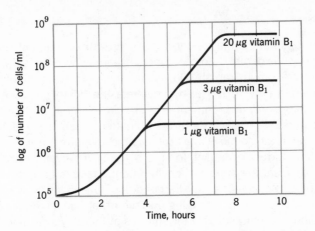

Fig. 6.3. The growth curves of three bacterial cultures with excess of all required nutrients except vitamin B₁. The culture with one microgram (μg) of the vitamin grows just as rapidly as the other cultures until the vitamin content of the broth is exhausted; then it enters the maximum stationary phase.

and the rate and extent of their growth. In waters or soil where organic matter is in short supply, **autotrophic** (sometimes called lithotrophic) **organisms** thrive since they can utilize gaseous CO_2 (carbon dioxide) as a source of carbon for building cell materials. Such organisms need as a source of energy an oxidizable inorganic material such as hydrogen sulfide, ammonia, hydrogen gas, or iron in the ferrous state. Or the organisms, depending on CO_2 as a source of carbon, may contain photosynthetic pigments and secure energy from sunlight. With the exception of algae, most microbes are **heterotrophic** (sometimes called organotrophic), using organic carbon compounds for both energy and building blocks. Some are omnivorous, using a wide variety of compounds ranging from the large molecules of starch and protein to two-carbon compounds such as alcohols or acetic acid. Many molds and the bacteria of the genus *Pseudomonas* are examples of organisms having a wide range of carbon substrates. Other microbes may be limited to relatively few sugars as sources of carbon. Nitrogen nutrition is equally diverse; some organisms can build proteins from ammonia or nitrate while others must have preformed proteinaceous material such as peptides or amino acids, preformed vitamins, and a few microbial species may even require nucleotides and coenzymes.

Laboratory media containing all of the materials necessary for growth have not been devised for some organisms; such organisms which grow only on other living cells are called **obligate parasites**. These include all viruses and rickettsiae. Among the disease-producing bacteria in man, they include the bacteria that cause leprosy and syphilis. The nutritional complexity of microorganisms ranges between the extremes represented by the obligate parasites on one hand and the autotrophs on the other. The pus-forming staphylococcus of the skin can grow in a medium containing sugar, mineral salts, eight of the amino acids, and three of the vitamins.

Analyses of the different nutritional groups indicate that all microorganisms contain the same complex chemical substances; those requiring vitamins or other special substances in the medium simply are unable to synthesize these materials for themselves. Biosynthetically, the autotrophic organisms are the most complex. They must contain not only enzymes to synthesize everything made by heterotrophs but also the additional substances which heterotrophs obtain prefabricated from the medium.

The organic food material brought into the heterotrophic bacterial cells is broken down into small molecules containing one-, two-, or three-carbon atoms. Specific adding, transferring, isomerizing, and oxidizing enzymes incorporate the food material with nitrogen, sulfur, phosphorus, and other mineral elements to produce new cell material. The biosynthetic steps involved in the conversion of sugar and simple nitrogen compounds into the complex molecules of the cell are well understood. When readily utilizable carbon compounds are present in high concentrations, the cell may synthesize large amounts of capsular material, fat, or other storage products. Since biosynthesis requires energy, biosynthetic reactions must be coupled with energy-yielding reactions in the cell. Heterotrophic organisms secure this energy from oxidizing organic matter, while autotrophic organisms secure the energy from oxidizing inorganic substances or from trapping sunlight.

MOISTURE

Foods are transported into the microbial cell, and wastes are removed from the cell in water solution. Water also serves as a building material in cell synthesis. Oxygen generally is introduced into a biologically synthesized molecule by the addition of water (H_2O), followed by the removal of the two hydrogen atoms; these hydrogens are transferred to other molecules in the cell, and thus the hydrogen from water also becomes a part of the living organism and its products. A medium which is high in colloids will hold much of the water as **bound water,** that is, water which is tied to the colloid in such a way as to be unavailable for the microorganism. The addition of 10% agar to nutrient broth instead of the usual 1.5% will result in a medium on which most organisms will not flourish because much of the water is bound by the colloid, and little free water is available. Cereal products such as flour may contain as much as 12% total water, but microbial spoilage will not occur because of the low **free water** content. Yeasts require less moisture than bacteria, and molds can grow on even dryer materials.

The addition of dissolved materials affects the availability of moisture to the cells. The **osmotic pressure** of a solution is a measure of this availability. When a solution has an osmotic pressure equal to that inside the cell, it is said to be **isotonic** with the cell. A solution of 0.85% sodium chloride is isotonic with red blood corpuscles, and since such solutions are widely available in medical laboratories, they are often used for suspending and diluting bacteria; actually three times that concentration of sodium chloride would be more nearly isotonic with most bacteria. A **hypotonic** solution exerts an osmotic pressure much less than that inside the cells. When a cell is placed in a **hypertonic** solution (one of higher concentration of dissolved materials), water is drawn out of the cell, and the solute particles tend to pass into the cell to establish an equilibrium on both sides of the cell membrane. Since the living membrane is differentially permeable, the water moves out faster than the dissolved substance can move in. In highly concentrated solutions this water loss proceeds to such an extent that the protoplast shrinks away from the cell wall, and the cell dies. This process is called **plasmolysis.** Organisms that live and grow best in high concentrations of dissolved substances are called **osmophiles**; a special subgroup is composed of organisms called **halophiles,** which grow preferentially in concentrated salt solutions. They are found in salt environments and are important spoilage organisms in salted foods.

If halophiles are removed from a salt solution and placed in distilled water, the high salt concentration inside the cell causes a reverse of the plasmolysis phenomenon. Water entering the cell to equalize the osmotic pressure will distend the cell membrane and may cause it to burst; this is called **plasmoptysis.**

Bacteria, yeasts, and molds can endure great variation in osmotic pressure because of the corset-type action exhibited by the cell wall. When cell walls are weakened or removed by enzymic digestion or chemical poisoning, the cells take up water and expand until the cell membrane ruptures. Bacterial cells with the cell walls removed are called **protoplasts** and **spheroplasts.** Keeping them in a solution of 5 to 20% sucrose will prevent their bursting.

Water also participates in the death phase. More vigorous heating is required to kill organisms in dry material than in wet material; hence, the steam autoclave is a more efficient sterilizer than the hot air oven. In fumigation, microbes are killed more readily if the air is humid. Pure ethyl alcohol does not kill bacteria nearly as rapidly as does a mixture composed of 70% alcohol and 30% water.

TEMPERATURE

Three points on the temperature scale—the minimum, optimum, and maximum temperatures—delineate the effect of temperature on microbial growth (Fig. 6.4). For some species the temperature range is very narrow; for example, the optimum temperature for growth of the gonococcus is about 35 C, the minimum temperature a few degrees lower, and the maximum about 35.5 C. Other species may have a broad optimum temperature spectrum, the range between the maximum and the minimum extending over 40 centigrade degrees. The optimum temperature for an organism varies with the characteristic that is measured. Some organisms grow fastest at 40 C, produce the biggest crop of cells at 35 C, produce the most acid at 30 C, and form acid at the fastest rate at 45 C. Usually the total number of cells produced is the criterion for determining the optimum temperature; although in a practical problem such as penicillin production, the optimum temperature is that at which the highest yield of penicillin is obtained in the shortest time.

Since biological reactions depend on the chemical activity of the cell, the temperature response of chemical reactions determines total cell activity (reaction rates approximately double with a rise in temperature of 10 C). Obviously, biological reactions cannot proceed when the cell water is frozen, nor can they operate if the temperature is raised to a point that cooks (denatures) the cell proteins. But between these values, the maximum and minimum temperatures are determined by the variation in the effect of temperature on the individual reactions which govern cell growth. A change in temperature will increase or decrease the rate of some chemical reactions in the cell to a greater extent than others, producing an excess of some cellular substances and, consequently, an insufficiency of others. Optimum growth occurs when the **best balance** is attained. The anthrax bacillus is modified in a peculiar manner by growth at temperatures higher than the optimum; such cultures lose their ability to produce disease in animals, and the ability to

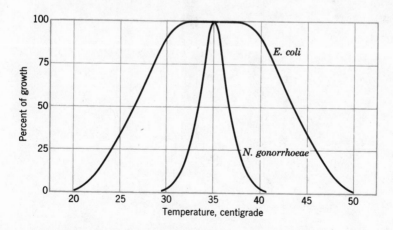

Fig. 6.4. The temperature range of some organisms show a broad optimum with a widely spread maximum and minimum. For other organisms it may be extremely narrow.

form spores. In other organisms, pigmentation and antibiotic production are markedly influenced by incubation temperatures.

The **psychrophiles** or cold-loving group of microbes grow slowly, but they thrive at refrigerator temperatures (below 20 C) and make some growth at 0 C (Fig. 6.5). They take part in spoilage of refrigerated foods; however, they may produce desirable changes in the flavor and texture of beef as it ages in cold storage. The **mesophiles,** which thrive in the moderate temperatures, are the largest group and include all of the animal pathogens. The **thermophiles,** which thrive in high temperatures (above 45 C), are of economic importance in food processing industries. Their growth in hot syrup may cause

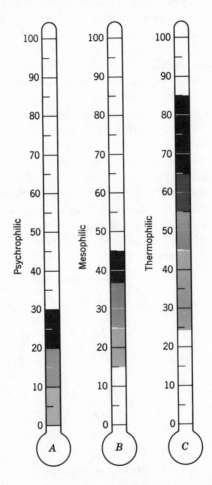

Fig. 6.5. The growth ranges of psychrophiles, mesophiles, and thermophiles are indicated on the centigrade thermometers. Convenient optimum temperatures to remember are: psychrophiles, 10C; mesophiles, room temperature (25C) or body temperature (37C); and thermophiles, 50C.

Microorganisms and Man

undesirable slimes in sugar factories; they are important spoilage organisms of canned foods. In the dairy industry, some species of thermophiles grow at pasteurization temperatures; in the production of Swiss cheese, the milk is held at a high temperature to favor the growth of certain thermophiles that produce desirable flavors. Thermophiles thrive in hot springs and in the centers of piles of decomposing organic matter such as composts or manures. Some organisms are **obligate** thermophiles, requiring high temperatures for growth; others are **facultative** thermophiles that are able to grow at elevated temperatures but are capable of growing in the mesophilic range. The adjectives obligate and facultative are also used to describe other microbial characteristics. The thermophiles are biological oddities because they grow best at temperatures which denature proteins (destroy protoplasm) and thrive under conditions generally regarded as incompatible with life. Constituents of thermophilic bacteria such as flagella and enzymes, which have been studied apart from the living organisms, are much more resistant to heat than are the analogous substances isolated from mesophilic species.

Increased temperatures give a steeper slope to the death phase of the bacterial growth curve. Even under very high temperatures the organisms do not all die at once, but the survivors follow the same type of exponential curve, that is, a straight line usually results when the logarithms of survivors are plotted against time. Many vegetative cells are killed by heating them for only 10 minutes at 55 C, and most are killed in less than a minute at 70 C. However, some mesophiles will survive long exposures to temperatures above the maximum for growth. Such organisms are described as **thermoduric,** and they are found, among other places, in pasteurized beverages and on utensils washed with very hot water. But the heat resistance of thermoduric vegetative cells does not approach that usually exhibited by endospores. Some of the latter may be boiled for an hour or more without being killed.

Only small ice crystals are formed if cultures are frozen rapidly. Microbial cultures can be stored for years in liquid nitrogen or dry ice without significant killing.

HYDROGEN-ION CONCENTRATION

The acidity or alkalinity (basicity) of the medium in which microorganisms grow is determined by the concentration of hydrogen ions. Most organisms have an optimum pH for growth at or near neutrality, pH 7.0. Those pathogenic for man or animals have an optimum pH of about 7.26, the pH of blood. With few exceptions, the range for bacterial growth fits somewhere between pH 4 and pH 10; some organisms cover the entire range, whereas others are restricted to less than 1 pH unit (Fig. 6.6). Most fruit juices are too acid for bacteria, but yeasts and molds grow well in media as acid as pH 2 or

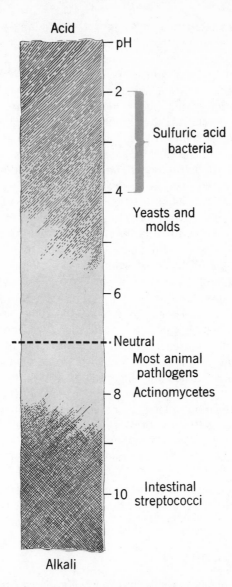

Acid

pH

2

Sulfuric acid
bacteria

4

Yeasts and
molds

6

Neutral
Most animal
pathlogens

8   Actinomycetes

10   Intestinal
streptococci

Alkali

Fig. 6.6. One end of a piece of absorbent paper is dipped into acid, the other into alkali, and placed on an agar surface. A pH gradient is set up, and the optimum pH values showing growth of mixed populations range from pH 2 to pH 10.

3. Plate counts of mold or yeast may be made without interference from bacteria by employing a medium adjusted to a low pH. Unlike the molds and yeasts, the actinomycetes do not tolerate acid conditions. Many bacteria produce organic acids from sugar; the organic acids lower the pH to an unfa-

vorable value before maximum growth is attained unless the medium contains effective **buffers.**

Buffers are substances which resist changes in pH, usually by accepting or releasing hydrogen ions. Phosphate, carbonate, acetate, and citrate are commonly used buffers in culture media. Amino acids and peptones are excellent buffers since they are amphoteric substances existing in solution both as weak acids and weak bases.

Strong alkalies are incorporated into washing powders not only because of their cleansing action but also because they kill microorganisms. Strong acids are also powerful killing agents, but they are seldom used because of their corrosive action. Milder acids such as citric, acetic, and lactic have a preservative action and are employed for retarding microbial action in certain foods and beverages.

THE GASEOUS ENVIRONMENT

Although there are species of microorganisms involved in the production and utilization of methane, hydrogen, carbon monoxide, hydrogen sulfide, ammonia, and nitrogen gases, these are special instances and will be considered in other chapters. The bacteriologist generally considers the gaseous environment in terms of carbon dioxide and oxygen, since these gases affect the growth of all microorganisms.

Carbon dioxide is indispensable to all living cells, and, except in the early lag phase of growth, most microorganisms are able to produce all they require. For certain organisms, such as freshly isolated strains of gonococcus or brucella, incubation of the inoculated medium under 10% CO_2 is necessary for growth. With other organisms, growth from a small inoculum is stimulated by an atmosphere enriched in CO_2. All organisms require special compounds made of four-carbon atoms plus hydrogen and oxygen, as catalysts in their metabolism. These compounds are very reactive and are easily lost from the cell through use as structural materials. The organism has an effective means of replenishing these catalysts by attaching gaseous CO_2 to compounds containing three carbons—the common breakdown products of sugars and peptones. The addition of CO_2 to a three-carbon compound by heterotrophic organisms is called the Wood-Werkman reaction, and, although it was first discovered in bacteria, it probably occurs in all cells. This is not to be confused with the autotrophic uptake of CO_2 which furnishes all the carbon used by the organisms. The Wood-Werkman reaction in heterotrophic organisms supplies only a fraction of one percent of the carbon of the cell. An excess of CO_2 in the atmosphere inhibits some microorganisms. In some cases the inhibition is due to the lowering of the pH of the medium; in others it is considerably more complicated. With the autotrophs, CO_2 is required as their sole source of carbon.

Free oxygen was regarded as an essential for all life until Pasteur demonstrated that, although some microorganisms require it, other species not only grow without it but are actually poisoned by free oxygen. These organisms are called **anaerobes,** and the organisms requiring gaseous oxygen are called **aerobes.** Anaerobic life is similar to aerobic life, in that energy-yielding processes involve oxidations which occur by the addition of water to the molecule being oxidized, followed by the removal of hydrogen atoms. However, in aerobes, hydrogens are ultimately transferred to atmospheric oxygen to yield H_2O as the end product while oxygen is used up from the gaseous environment. In anaerobes, hydrogens are ultimately transferred to some other molecule and become a part of ethyl alcohol, lactic acid, ammonia, or other organic or inorganic end products. Some organisms are capable of both aerobic and anaerobic life; they are termed **facultative** and are thus distinguished from the **obligate** aerobes and anaerobes. Other organisms require gaseous oxygen but grow best at a concentration less than that which exists in air. Such organisms are called **microaerophiles.**

A medium containing such oxidized substances as nitrate or permanganate will fail to support the growth of anaerobic organisms even if oxygen is excluded. If a crystal of sodium sulfide or some other highly reducing substance is placed in the medium, anaerobic organisms will grow in the immediate vicinity of the crystal and microaerophilic organisms will grow at some slight distance, while aerobic organisms will not grow in the medium anywhere near the reducing substance (Fig. 6.7). The medium varies with the distance from the crystal, and the measurement of this variation is expressed as the oxidation-reduction (O-R or redox) potential (Fig. 6.8). In electrometric measurements a platinum electrode is placed in a culture medium saturated with oxygen at pH 7.0. The resulting reading is about 800 millivolts more positive than that given by a standard hydrogen electrode. If oxygen is excluded so that anaerobes will grow, the reading may be about -400 millivolts. By varying the amount of oxygen or the amount of reducing substance, intermediate values are obtained. There are colorimetric indicators which show the O-R potential; thus methylene blue is deep blue at about $+50$ millivolts and colorless at about -100 millivolts on the O-R scale.

Desirable O-R potentials are more easily obtained than are desirable pH values. Since the common procedures used in the preparation of culture media result in satisfactory values, adjustment is generally unnecessary. High potentials for aerobes are obtained by growing them on an agar surface, while low potentials for anaerobes are found at the bottom of a deep tube of agar. For optimum growth of aerobic organisms, the amount of oxygen required may be very large, and the medium must either be agitated or dispersed in shallow layers. To secure the low O-R potentials necessary for anaerobic growth, oxygen is excluded.

Fig. 6.7. On plates of nitrate agar, aerobes were streaked upward to the left; anaerobes were streaked upward to the right, and microaerophiles were streaked horizontally as indicated by the three arrows on the upper plate. If all had grown (which they would not) the plate would have appeared as indicated. When a crystal of the highly reducing substance, sodium sulfide, was placed in the center, the anaerobes (wavy lines) grew only near the crystal, the aerobes (straight lines) grew near the edge of the plate, and the microaerophiles grew in between. Differences in oxidation-reduction potential for the three areas with a platinum electrode can be measured.

SURFACE TENSION

Surface tension is a measure of the cohesive forces between molecules at the interface between liquid and air. The surface tension of water is about 72 dynes per centimeter at room temperature. Certain chemicals, such as salt, added to a medium will raise the surface tension slightly; others such as

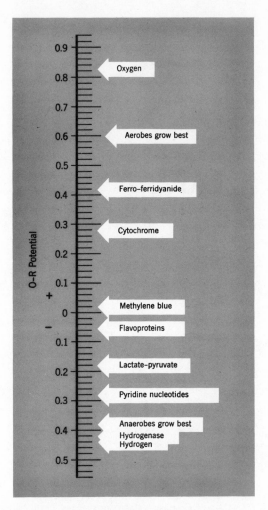

Fig. 6.8. At pH 7.0 the oxidation-reduction potential ranges from −0.4 to +0.8 volt. Methylene blue becomes colorless at values below 0. Some important biological chemicals are indicated at the potential where they exist as half oxidized and half reduced.

soaps lower the surface tension markedly. Gram-positive organisms generally prefer high surface tension. Gram-negative organisms, especially those capable of growth in the intestinal tract of animals, do very well in a medium with low surface tension. A medium with low surface tension favors the growth of intestinal bacteria by inhibiting or killing other forms. The tubercle bacilli carry a waxy surface coat and float on the surface of a medium. More vigorous growth is achieved in a medium containing a chemical which lowers the surface tension, because under such conditions the cells sink into the medium, are wetted on all sides, and thus present more surface for absorbing food materials and excreting wastes.

Variations in the atmospheric pressure of the earth's surface have no measurable effect on the life and death of microbes. Measurable killing does not occur until pressures of at least 5000 lb/in.^{2} are applied. However, microbes from ocean sediments often fail to grow when brought to the surface, as the pressure to which they are accustomed may exceed 12,000 lb/in.^{2}. These are **barophilic** organisms and can be cultured, if the medium is incubated in a high pressure chamber. This pressure effect must not be confused with the autoclave pressures of 15 lb of steam per square inch, where the effect is due entirely to the higher temperature of steam under pressure.

TOXIC MATERIALS

The presence of toxic materials often has a predominant influence on the growth of microorganisms. These toxic materials may be such ordinary substances as mineral salts, vitamins, or amino acids if they are present in unusually high concentrations. Or they may be normal constituents of fresh milk, eggs, blood, soil, and natural waters. Even the broths prepared by the bacteriologist often contain chemicals inhibitory, or even lethal, to certain microorganisms. Some of these substances can be destroyed by heating or may be bound in an inactive form by the addition of an adsorptive colloid. Such addition of colloidal material to a culture medium will improve microbial growth by removing or neutralizing some toxic element rather than by supplying new food. Increased growth resulting from the addition of starch, protein suspensions, and extracts often results from adsorption of soluble inhibitory substances by the particulate material.

Some toxic materials are byproducts of microbial metabolism. Accumulated acids or alcohols, formed from sugar by some organisms, and the protein decomposition products from dead and dying organisms effectively inhibit microbial growth. When organisms produce hydrogen peroxide and do not form the peroxide-destroying enzyme, **catalase,** the curve cf the death phase descends abruptly.

Sources of Energy

PHOTOSYNTHESIS

The green- and the purple-pigmented bacteria use light as a source of part or all of their energy. The autotrophic bacteria that are photosynthetic and independent of organic matter are termed **photolithotrophic,** which implies feeding by light and minerals. The rest of the photosynthetic bacteria require organic matter; these are **photoorganotrophic.** Photosynthesis involves splitting of electrons from the chlorophyll and attachment to a coenzyme together with a hydrogen ion. The resulting hydrogen atom serves as fuel for the cell, but the electron removed from the chlorophyll must be replaced. The algae and higher plants secure the electron by splitting water and thus releasing oxygen gas. The photosynthetic bacteria employ a more primitive and less efficient method which makes them dependent either on organic matter or hydrogen sulfide. During the transfer of the electron, energy is released to the cell and is stored by hooking on additional phosphate to the coenzyme-type compound **adenosine diphosphate** (ADP) to make the versatile energy-rich **adenosine triphosphate** (ATP). (Fig. 6.9).

Bacterial photosynthesis differs from that of the algae and the green plants in three important ways. First, in bacterial photosynthesis, oxygen gas is never produced; in fact, green and purple bacteria are usually obligate anaerobes. Second, photolithotrophs use hydrogen sulfide and photoorganotrophs use organic matter to secure electrons. The algae and green plants substitute water for this purpose. For this, two photons of light are required in contrast to the single photon needed for each photosynthetic event by the bacteria. Third, the active pigments differ chemically, and therefore the bacterial chlorophyll absorbs at the very long wavelengths of light while the chlorophyll of algae and other green plants absorbs at two wavelengths. One of these is shorter than that encountered in bacterial chlorophyll and the other is moderately long. Consequently, the photosynthetic bacteria can grow deeper in pond water than the algae, since they are able to use wavelengths of light that pass through the pond scum. All of these bacteria contain green chlorophyll pigments but, in addition, have red, orange, and brown pigments that sometimes mask the green color.

The CO_2 taken up during photosynthesis is added to organic molecules in the cell. The hydrogen atoms produced in photosynthesis by the action of light on chlorophyll reduce the combined CO_2 to stable organic molecules, which can be used for building blocks or stored as sugars or polymers of sugar such as starch or cellulose. The newly formed organic molecules now contain part of the energy captured from the light, energy that will be released when heterotrophic organisms break up the molecules.

ANAEROBIC HETEROTROPHS

During breakdown reactions, molecules characteristically release chemical bond energy. Living cells secure this energy by shifting the chemical bonds and rearranging the molecules. Although isomerizing, splitting, adding, and

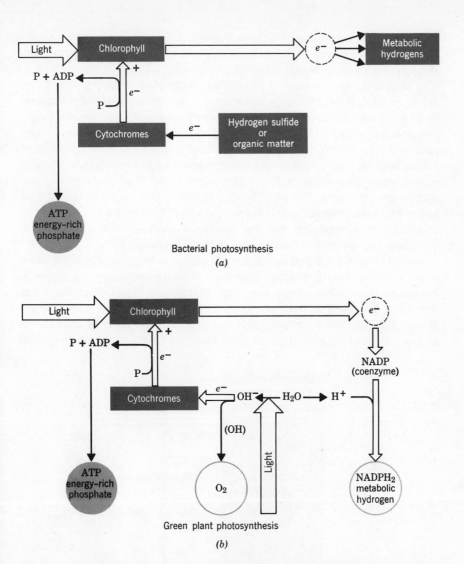

Fig. 6.9. A concept of photosynthesis. (a) In photosynthetic bacteria the light displaces an electron (e^-) from chlorophyll. Electrons can unite with hydrogen ions (H^+) to form the metabolic hydrogens (as reduced NADP or NAD) needed to reduce CO_2 to carbohydrate. The electron is replaced in the chlorophyll from hydrogen sulfide (in photolithotrophs) or organic matter (in photoorganotrophs). This electron passes through the cytochrome-enzyme system and in doing so generates an energy-rich phosphate bond. (b) In algae and other green plants the electron displaced from the chlorophyll is replaced by one secured from splitting a water molecule. Whether light is needed to split the water molecule is not certain. But the OH radical, having given up the electron, combines with itself to release gaseous oxygen thus: $4\ OH \rightarrow 2\ H_2O + O_2$. As with the bacterial photosynthesis, the electron removed from the chlorophyll unites with a hydrogen ion to yield metabolic hydrogens, which serve to drive forward the synthetic activities of the cell.

transferring enzymes may participate in the process, the energy release step is dependent primarily on oxidizing and reducing enzymes. Energy release through oxidation can occur aerobically or anaerobically.

In industrial terminology, **fermentation** implies any change in organic matter produced by microorganisms. Strictly, fermentation is the energy release from organic molecules under anaerobic conditions where the oxidation and the reduction involves removing hydrogen from one organic molecule and transferring it to another. A yeast ferments one molecule of a simple sugar ($C_6H_{12}O_6$) to produce two molecules of alcohol (C_2H_6O) and two molecules of carbon dioxide (CO_2). The total number of carbons, hydrogens, and oxygens in the alcohol plus the CO_2 is exactly that found in the original sugar. The yeast has extracted some of the energy. A consideration of the structural formulas shows how the process involves oxidations and reductions even though oxygen from the air does not participate (Fig. 6.10). As can be seen in Fig. 6.10, some of the carbon atoms in the products have more oxygen and less hydrogen attached to them than the corresponding atoms in the glucose from which they came; these atoms have been oxidized with the release of energy, and the hydrogens have been placed on other carbon atoms. Although the equation shows a single step, the process actually requires a number of distinct steps, each involving isomerizing, splitting, add-

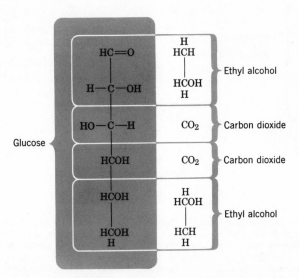

Fig. 6.10. The 6-carbon glucose molecule is split into four pieces in the alcohol fermentation. The hydrogens are removed from those carbons that end up as CO_2 and are accepted by those fragments that end up as ethyl alcohol. These oxidations and reductions net the cell energy sufficient to make two energy-rich phosphate bonds. The ethyl alcohols still contain most of the chemical bond energy which could be released to the cell by aerobic mechanisms.

ing, transferring, or oxidizing enzymes. This involves a sequence called glycolysis or the Embden-Myerhoff pathway. Some of the bond energy is lost as heat. The rest is transferred to energy-rich phosphate bonds (as in ATP) by special enzymes and processes which involve the transfer of electrons in the same manner as occurs in photosynthesis. The energy-rich phosphate bonds are energy packets of a convenient size for use in the energy-requiring reactions of biosynthesis and, like the genetic code, are universal throughout biology.

The term **intermediary metabolism** summarizes the enzyme reactions going on in conversion of sugar to end products. Many of the steps serve a double purpose in yielding fragments readily used by the biosynthetic enzymes. Ethyl alcohol and carbon dioxide are fermentation end products of most yeasts and a few bacteria. Instead, or in addition, other bacteria may yield lactic acid, formic acid, acetic acid, propionic acid, butyric acid, acetone, butyl alcohol, isopropyl alcohol, acetoin, and hydrogen gas. When proteinaceous material is fermented, ammonia and hydrogen sulfide are produced in addition to acids and alcohols. Anaerobic decomposition of protein is termed **putrefaction.** The important feature of any anaerobic process is that a large part of the organic compound which was oxidized must remain in the medium to accept the hydrogen. Since only a small part of the energy of an organic compound is extracted by this process, the anaerobic organism must work over tremendous amounts of material and leave large amounts of fermentation products in the medium. A microorganism may ferment its own weight in sugar in less than the time for one cell division.

Some anaerobic organisms are able to oxidize organic compounds and transfer the hydrogen to inorganic compounds such as nitrate or sulfate. When inorganic substances other than oxygen are employed as the final hydrogen acceptors, the process is called **anaerobic respiration.** Since nitrate is reduced to nitrite, ammonia, or nitrogen gas, and sulfate is reduced to hydrogen sulfide, these processes are also termed **nitrate reduction** and **sulfate reduction.**

AEROBIC HETEROTROPHS

In microbiology the term **aerobic respiration** describes the energy-yielding process in which oxygen gas participates. Hydrogens (and electrons) removed from the substrate are transferred to atmospheric oxygen through a series of enzymes and coenzymes. Each pair of hydrogens transported in this way in the aerobic system may yield up to three energy-rich phosphate bonds. In contrast, the anaerobic system yields only one energy-rich bond when a pair of hydrogens is taken from one molecule and transferred to another. In addition to giving greater energy yield for each pair of hydrogens, the aerobic system permits the entire molecule to be oxidized since, because of the participation of atmospheric oxygen, no molecules need be left in the medium to accept hydrogens, as is required in anaerobic systems. At least twenty different enzymes of all the functional types are involved in aerobic respiration. The process includes adding water, removing and transferring hydrogens, splitting out carbons in the most oxidized form, CO_2, and producing energy-rich phosphate bonds. Several aerobic pathways exist; most start with an anaerobic fragmentation of the sugar to organic acids, upon which is superimposed an aerobic pathway such as the tricarboxylic acid cycle (also called the citric acid cycle). The relative efficiency of aerobic versus anaerobic processes is well demonstrated in yeast, one of the many facultative microorganisms. Aerobically, a large amount of yeast is produced from respiration of a small amount of sugar or other organic food; anaerobically, a small number of cells is produced from the fermentation of prodigious quantities of sugar.

The ideal aerobic system found in higher animal tissue and many aerobic microorganisms oxidizes an organic compound completely, extracting all available energy and converting all of the carbons to CO_2. The hydrogens that are removed and their accompanying electrons are passed by enzyme systems to oxygen through a series of three hydrogen or electron carriers. This breaks up the 50 calories of energy released in the reaction $2H + O = H_2O$ into three parts. Each part is enough to make one energy-rich phosphate bond. Many aerobic microorganisms have less than perfect enzyme systems and carry out incomplete oxidations. Such aerobic organisms may oxidize a sugar partially and return to the medium such acids as gluconic, acetic, citric, and oxalic. For example, vinegar bacteria are aerobic organisms useful to man because they produce acetic acid by incomplete oxidation of alcohol. Other microbes oxidize other substances incompletely either because they are missing some part of the total ideal aerobic system or because they are cultivated on a medium that makes part of the system nonfunctional.

ENERGY FROM RESPIRATION OF INORGANIC SUBSTRATES

Autotrophic microorganisms that are unable to use light as an energy source must oxidize inorganic substances to secure the energy needed to build their organic structures entirely from CO_2. These bacteria have been termed **chemolithotrophs.** While their numbers are not large, they carry out certain very important steps in the cycles of the elements. These include the oxidation of inorganic sulfur compounds, the oxidation of inorganic nitrogen compounds, the oxidation of hydrogen gas, and the oxidation of the reduced forms of iron and manganese.

In the oxidation of ammonia the genus of bacteria, *Nitrosomonas,* produces nitrite ($-NO_2$) or nitrous acid (HNO_2). Another bacterial genus, *Nitrobacter,* oxidizes the nitrite to nitrate or nitric acid (HNO_3), the most oxidized form of nitrogen. These energy processes are similar to those used by heterotrophic organisms (chemoorganotrophs) in the oxidation of organic materials. Hydrogen atoms are removed by enzyme systems and are transferred to oxygen with the release of energy; at the same time CO_2 is taken up by the autotroph, built into a more complex carbon compound, and stabilized by reduction with hydrogen atoms.

Some sulfur-utilizing autotrophs start with hydrogen sulfide, others with sulfur itself, and still others with thiosulfate. The end product is usually sulfuric acid (H_2SO_4), the most oxidized form of sulfur. The hydrogen bacteria convert gaseous hydrogen into metabolic hydrogens and transfer them to oxygen gas, securing energy from the process. The iron bacteria oxidize soluble ferrous ions (Fe^{2+}) to ferric (Fe^{3+}) ions. This oxidation is also accomplished by the removal of hydrogen:

$$Fe\,(OH)_2 \quad + \quad H_2O \quad - \quad H \quad = \quad Fe\,(OH)_3$$
$$\text{ferrous hydroxide} + \text{water} - \text{hydrogen} = \text{ferric hydroxide}$$

This hydrogen is used to produce energy-rich bonds necessary for biosynthesis in the organism.

Summary

The cell has no fuel but electrons; often "hydrogen" is substituted for electrons in this statement because electrons are usually transferred as hydrogen atoms, that is, electrons attached to protons (H^+). Photosynthetic organisms use light energy and chlorophyll to secure the electrons. The heterotrophic organisms split electrons (with attached hydrogen ions) from organic molecules, and the chemoautotrophs extract them from inorganic molecules. Anaerobically, hydrogens are transferred to some organic or inorganic hydrogen acceptor, which then appears in the medium as the product of the fermentation or anaerobic respiration; a relatively small amount of energy is thus yielded to the cell. Aerobically, the electrons are transferred to oxygen in a three-step process, yielding much larger quantities of energy. Some of the energy is dissipated as heat; the rest is used by coupling energy-yielding with energy-consuming reactions. Energy-rich phosphate bonds are important intermediates. Energy is also stored by the organisms in the chemical bonds of the molecules making up the fats, carbohydrates, and proteins of its structure.

The division and growth of bacteria are useful models for studying cell division and growth in other forms. The environmental factors that control the extent and rate of growth can be studied as independent parameters. Microbial life is immensely variable as compared with the restrictions that apply to the cells of higher forms and many processes that have been adequate for sustaining some types of microbes for billions of years were rejected in the evolutionary development of higher forms. On the other hand, most biological processes that are found in higher forms are also encountered in some of the bacteria.

SEVEN

Classification of Bacteria

Scholars characteristically arrange the objects of their study in an orderly fashion, grouping related forms and assigning descriptive names. In the case of a plant or an animal, the anatomy or morphology indicates relationships to other plants or animals. Therefore, an understanding of similarities between individual organisms allows the scientist to form groups, to write descriptions, and to assign names on the basis of evolutionary relationships. However, morphology is not sufficiently varied among bacteria to afford very extensive groupings. Furthermore, in bacteria, similarities in form do not necessarily indicate natural relationships. In classifying these simple, unicellular organisms, such nonmorphological characteristics as physiological, biochemical, and, in some instances, serological differences must be considered. Such "serological" differences are detected by inoculating the organisms into laboratory animals and observing changes in the chemistry of the blood **serum** components which are distinctive of the organisms inoculated. Consequently, the schemes of classification suggested for bacteria are not based on evolutionary criteria, are subject to frequent revision, and probably will continue to change as new information becomes available.

Bacteria were first considered animals. Leeuwenhoek spoke of them as "little beasties." To some workers they constituted a third kingdom of living things—the *Protista*. It was early recognized that they had characteristics which made them more plantlike than animallike, so they were placed in the plant kingdom. Most microbiologists were in accord with this view but current sentiment has shifted in the direction of a distinct kingdom for the microbes.

Even though bacteria cannot be classified satisfactorily on the basis of their natural relationships, the same general principles laid down for the classification and naming of plants and animals are followed. That branch of biological science which deals with classification is termed **taxonomy;** the system of names is called **nomenclature.**

Descriptions and names of various plants, animals, and microorganisms have been assembled over the years into a volume on taxonomy with a key which permits tracing an unknown to its proper place in the scheme. There is no official system of taxonomy for bacteria, but a semiofficial system, presented in Bergey's *Manual of Determinative Bacteriology* is very widely accepted. Descriptions of almost 2000 organisms are included in this publication. It should be noted that some bacteriologists do not accept the Bergey system in all its details.

The Bergey *Manual* suggests that the blue-green algae, the bacteria, the viruses, and rickettsiae be placed in a division known as *Protophyta,* or primitive plants. The blue-green algae, placed in a class known as the *Schizophyceae,* are studied along with the other types of algae and the higher fungi in cryptogamic botany. Those organisms which are normally regarded as bacteria, the so-called higher bacteria, and a number of less familiar types have been placed in the class *Schizomycetes.* Organisms, so small that they will pass through filters which retain the more familiar bacterial types, have been allocated to a class designated *Microtatobiotes.*

Bacteria occur in nature in mixed cultures; pure cultures are uncommon. In order to establish the characteristics of an unknown organism and assign it a name, a study of a pure culture is necessary. Pure cultures may be obtained by standard procedures (Chapter 3), and the culture so obtained, prior to identification, is referred to as an **isolate,** or, since the methods of isolation separate the descendants of one individual cell, it is also called a **clone.** If there is some reason why a particular isolate should be differentiated from other isolates, it may be designated as a **strain.** This designation may be a number, as putrefactive anaerobe No. 3679; it may be a letter, or letters, or the name of a place or person, as *Staphylococcus aureus* H, or *Bacillus subtilis* Marburg.

Often the isolate is identified in a taxonomic study by determining its characteristics, organizing the data obtained, and comparing the data with that given in the taxonomic key. The bacterial characteristics commonly studied include some or all of the following:

1. **Cell morphology,** including, in addition to cell size, cell shape and arrangement: staining reaction, size, flagella (if present, their distribution on the cells), the presence or absence of capsule, spores, and cell inclusions.

2. **Culture morphology,** including the size and shape of colonies, the appearance of the growth in liquid, solid, and special media.

3. **Physiological and biochemical reactions,** including the temperature range of growth, oxygen requirements (aerobic, anaerobic, or microaerophilic), utilization of various carbon and nitrogen compounds incorporated into the medium, and formation of end products of metabolism, such as acids, gases, sulfide, ammonia, and so forth.

4. **Disease production** in either animals or plants in the cases where there is some indication that the organism under study is pathogenic.

5. **Serological reactions** (serology: the study of the properties and reactions of blood serum) sometimes may be a determining factor in identification of an unknown. The killed organism is injected into an animal, in which it stimulates the formation of specific chemical substances called **antibodies.** The antibodies, which are contained in the fluid serum portion of the blood, will react specifically in laboratory test systems with all cultures of this species or variety (**serotype**), but not with unrelated species and serotypes.

Since sexual reproduction is not common in bacteria, a species cannot be defined in the terms used for plants and animals. Modern microbiological research indicates that the results of genetic recombination usually confirm the relationships defined by other taxonomic criteria. Nevertheless, some adjustments are necessary. Recent experiments reveal that when organisms are closely related, the melting points of their DNA are almost identical because the molecular constitution of these substances are very nearly identical. Such differences can also be detected by measuring the densities of the DNA in the ultracentrifuge and can be confirmed by the more tedious methods of chemical analysis.

The DNA, isolated from closely related organisms, will intertwine when a mixed aqueous solution containing the chemicals is heated and cooled. Unfortunately, differences in the properties between very closely related organisms are small and the identification tests of the microbiologist are still essential. Until research on DNA progresses much further, bacterial species must be regarded as a group of isolates which the classical study, using morphology, physiology, etc., characterize as more like each other than like any other group. In some instances individuals within a species may differ, but not sufficiently to justify creation of another species designation. For such minor variations, designations of **subspecies, serotypes,** or **variety** may be added as a part of the species name.

The isolate or clone is the smallest unit of bacterial taxonomy. The possible number of isolates is infinite, since it is limited only by the amount of work done in securing additional pure cultures. To prepare a classification scheme identical or nearly identical, isolates are grouped into species. More commonly, the problem confronting the microbiologist is that of classifying a newly isolated organism. Once the characteristics of an isolate have been determined and organized, the organism is traced to its proper place in a descriptive key such as Bergey's *Manual*. Bacteria are members of the phylum *Protophyta* and the class *Schizomycetes*. The sequence of divisions, or categories, remaining which must be followed in tracing through the key are:

Order. The *Schizomycetes* are divided into ten orders, designated by the suffix **-ales.** The largest order is *Eubacteriales*, with 607 named organisms and about 350 serotypes. The smallest order is *Hyphomicrobiales*, with only four organisms included.

Suborder. Some of the orders are divided into suborders, representing a slight narrowing of the characteristics of a broadly related group of organisms. The suborder consists of organisms which share certain significant characters not shown by other suborders of the parent order. The suffix which designates a suborder is **-ineae,** as in *Pseudomonadineae*.

Family. Further narrowing of shared characteristics results in the grouping known as the family. The suffix which designates the family is **-aceae,** as in *Enterobacteriaceae*.

Tribe. Sometimes the family constitutes such a broad group that it must be subdivided into tribes in order to make the descriptive key readily usable. The ending which denotes a tribe is **-eae,** as in *Salmonelleae*.

Genus. The tribe (or family, if there is no division into tribes) is subdivided still further to give the genus for which there is no special ending.

Species. The final narrowing of characters gives the species, which like the genus, has no characteristic ending (Table 7.1).

The first described member of a species is the **type species.** The type species should be thoroughly characterized in order that comparisons will readily identify later isolations. Such species are kept in stock culture collections where they are available for comparison, like animal specimens in a museum.

If a previously undescribed organism is isolated, certain principles of nomenclature must be followed in assigning a name. The name should be a binomial, the first word designating the genus; the second, the species. These names are Latin, and though they may be derived from some other language, they must be Latinized. When written in English, Latin and other foreign words are italicized or underlined. The generic name is a noun in the singular and is written with a capital letter. It may be descriptive of some characteristic of the organism, such as *Aeromonas*, a gas-producing bacterium. The generic name should represent some important feature of the organism, but it may be given in honor of an individual, usually a scientist, who has studied the organism. *Escherichia* is named for the bacteriologist, Escherich, who first described the organism.

The second word of the binomial is the specific epithet, which is not capitalized. It may be an adjective, in which case it must agree in gender with the

Table 7.1 Genus and Species Names

Name	Meaning
Bacillus (generic name)	Rod shaped
subtilis (species name)*	Slender
mycoides (species name)	Fungus like
cereus (species name)	Wax colored or waxen
Serratia (generic name)	Honoring the Italian scientist Serrati
marcescens (species name)	Decaying
Escherichia (generic name)	Named for the bacteriologist, Escherich
coli (species name)	Living in the colon
Streptococcus (generic name)	Flexible chain of cocci
pyogenes (species name)	Pus producing
equi (species name)	Causing a disease of horses
agalactiae (species name)	Causing a disease which stops milk production
Staphylococcus (generic name)	Irregularly clustered cocci
aureus (species name)	Golden pigment
Proteus (generic name)	Referring to marine god who could assume any shape
vulgaris (species name)	Common
Clostridium (generic name)	Spindle shaped (as a common name with a small c)
botulinum (species name)	Sausage-shaped
perfringens (species name)	Breaking through
Pseudomonas (generic name)	False (pseudo) and a unit (monad)
fluorescens (species name)	Producing a fluorescing pigment
synxantha (species name)	Producing a yellow pigment
syncyanea (species name)	Producing a blue pigment
mephitica (species name)	Producing a skunk-like odor

*Used alone this is the specific epithet; the species name results when used in combination with the generic name, as *Bacillus subtilis.*

noun it modifies, as *Bacillus albus,* the white bacillus, and *Sarcina alba,* the white sarcina. It may be an adjective in the form of a present participle, in which even the ending will be the same for all genders, as *Clostridium putrefaciens* or *Bacillus putrefaciens.* It may be a noun in the genitive case modifying the generic name, as *Clostridium belfantii,* the clostridium of Belfant, who was an Italian bacteriologist.

Taxonomic studies of a number of strains of a species may establish more precisely the characteristics significant for classification. Variability is a property of all living matter, hence comparative studies often reveal variations in characteristics. Theoretically, each test applied to the group permits division into two groups, those which react positively and those which react negatively. Some judgment is, therefore, essential in choosing which characteristics have taxonomic significance. The number of types possible increases at the rate of two raised to the power indicating the number of tests. This is the 2^n principle in taxonomy. If a single test is applied, for example, the gram reaction, only two groups are possible—one positive and one negative. If a second test is applied, for example, motility, a positive and a negative group is possible for each of the original two groups. For three tests the number of groups becomes 2^3 or 8 and for n tests it is 2^n. Reasonable and workable limits demand that only tests which yield significant data should be considered. The use of automated devices for collecting large volumes of data and of computers for processing it has stimulated study of "numerical taxonomy" which may eventually replace the standard nomenclature in the bacteria.

Summary

Taxonomy, or descriptive bacteriology, is not a glamorous or exciting subject. However, it is of great importance in every aspect of the science, and students should become familiar with it early in their study of bacteria.

The algae, fungi, and protozoa show greater complexity, and their taxonomic position follows precisely that used for higher plants and animals. Since these "higher" organisms usually have a sexual cycle, major emphasis in classification is attributed to the nature of the sexual stages.

EIGHT

Genetics of Bacteria

The study of bacterial genetics has supplied a better understanding of the operation of genetics on the molecular level. However, bacterial genetics is a relatively new science. With the development of pure culture methods, especially by Koch and his associates, the concept of **monomorphism** (the stability of single types) became so firmly established that variability of bacterial forms was derisively ascribed to culture contamination, or possibly, to degeneration of the cells of a pure culture. Eventually the doctrine of strict monomorphism died out and was succeeded by what has become an extremely active and productive field of research — the recognition and investigation of true heritable or genetic forces in these relatively simple cells. Genetic studies on higher organisms have revealed much about the cytology and the overall mechanics of inheritance.

Despite the relative deficiency of cytological data for microorganisms compared to the available information for higher plants and animals, the genetic terminology and interpretation of observed phenomena as seen in the higher forms have been largely transferred to the field of bacterial genetics although these terms do not always carry the same meaning. For example, the bacterial nucleus, although identical in function, does not have the same physical organization as the nucleus of higher organisms (Fig. 8.1). Thus, the term **nucleoid** was recently coined for the bacterial nucleus to make the distinction between the two types. Physical structures, the chromosomes, composed of deoxyribonucleic acid (DNA) and protein, carry the genetic information in eucaryotic cells. Genetic and cytological studies on plants and animals have established that genes (the DNA component), lo-

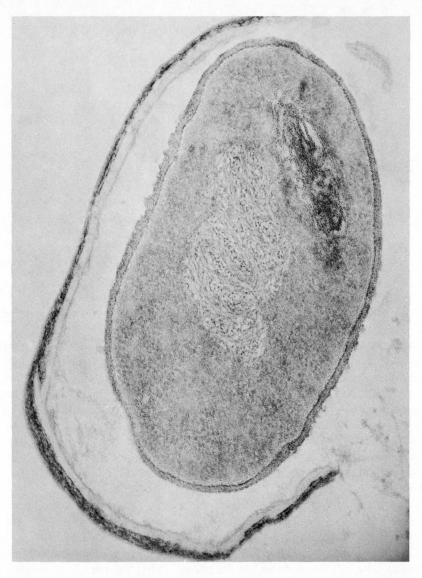

Fig. 8.1. The coiled nuclear thread of DNA is visible in this electron micrograph of a thin section of a germinating spore of Bacillus subtilis. Photo courtesy of C. F. Robinow.

cated along the threadlike chromosomes in the cell nucleus, are the determiners of the heritable characteristics. The genes not only carry the genetic code for the direction of the multitude of activities of the cell, but also possess the ability to replicate themselves. However, there are no chromosomes *per se* in the bacterial cell but only a double strand of DNA (Fig. 8.2). New terms have been introduced to describe cytological and functional situations that occur only with the microbes or that can be studied more exhaustively and precisely with these organisms.

When a cell of a higher organism divides by mitosis, the chromosomes also divide, and an identical set of genes goes to each daughter cell. Unless one or more of the genes changes, the two cells will be identical in all respects. If one of the genes changes, or mutates, the characteristic governed by the altered gene is different from the comparable characteristic in the "wild type," the cell with the unchanged gene.

Each chromosome carries a number of genes and, since chromosomes are usually transmitted as a unit, the genes of any single chromosome are likewise transferred as a unit and are spoken of as **linked** genes. In higher organisms, the sex cells carry only one chromosome of each linkage group. Such cells are called **haploid.** When fertilization takes place, the number of chromosomes found in the zygote will be twice that found in the sex cells. Such a cell is termed **diploid.** If each chromosome is represented more than twice, such a cell would be **polyploid.**

In diploid organisms, each chromosome, and likewise each gene, is paired. A gene located at a particular site on one chromosome is matched by a corresponding gene at a similar location in the homologous chromosome. Frequently, the members of a gene pair produce different effects. Such genes are called **alleles.** If one dominates, covering the detectable effects of the

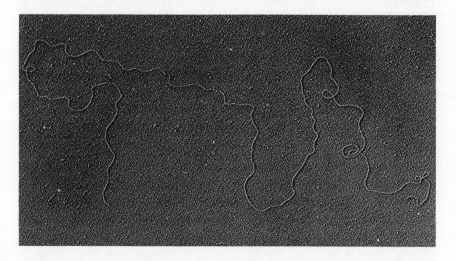

Fig. 8.2.   Electron micrograph of a DNA molecule. This DNA was extracted from E. coli **bacteriophage T7, shadowed with platinum. Magnification (21,000×) is still not sufficient to see the double helix.** Courtesy of D. Lang.

other pair member, the gene is said to be **dominant.** The opposite member of the pair is called **recessive.** If the gene pairs are identical, the cell is said to be **homozygous.** A cell containing a dominant and a recessive gene is **heterozygous.** Since bacteria are haploid, they ordinarily behave as homozygous cells, although a cell with two nuclei can behave as if it were a heterozygous diploid if any of the gene pairs of the two nuclei are unlike in the effects produced.

Mitosis does not occur in bacterial cells, but cell division occurs through the process of binary fission or amitosis. Cytological evidence of bacterial chromosomes is moot, although the genetic evidence definitely demonstrates true heredity-governing units that have an orderly arrangement and are precisely proportional between the two daughter nuclei upon cell division.

The sum total of the genes of an individual is its **genotype,** and is heritable. The morphological appearance and physiological behavior of a genotype or in conjunction with the environment is the **phenotype** or outward characteristic of the organism. The phenotypic differences are frequently termed modifications, adaptations, or environmentally-induced variations. The idea that phenotypic differences can be transmitted has been very attractive at times to inadequately trained individuals, but the idea has not survived the test of scientific analysis.

There are two types of bacterial variation, **environmentally-induced variations** and **mutations.** The environmentally-induced variation involves only a phenotypic change. Therefore, the expressed phenotype is not inherited by the progeny cells since there is no alteration in the genotype. All of the cells within the population show the new characteristic. Once the environment inducing the new phenotypic expression is removed, reversion to the original phenotype occurs. Genotypically motile bacteria may lose their flagella and become nonmotile when grown in broth containing 0.1% phenol. They may be repeatedly transferred in such a medium wherein for months or even years they possess no flagella; but when transferred to suitable phenol-free medium, they develop flagella and resume motility. The bacterium, *Serratia marcescens,* produces a red pigment during colonial growth at 25 C, but the cells are unable to produce the pigment when grown at an elevated temperature. The higher temperature appears to inactivate one or more of the enzymes which are functional only at the lower temperature.

The second type of bacterial variation occurs through the process of mutation. The genotypes resulting from the division of a cell are duplicates in morphology and physiology of the parent cell within the limits imposed by the environment. In a very low percentage of cases, a gene may change or mutate. That is, a chemical change has occurred in the genetic code of the cell, and therefore, the mutant genes give rise to new genotypes. If the mutation occurs in a diploid cell and is recessive, it may not be detected for several generations. In uninucleate haploid organisms a mutant gene, unless lethal, expresses itself promptly.

Mutations may occur spontaneously and may be induced experimentally. **Spontaneous mutations** are those for which no definite cause is known; probably factors such as radiation and chemicals produced during growth influence mutations. The rate of mutation is usually expressed as the rate per bacterium per cell division. The spontaneous rate is fairly constant for any one gene controlling a particular characteristic, but it varies among different genes, as can be illustrated with a few of the rates which have been determined, such as sulfathiazole resistance for *Staphylococcus aureus,* 1×10^{-9}, and penicillin resistance, 1×10^{-7}, and streptomycin resistance (1000 micrograms) of *Escherichia coli,* 1×10^{-10} (Fig. 8.3).

Mutations of all types occur constantly in a bacterial population. For exam-

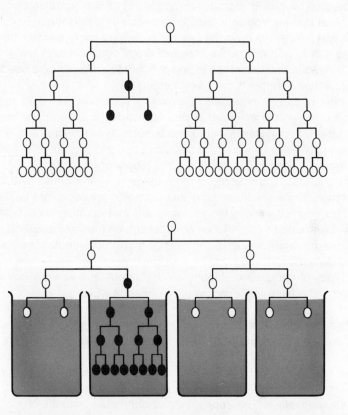

Fig. 8.3. A mutation (black dots) occurs in the bacterial colony in the top diagram. Under usual circumstances, the mutant divide more slowly than wild type (open circles), are soon overgrown, and make only an insignificant contribution to the total population. In the bottom diagram, the cells are placed in a selective environment, e.g., in broth containing an antibiotic. There the wild type do not grow at all, and the mutant dominate the final population.

ple, drug-resistant mutants are appearing in populations that have never been in contact with a drug. In an ordinary medium, such mutant organisms are outgrown by the wild-type population; only when the drug is present and the wild type is inhibited or killed do the mutants dominate the population. Consequently, such a mutation to resistance is a survival factor.

The spontaneous occurrence of mutants, prior to contact with the selecting environment, can be demonstrated by the indirect selection test using replica plates. In this procedure, the colonies are transferred en masse to other plates by the use of a velvet-covered wood or metal disc. The fabric is pressed gently onto the surface of the culture plate, and the threads of the fabric pile pick up numerous organisms from each colony. The velvet is stamped on several plates containing different kinds of selective media designed to select mutants from the wild-type population. The colonies which grow develop in the same spatial arrangement on the stamped plates as on the original culture, and thus their origins can be determined. Mutants that appear can be identified on the original plate, and every transfer from such colonies yields mutants, while transfers from other colonies yield nonmutants. The replica plate method is one of many techniques developed for the rapid detection and identification of mutants.

Mutation rates are increased by exposing populations of cells to various physical or chemical agents that interfere with the DNA of the cell. Such agents are called **mutagens.** Radiations such as ultraviolet and X-rays or chemical substances such as peroxide, mustard gas, nitrous acid, acriflavine are frequently used in the experimental laboratory to obtain mutants with higher frequency. The majority of the mutagenic agents cause death as well as mutation; some scientists hold that such killings constitute lethal mutations. Those which survive the mutagen will mutate at a much higher rate than will untreated cells. Proper applications of mutagenic agents can increase the mutagenic rate 1000 times or more. Such mutations are called **induced mutants.**

The types of bacterial mutations are numerous and varied. They may be included in the following representative mutation types.

1. **Pigmentation.** Since color variation is affected by environment, standard conditions are essential in the study of color mutants. But when a culture shows a sudden inherited change in pigment production, the gene controlling this trait has probably mutated, and an enzyme involved in the synthesis of the pigment is absent or modified.

2. **Cell morphology.** The appearance of nonmotile mutants of motile species, variations in cell size, and the loss of the ability to form capsules or spores are representative changes in cell morphology that result from mutations. One of the very important manifestations of this type of mutation is seen in the somatic (body) **O** and flagellar **H** variation in colonies. Colonies of

motile *Proteus vulgaris* spread to cover the surface of a moist, solid plating medium in a form suggestive of the film produced by breathing on glass. Nonmotile variants remain as discrete colonies. These two colony types were designated as **H** (Hauch = breath) and **0** (Ohne Hauch, Ohne = without). The terminology has been adopted as laboratory shorthand and is used to describe flagellar antigens (H) and cell or somatic antigens (0) regardless of variation.

3. **Colony morphology.** Differences in cell wall constituents, capsules, and arrangement of cells during cell division result in differences in colonial morphology between the wild type and mutant strains. For example, members of the *Salmonella-Shigella* groups form both smooth and rough colonies (**S** and **R**). The normal parental type develops convex, smooth, glossy colonies with a smooth or entire margin on solid media. In contrast, the mutant colonies have an irregular margin and are dull and rough in appearance. The smooth type gives a diffuse growth in broth and is stable in physiological saline solution. The rough type, however, shows a granular growth in broth and forms clumps of cells in saline solution. **S** cultures tend to yield spontaneously **R** mutants which usually overgrow the wild type, especially as the culture ages. It is now known that the smooth wild-type organism loses the ability to synthesize certain cell wall constituents, therefore resulting in the change in colony morphology formed on an agar plate as well as the previously mentioned characteristics.

In addition to **S** and **R** colonies, some organisms may produce colonies which appear mucoid which is attributed to the presence of a capsule. The loss of the ability to synthesize capsule material is reflected by a loss in the mucoid nature of the colony (Fig. 8.4).

4. **Antigenic variation.** Antigens are certain chemical substances that stimulate the formation of antibodies in the animal body under suitable conditions. The antibodies will react specifically with the antigens that stimulate their production. The somatic or **0** antigens of a bacterial group, especially exemplified with the gram-negative bacteria, can be divided into a number of different components or antigens present in mosaics and in cell wall layers. Cells with a complement of surface antigens give **S** type colonies. A mutant cell can develop that no longer produces the surface antigens, exposing the next layer of the cell wall which is then revealed and able to manifest itself. The cell is roughly similar to an onion; peeling off one layer (antigen surface) reveals another layer. The encapsulated pneumococci cells show a variety of antigenic types depending upon the chemical nature of the capsule. However, the loss of the capsule reveals the somatic antigen (**0**) which is identical for all pneumococci.

Variation in H antigen is commonly referred to as **phase variation.** Many of the species of motile, enteric bacteria are diphasic, that is, the flagella may assume two antigenic types, each of which is an antigenic mosaic, or mix-

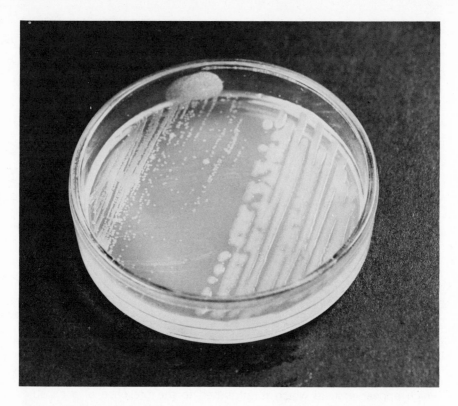

Fig. 8.4. A wild-type mucoid culture of Azotobacter agilis (right) and a rough mutant isolated from it (left) are streaked on the same plate.

ture of antigens. Originally, these were termed the specific and nonspecific phases, but they are now designated as phase 1 and phase 2. A culture in one phase is capable of giving rise to descendants in the other phase. The individual cells may have flagella of either phase, but not of both, as can be established by streaking and testing individual colonies. The phase is maintained by culture from a "pure" colony for a time but, after phase variation occurs, the culture contains both phases.

5. **Virulence and toxicogenicity.** For most pathogenic organisms, the virulent form is the **S** (smooth) or **M** (matt) form. A change from **S** to **R** (rough) or **M** to **S** is accompanied by a reduction in virulence or, in some cases, even a complete loss. Variation resulting in a loss in the capacity to produce a toxin by toxicogenic bacteria also appears to be dependent on **S-R** variation.

6. **Fermentation mutants.** Differences in the ability to ferment lactose by a strain of *Escherichia coli* were described by Massin in 1907. A mutant strain streaked on a differential lactose medium, such as eosin methylene blue

agar, may show only colorless nonlactose fermenting colonies (lac⁻). However, lac⁺ purple papillae appear on the surface of the colonies after several days of incubation (Fig. 8.5). Subcultures from the lac⁺ sector of the colony are prompt fermenters of lactose. Isolates from the lac⁻ portion of the colony repeat the pattern of the parent strain. Positive mutants that have gained or lost the ability to ferment other sugars are easily obtained and are useful tools for genetic studies.

7. **Nutritional mutants.** In the process of a mutation, some organisms gain or lose a nutritional requirement. Organisms which require additional growth factors (low molecular weight compounds such as amino acids, vitamins, nucleotides), not required by the wild-type strain (**prototroph**), are called **auxotrophs.** A biotinless auxotroph (biotin⁻) is a mutant strain that needs the vitamin biotin added to the minimal salt medium on which the prototroph (biotin⁺) can grow; a methionineless auxotroph (meth⁻) requires methionine added to a medium adequate for the prototroph (meth⁺). These nutritional mutants have been especially useful in the studies of genetic exchange between bacteria.

8. **Drug-resistant mutants.** The development of drug resistance, particularly to antibiotics and sulfa drugs, has become a problem of obvious major interest to clinicians as well as to the bacterial geneticist.

9. **Phage-resistant mutants.** Surprising as it seems, bacteria are subject to infection by a group of viruses called bacteriophages or phages. Bacteria can become resistant to phage infection by a mutation within its genetic code that results in a modification of the cell wall. Consequently, the neces-

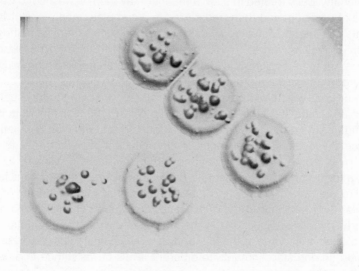

Fig. 8.5. Lac⁺ papillae on the surface of lac⁻ colonies of E. coli.

sary receptor sites required for phage adsorption are absent; thus, infection is impossible.

The detection of mutants in a culture, even when present in very small numbers (e.g., one mutant per 10^8 parent or wild-type cells), is a fairly simple matter when a very large proportion of the culture is plated on a medium sufficiently selective to allow only the mutant cells to form colonies. For example, drug-resistant mutants may be selected from a culture when plated on a medium containing an adequate quantity of the inhibitory drug sufficient to prevent growth of the sensitive wild-type strain. Mutants capable of synthesizing an amino acid required by the wild-type culture may be selected on a medium deficient in this amino acid. The detection of antigenic mutants, though more difficult, has been accomplished by repeated single cell isolations. When the bacteria are cultivated in medium containing specific antiserum for one of the antigenic types, colonial differences result between parent and mutant colony types. Nutritional mutants may be easily obtained by first exposing the wild-type culture to a mutagenic agent such as ultraviolet light. The culture is then transferred to a nutritionally-deficient medium such as a glucose-mineral salts broth containing penicillin. The surviving prototrophic cells, which are capable of synthesizing their nutritional requirements from the supplied ingredients, begin to metabolize and grow. However, the penicillin causes the cells to become osmotically-sensitive due to its effect upon the formation of new cell wall material. Auxotrophic mutants, which are unable to make all of their growth requirements from the food supplied, remain dormant. Thus, penicillin has no effect upon their viability. Cells treated in such a manner may then be washed to free the culture of penicillin and then transferred to a complete medium. Nutritional mutants can then be detected by using a method such as replica plating.

If an organism varies in response to an environment, the observed changes are often spoken of as an **adaptation** or training. Training is usually a procedure for selecting mutants. Cultivation of an organism on a concentration of drug which will prevent growth of the wild-type strain usually results in an isolate that is drug-resistant. The cells in the resistant culture are descendants of one or a few resistant mutants in the wild-type population at the time of inoculation. Resistance to high concentration of drugs, in most cases, may require several mutations within the same organism. In such cases, a low level of drugs selects a population of mutants resistant to that level. Transfer to a higher drug concentration selects mutants of the original mutants which are more highly resistant. Observation of this phenomenon led early microbiologists to assume mistakenly that the population was being trained or adapted to the higher drug level. In limited cases, organisms can sport, in one step, mutants to different levels of resistance to a certain antibiotic such as streptomycin. Since mutations to drug resistance can occur in the body of infected individuals, the importance of an adequate initial dosage of the chemotherapeutic agent is obvious.

Genetic Transfer Between Bacteria

Although no mutual exchange of genetic material between bacteria has been discovered, certain novel and amazing mechanisms are now known that involve the transfer of genetic material between the same or very closely related organisms. Certain general features characterize and limit the exchange of genetic material. The transfer is always unidirectional; that is, the genetic material passes from a donor to a recipient cell. The transfer is incomplete since only part of the donor's genetic material is transferred. In most cases, only one gene or closely-linked genes can be transferred at any one time. Consequently, the recipient cell is not considered a true zygote due to the incomplete transfer and, therefore, is considered only a partial diploid cell or **merozygote.**

At the present time, there are three known mechanisms for genetic transfer: **transformation, transduction,** and **conjugation** (Fig. 8.6). The principle difference between these mechanisms is their mode of transfer.

TRANSFORMATION

The simplest method of genetic exchange and, therefore, the most primitive evolutionary development in the direction of sexuality, is called transformation. It involves a mechanical feeding of DNA from one strain of bacteria to a second. If the recipient organism incorporates a gene from the DNA of the donor by genetic recombination, the difference will become a stable inherited trait in the progeny of the recipient organism (Fig. 8.7). Usually only a single gene is introduced by a single transformation. Much of the early experimentation in transformation was done with *Diplococcus pneumoniae* wherein the antigenic type of the capsule of the donor was transformed into the recipient. Since this fundamental work, many different characteristics have been transmitted from a donor strain to a recipient by DNA extracted from the donor bacterium. Furthermore, transformation has been discovered in a variety of other species of bacteria. Transformation offered the first convincing evidence that pure DNA carries the genetic information in the hereditary apparatus of the cell.

TRANSDUCTION

The second method involving the transfer of genetic characteristics is the accidental incorporation of a small, random piece of bacterial host DNA in a

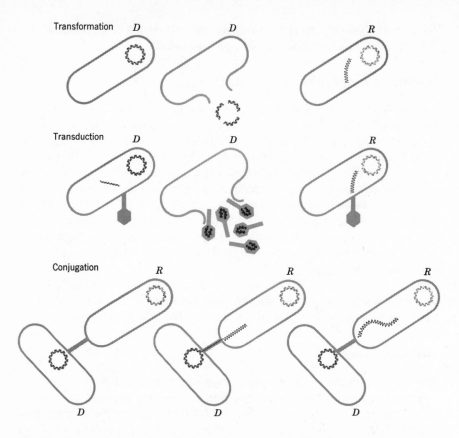

Fig. 8.6. Three methods of genetic exchange between bacteria are illustrated diagrammatically emphasizing only the mode of transfer. Genetic recombination is not indicated. D = donor cell; R = recipient cell.

temperate phage during maturation in the infected donor cell. When such a temperate phage infects another bacterium, the gene from the donor bacterium is introduced. The size of the genetic fragment is limited by the small size of the phage; therefore, usually only one gene is ever transferred, although occasionally several genes located close together on the donor's chromosome may be included. Only one bacterial phage particle per million transduces any given character. Each temperate phage may include some bacterial DNA, but since the bacterial chromosome is so large relative to the phage particle, there is only one chance in a million that the included genetic piece will carry the gene that is under study. Only certain temperate phages can transduce, and the recipient cells are not generally lysogenized by the transduction process. Of course, if the recipient bacterium is lysed by the transducing phage, the effect of the incorporated genetic material could not be observed.

A third type of genetic exchange in bacteria is the process of conjugation in which the DNA is passed directly from the donor cell to the recipient cell. When certain strains of *Escherichia coli* are mixed, fusion tubes form between individuals of the two strains, and genetic material from the donor cells passes through the tube into the recipient cells.

The phenomenon involves the possession by the cell of a **fertility** or "**F**" **factor** which confers upon the possessor the ability to initiate conjugation. The cell serves as the donor cell and is referred to as the male. Cells void of the F factor are the recipient or female cells.

The F factor behaves as a genetic determinant in addition to the usual chromosome of the cell. It is, therefore, an **episome.** It may exist as a cytoplasmic particle capable of autonomous replication and carries the genetic information necessary for the synthesis of the fusion tube or specialized pilus required for this type of genetic exchange.

When the F factor is not associated with the circular chromosome, the male is designated F$^+$. The bacterial genes of an F$^-$ female strain are arranged in a linear order on a circular chromosome. During conjugation with an F$^-$ cell, one of the replicated F factors from the F$^+$ cell is passed to the F$^-$

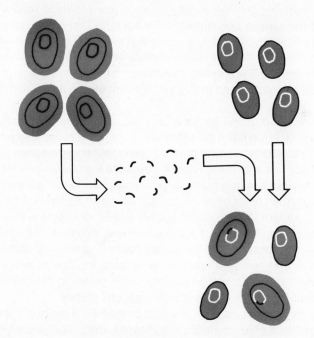

Fig. 8.7. Transformation involves chemicals or physical extraction of the DNA from a donor culture and feeding it to a recipient culture. In this illustration, the donors' gene for capsule formation becomes integrated into some of the nonencapsulated recipients' chromosomes.

cell, the F⁻ then becoming an F⁺ strain. During growth on unfavorable media, some male cells may lose the F factor; therefore F⁻ strains can be isolated readily from F⁺ strains.

In certain cases, the F factor in the male strain may associate itself with the circular chromosome. The male cell is then designated as an Hfr male (high frequency recombining). The bacterial chromosome breaks at the attachment point. Instead of the F factor passing first through the conjugation tube, the chromosomal thread with the F factor on the trailing end passes into the F⁻ cell. The amount of donor's genetic material transferred depends on the length of time conjugation takes place. The male chromosome is fragile and usually breaks before the entire chromosome enters the female. The female remains as such unless complete genetic transfer occurs allowing the receipt of the F factor.

The chromosome of the male is replicated as the genetic material is being transferred to the female so the male cell, following conjugation, is still genetically complete and viable.

The phenotypic characteristics of male cells will not have changed. Descendants of the female cell will exhibit genetic characteristics inherited from both male and female strains depending upon the frequency of genetic recombination occurring within the female cell. If, for example, the male was phage resistant and the female was streptomycin resistant, then among the progeny of the female line will be some bacteria that are resistant to both phage and streptomycin and some cells that are resistant to neither.

Although the transfer of the entire male genetic material is rare in this type of exchange, the proportion transferred contains hundreds of times as many genes as are involved in any single event of transformation and transduction.

Another variation of the conjugation process is known as **sexduction.** At times, the F factor breaks off the male (F′) chromosome carrying one or more closely-linked genes with it. When the F′ cell transfers the F particle to the female cell, not only is the F factor passed to the female cell, thus conferring maleness, but also the bacterial genes attached to it. If genetic recombination occurs, then progeny of the cell will demonstrate the new phenotype.

Dozens of mutant strains of both the Hfr and F⁻ strains have been studied, and various recombinations of these have been obtained. The relative location on the chromosomes of the genes controlling various bacterial traits have been determined, and maps showing gene location have been drawn. Such procedures have relied on analogy in the determination of the linkage groups of fruit flies, corn, etc. by classical genetic studies.

Genetic analysis is based on the fact that when individuals differing in several hereditary characteristics are crossed, the progeny contain a reassortment of genes. The assumption is made that when an Hfr chromosomal segment enters the F⁻ cell, it lines up beside that portion of the F⁻ chromosome bearing the corresponding genes. When the new chromosomes are synthesized, preparatory to cell division, some information is taken from each of the parental structures. When genes are close together a "switch-over" or copy choice probably will not occur when the chromosome copy is being made for the daughter cells. The frequency of such "copy choice" is a function of the distance between the genes, and studies of these distances aid in mapping the gene location. Because of the ease in handling bacteria, great numbers of observations can be made, and details on short chromosomal segments can be studied which would not be resolved in recombination studies using higher animals or plants. For example, the lactose fermentation region of the chromosome of *E. coli* can be resolved into three distinct parts. One controls the synthesis of the enzyme β-galactosidase; one determines the enzyme as being inducible or constitutive; and one controls the synthesis of a permease which brings lactose into the cell.

A second technique for mapping the bacterial chromosome entails the measuring of the length of time required for penetration of the donor's chromosome into the recipient cell. When an Hfr male culture is mixed with an F⁻ strain carrying many mutant genes, conjugation occurs, and the penetration of the Hfr chromosome begins. Aliquots of the conjugating cells are removed every few minutes thereafter and shaken violently to break the connecting conjugation tubes. Analysis of the genetic constitution of the progeny will reveal the time of penetration for each gene (Fig. 8.8). With one Hfr strain, the threonine-synthesizing gene (thr) enters at 8 minutes, the lactose-fermenting gene (lac) at 18 minutes, the streptomycin-resistance (str) at 90 minutes, and the vitamin B_1 gene (B_1) at 115 minutes. Therefore, thr is located on the chromosome 10 units (minutes) from lac, 82 units from str, and 107 units from B_1.

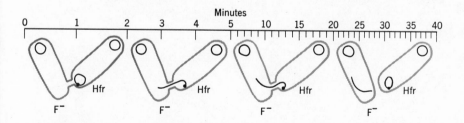

Fig. 8.8. **The location of various genes along the bacterial chromosome can be mapped by interrupting the mating process by vigorous shaking and then determining which genes had been transferred.**

Summary

In bacteria, as in higher animals and plants, errors in the replication of DNA, called mutations, introduce the variability necessary for survival and development in the ever-changing environment. Bacteria are able to colonize new environments more readily than multicellular forms for several reasons. Since bacteria are unicellular, each cell is a forerunner to an ancestral line; thus, mutations are passed to the next generation. Mutations are expressed immediately, the environment permitting, since the bacterial cell is haploid. The appearance and selection of mutations are more rapid since bacteria have faster growth rates and higher population densities.

The genetic code is a linear arrangement of genes which directs the activities of the cell and divides prior to cell division. New combinations of mutant and wild-type genes can arise in an individual bacterium as a result of genetic exchange between a donor cell and a recipient cell. The new genotype arises if genetic recombination takes place in the recipient cell. Due to the simple nature of the genetic apparatus of bacteria and to the existence of several types of genetic transfer, bacteria have been extensively employed as tools to elucidate genetic information at the molecular level. Work with bacterial genetics has uncovered the mechanism of the code and its errors.

NINE

Inhibiting and Killing of Microorganisms

Man has developed many procedures to control and destroy microorganisms in order to protect himself, his animals, and his plants from disease, and his food, clothing and other equipment from microbiological deterioration. For this purpose he has used a wide variety of physical and chemical agents.

The destruction of all microorganisms in an environment is referred to as **sterilization,** and the term sterility implies freedom from seeds of reproduction. The removal of disease-producing organisms (disinfection) implies that the infectious material has been removed, but sterility has not necessarily been attained. Some people would restrict the term disinfectant to those chemicals used in destroying bacteria on nonliving materials. Antiseptics are milder agents and are applied to the living body. Some chemicals are used in high concentrations for disinfection and in lower concentrations for antisepsis. Large volumes of such chemicals are used in the United States to lower the bacterial count and thus minimize odor problems both on the human body and in the environment. Sanitizers are chemicals used to lower the microbial count on utensils, and so forth. A chemical taken or injected into the living body for the purpose of inhibiting or destroying microbial infections in cells or tissues is called a **chemotherapeutic agent.** Such substances are generally used for specific organisms and must have a favorable **therapeutic index,** that is, they must be much more toxic to the invading organisms than to the host cells.

Agents that kill bacteria are called **bactericidal agents.** If they merely inhibit bacterial growth by their presence, they are referred to as **bacteriostatic agents. Bacteriostasis** is a synonym for bacterial inhibition. **Fungicides** kill fungi, and **viricides** kill viruses. The term **germicide** is the more inclusive term.

Nine factors markedly affect any attempt to kill or inhibit organisms.

1. **Intensity or concentration and time of exposure.** Time and intensity are intimately bound together in any sterilization process; low concentrations require longer exposure times and vice-versa. Penicillin in a massive dose may cure syphilis in one day; a lower dose rate must be continued for weeks to achieve the same response. A still lower rate may require infinite time; that is, it will fail to destroy the microorganism causing the disease. At very low intensities or concentrations, many useful antibacterial agents will actually stimulate the growth of the same organisms which higher concentrations will inhibit and still higher concentrations will kill (Fig. 9.1).

2. **Number of microorganisms or the contamination load.** Dying microorganisms generally exhibit a straight-line death curve when the logarithms of the numbers of survivors are plotted against time. A death-phase curve

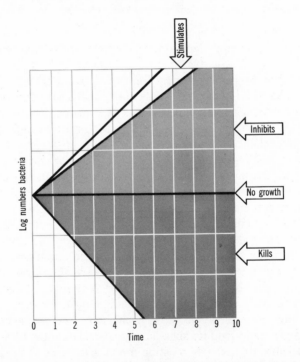

Fig. 9.1. An infinite number of curves can be obtained with a single antibacterial chemical used at varying concentrations. Very low concentrations have no effect; a slight increase results in stimulation. Further increase in concentration will give a decreased growth until a level is reached where killing occurs. The rate of killing is a function of the concentration.

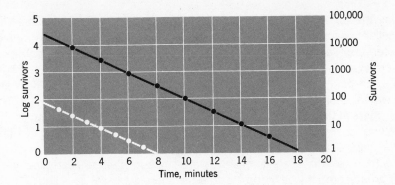

Fig. 9.2. Death rate of bacteria where the same disinfectant was applied at two levels of contamination. The curves are parallel, so the lower curve reaches the bottom of the graph 10 minutes before the upper one.

starting with high numbers will, therefore, take longer to descend to the bottom of the graph than will a curve of identical slope that starts lower down on the graph (Fig. 9.2). A resistant variant (or in mixed populations, a resistant species) is more apt to be present if the inoculum or the initial load of contaminants is large (Fig. 9.3).

3. **Type of microorganism.** While certain types of chemicals are toxic to all

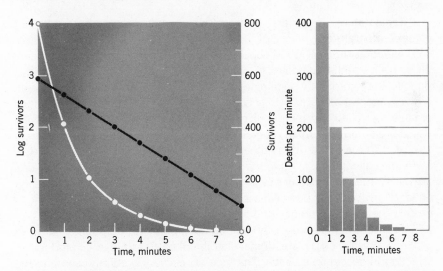

Fig. 9.3. In a typical disinfection experiment, a straight line usually results when the logarithms of the survivors are plotted against the time of exposure to the killing agent. If logarithms are not used, the curved line results. In microbiology it is usually impossible to count dead organisms but, if we could, the bar graph on the right would show a typical result.

living things, most of the useful inhibitory agents act selectively. Some are excellent against fungi and useless against bacteria; some destroy or inhibit bacteria but fail against viruses; some are more effective against gram-positive than gram-negative bacteria. In general, spore formation, capsule formation, and acid fastness help the organism to resist killing but do not deter growth inhibition.

4. **State of the culture.** Even a single culture will vary in its reaction to antibacterial agents. A synchronized culture is usually most sensitive to killing while the nuclei are dividing. Sensitivity varies in different phases of the growth curve with greatest sensitivity often occurring in the early log phase and greatest resistance in the maximum stationary phase.

5. **Moisture.** Wet organisms are more easily killed than dry organisms.

6. **Extraneous material.** Usually microorganisms suspended in pure water are easiest to kill. Dissolved and suspended solids protect them from immediate contact with the antibacterial agent. Extraneous material such as blood or pus may absorb or otherwise react with the disinfectant, thus wasting it and lowering its effective concentration.

7. **pH.** Acidity and basicity affect most antibacterial activities. Cells are usually most resistant at neutrality and become more sensitive in both acid and alkaline substances. On the other hand, streptomycin and sulfonamides are less active at acid pH values, while phenol and chlorine are less active under alkaline conditions.

8. **Temperature.** In general, antibacterial agents are more effective at higher temperatures.

9. **Antidotes.** In addition to nonspecific interference by extraneous materials, certain specific antidotes in the medium can often restrict the damage done to the microbe by a killing agent, or interfere with the action of an inhibiting agent. The human body contains antidotes which may revive microorganisms "killed" by mercurial disinfectants. An infected lesion involving tissue breakdown (**necrosis**) contains enough of the vitamin, para-aminobenzoic acid (the antidote for sulfonamides), to render those drugs completely ineffective as inhibitors. When the bacteriologist attempts to isolate a microorganism from the blood or tissue of a patient under penicillin treatment, the antidote, in this case penicillinase, must be placed in the broth or agar to destroy the penicillin that is introduced with the inoculum.

"Death" of microorganisms is defined as inability to reproduce, that is, to produce a colony on a plate, to make a tube of broth turbid, or to produce an infection in susceptible tissue. The type of recovery medium used and the antidotes it contains may influence the "death" condition. A chemically treated polio virus may appear dead when tested for growth in a culture flask of monkey kidney tissue, but may still produce an infection in the brain of a living monkey.

Test Methods for Antimicrobial Action

While no single method of studying antibacterial action can be used with all agents, two methods are widely used to study bacteriostatic action. (1) Measured portions of the bacteriostatic agent in concentrations ranging from zero to the highest practical concentration are added to several tubes of broth. These are inoculated with the test organism. After incubation, the tube that contains just enough of the antiseptic or disinfectant to prevent visible growth contains the **minimum inhibitory concentration.** Such experiments compare different substances in order to determine which is superior as a bacteriostatic agent. (2) Nutrient agar plates are heavily seeded with the test organism, and a small piece of absorbent paper soaked with varying amounts of the test chemical is placed on each plate. After incubation, the plates will be covered by growths of the test organism except for clear zones around the inhibitor. The width of the inhibition zone is a measure of the inhibiting ability of the chemical concentration (Fig. 9.4).

When bacteriostatic agents fail to inhibit completely, it is because of individual mutants in the test culture that have a heritable resistance to the inhibitory action. These mutants will produce colonies in the zones of inhibition on agar plates used for assay of inhibitors. Such mutants occur even in the absence of inhibitors, but they remain as an insignificant part of the normal population. In the presence of a substance that inhibits all of the normal organisms, the resistant mutants will grow and the resulting culture will be composed exclusively of resistant individuals and the culture is said to be **drug resistant.** By transfer to media containing still higher inhibitor concentrations, more highly resistant mutants develop from the already partially resistant population. Fortunately, drug resistance is a very specific phenomenon; a population of microorganisms resistant to one agent such as sulfonamide is still as sensitive to penicillin as the normal or wild type. It is also fortunate that in the absence of inhibitor the wild-type organisms will grow faster than the mutants and soon become dominant again.

Several practical consequences result from these considerations of resistance to inhibitors.

1. A massive infection is more likely to contain drug resistant organisms than a limited one containing fewer organisms.

2. Partially resistant groups of organisms are likely to contain individuals that are completely or highly resistant. Inadequate treatment with small amounts of inhibitors can produce such partially resistant populations.

3. In hospitals and in other areas where inhibitors are used, the microorganisms found are more likely to be partially resistant or highly resistant

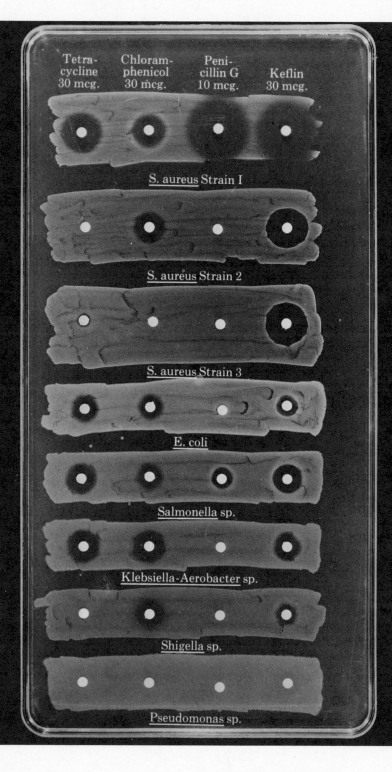

Tetra-
cycline
30 mcg.

Chloram-
phenicol
30 mcg.

Peni-
cillin G
10 mcg.

Keflin
30 mcg.

S. aureus Strain I

S. aureus Strain 2

S. aureus Strain 3

E. coli

Salmonella sp.

Klebsiella-Aerobacter sp.

Shigella sp.

Pseudomonas sp.

than the same types of microorganisms found in other areas. Patients may harbor drug-resistant populations but, as has been stated, when they leave the hospital environment the resistant populations are replaced by the normal wild type.

4. The use of combinations of inhibitors of different types minimizes the resistance problem since simultaneous mutation to resistance to two drugs is a rare event. This does not mean, however, that a population of bacteria already resistant to one drug will not yield mutants to another at the usual mutation rate.

BACTERICIDAL TESTS

Experiments determining the slope of bacterial death curves determine the basic principles of bactericidal action and provide maximum information about the killing process. A standardized inoculum of the test organism is subjected to the killing agent, samples are withdrawn at measured time intervals, and plate counts are made to determine the number of survivors. When data are plotted showing the relationship of the survivors to the time, a death curve is revealed for the total population; the slope of the curve is a measure of the killing. Repeated experiments with modifications in pH, inoculum size, type of organism, concentration, and all the other factors governing disinfectant action, reveal the effectiveness of the killing agent for the intended use in comparison with others already tested.

Disinfectants may be tested under conditions simulating those under which the agent will actually be used. Some rather bizarre tests have been designed that give useful information. For example, the infected skin from a mouse is painted with the test antiseptic; bits of skin are surgically removed and inserted in the abdominal cavity. If the antiseptic is ineffective, the mouse will succumb to the infection; if effective, he will survive and his wounds will heal. The efficiency of chemicals used to sanitize drinking glasses may be tested by placing typical organisms from the mouth or skin on glass beads. The time and the concentration of disinfectant required for sterilization of the glass surfaces measure the usefulness of the substance as a sanitizing agent.

The **phenol coefficient test** determines the amount of a disinfectant required to kill organisms in 10 but not in five minutes and compares this with the amount of phenol required for the same effect. Carefully standardized as to inoculum size, temperature, and so forth, this test employs both a gram-

---

Fig. 9.4. Clear zones around the antibiotic-impregnated discs indicate activity against the bacteria on the streak plate. Note the zones of stimulated growth which is especially marked with the chloramphenicol test on S. aureus, Strain I. Photo courtesy Eli Lilly and Company.

positive organism, *Staphylococcus aureus,* and a gram-negative organism, *Salmonella typhi.* The **phenol coefficient,** therefore, is a number which relates the effectiveness of the disinfectant to that of phenol; a compound is said to have a phenol coefficient of five if it takes only one-fifth of the concentration to do as well as phenol in the test described (Table 9.1). It is especially useful in comparing efficiency of substances used as sanitizers. Some microbiologists would limit its application to disinfectants of the phenolic type.

Table 9.1 Phenol Coefficient Test

| | 37 C | | |
	5 min	10 min	15 min
S. aureus			
Phenol:			
1 - 80	+	−	−
1 - 90	+	+	−
1 - 100	+	+	+
Compound X			
1 - 160	+	−	−
1 - 180	+	+	−
1 - 200	+	+	+

Staphylococcus aureus is killed in 10 minutes but not in 5 minutes by a 1:80 dilution of phenol. This same organism is killed by compound X at the same time by a dilution of 1:160. Compound X is therefore twice as good as phenol and therefore has a phenol coefficient of 2.

MAJOR APPLICATIONS OF DISINFECTANTS AND ANTISEPTICS

Thousands of chemical compounds have been promoted for use as disinfectants and antiseptics. Chlorine is used to kill bacteria in drinking water, in wash waters in food and dairy industries, in swimming pools, and even in treated sewage. Chlorine is an extremely active germicide; 0.1 part per million is adequate for rapid killing of vegetative bacteria in moderately pure water. It is a general protoplasmic poison, that oxidizes, and thus destroys, essential protoplasmic components in all types of cells. It reacts so vigorously with any organic material that its killing action is largely dissipated in highly impure water; of course, killing can still be obtained by greatly increasing the concentration. Chlorine leaves no toxic residue because it is converted to chloride, which is a normal harmless constituent of all foods containing salt.

For sanitizing desk tops, mop water, toilet bowls, and hospital and laboratory equipment, phenol and the related cresols are highly effective. This group, which includes the pine tar and coal tar distillates, destroys all types of organisms; extraneous organic matter does not interfere with disinfection to the same extent as with chlorine action. These substances leave an active toxic residue on the desk top or other treated surfaces. Soaps and detergents are effective for the same purposes, especially when supplemented with an alkali.

High detergent concentrations react with and denature many protoplasmic constituents, but when used in low concentrations, the initial damage to the microorganism appears in the cell membrane. Death results from leakage of protoplasmic constituents through the membrane damaged by the detergent-type chemical. Gram-positive organisms are killed by lower concentrations of detergents than are gram-negative organisms. In fact, natural detergents, the bile salts, inhibit many of the gram-positive bacteria in the intestine. Microbiologists make use of this fact by adding detergents to media when they wish to cultivate typical intestinal bacteria such as *Escherichia coli* or *Salmonella* species. Chemists have synthesized thousands of detergents, some of which have strong germicidal activity. Quaternary ammonium compounds (nicknamed "quats") are cationic detergents most commonly promoted and they have a wide application in a great variety of processes for microbial control.

Some physicians use only soap and water for removing and killing microorganisms on the surface of the skin, but most use germicidal soaps as well as antiseptics containing alcohols, iodine, mercurials, or cationic detergents. These substances may provide preventive treatment of minor skin injuries and may be used to disinfect medical and dental instruments when heat sterilization is not available. Ethyl and propyl alcohols are effective skin degerming agents and, properly used, are effective for thermometers and other hospital equipment; substances such as thymol and salicyclic acid may be dissolved in the alcohols to increase their germicidal efficiency. The attachment of atoms of chlorine to the phenol ring yields chlorophenols which are strong protoplasmic poisons even at low dilutions. Some of them, such as DDT, are powerful insecticides. Others kill plants and a number are powerful inhibitors of microbial growth. One of them, hexachlorophene, is used extensively in germicidal soap and is widely used as a skin antiseptic when mixed with detergent. Hexachlorophene persists for hours on the skin and restricts the bacterial population and minimizes the body odor that results from the bacterial decomposition of skin excretions.

Iodine, one of the most effective antiseptics, is usually used as a tincture, which means it is dissolved in alcohol. Because iodine is highly irritating to mucous membranes and certain sensitive skins, less irritating formulations have been developed. Mercurial compounds are sold under trade names such as Mercurochrome, Merthiolate, and Metaphen. While they have powerful bacteriostatic action, their bactericidal action is only moderate. Mercury compounds leave a toxic residue which is desirable for some purposes but not for others. Like other heavy metals they react with the sulfhydryl (—SH) groups of enzyme proteins and stop biological processes. Thus they are nonspecific, that is, toxic not only to all types of microorganisms but to all living cells. A germicidal silver preparation is widely used in treating infections of the eye; for many years state laws have required its application to the eyes of newborn infants to prevent congenital gonococcal infections. The ability of very small amounts of silver and a few other metals to inhibit microorganisms is designated as **oligodynamic** action.

The prevention of deterioration utilizes the greatest tonnage of antibacterial agents. Wood posts, poles, piling, and railroad ties can be preserved from fungal and bacterial deterioration for periods exceeding twenty years if treated with creosote, a distillate from tars. Also effective are copper salts and pentachlorophenol (penta) and its chemical relatives (Fig. 9.5). Penta is a comparatively insoluble material retained by wood for long periods of time. Penta and insoluble salts of copper preparations are employed to protect rope and tent fabric from antibacterial action and to preserve paints; most paints also contain zinc salts, and the modern water-soluble paints usually need some active preservative to prevent spoilage in the can. Paper not used for food cartons can be impregnated with mercurials such as phenyl mercuric nitrate, but often penta or related compounds are employed.

Formulations of copper and of sulfur have been used for hundreds of years to inhibit the microbial attack on growing plants. They are especially effective against fungal infections. Hundreds of organic compounds with high antimicrobial activity have been prepared, but their usefulness depends on several factors: completeness of spread and coverage over the leaf surface, the persistence on the leaf through rain and wind, and the cost. **Chelating compounds,** which tie up the nutritive mineral elements in chelate

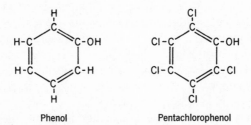

Phenol   Pentachlorophenol

Fig. 9.5. Pentachlorophenol is used where a highly active and long lasting inhibitor is required. It is too insoluble to give the rapid killing exhibited by phenol.

complexes and render the minerals unavailable for microbial growth, have diverse uses, from seed disinfection to inhibition of growth in water cooling towers. Oxine (8-hydroxy quinoline) is an example.

Edible chemicals are sometimes added to foods as preservatives to inhibit microbial growth. Salts, sugar, spices, and the edible organic acids (such as found in vinegar, sour milk, fruit juices, and cheese) inhibit the action of many microorganisms and therefore are acceptable food preservatives. Of course, most food is preserved by some physical means which leaves no toxic residue. The most important of the physical procedures are drying (cereals, cookies, etc.), heating in a closed container (canning), and refrigeration. In refrigeration, a chemical inhibitor is sometimes used as a supplementary preservative. Uncooked chicken is dipped into a solution of the antibiotic aureomycin (also called chlorotetracycline). Psychrophilic growth is thus inhibited, allowing the chicken to be refrigerated for a longer period without evidence of spoilage. Since all chicken is cooked before eating, and since cooking destroys the aureomycin, this type of additive has been approved by the U. S. Food and Drug Administration. Inhibitors such as benzoate are permitted in some foods which constitute only a minute part of the diet.

The gases currently used for disinfection and sterilization are formaldehyde and ethylene oxide. Formaldehyde, though in use for many years, is highly irritating, and the odor lingers even after prolonged airing. Ethylene oxide has much greater penetrating power and will kill the spores as well as the vegetative cells of all types of bacteria. Almost any material can be sterilized at room temperature if it is placed in a tight container, the air removed with a vacuum pump, and the container filled with ethylene oxide. Usually materials are sterilized overnight, but four hours will suffice. Pieces of human arteries to be used in surgical transplants, surgical bandages, and bulk foodstuffs, such as spices, are a few of the materials that have been sterilized in this way. Ethylene oxide reacts with chemical groups on bacterial proteins in an irreversible manner.

Chemotherapy

Chemotherapy involves the treatment of internal disease by chemicals which have a specific and toxic effect upon the microorganisms causing the disease without seriously poisoning the patient. Antiseptics that are used externally—on the skin or even on the mucous membranes of the throat—are not regarded as chemotherapeutic agents. Although these agents may be useful when applied topically, chemotherapy implies systemic treatment, that is, circulation of the drug throughout the body following injection or ingestion. Chemotherapy of malaria with quinine has an ancient history; Paul Ehrlich first used the term to describe his use of arsenic compounds in the treatment of syphilis.

Today, the useful chemotherapeutic agents include sulfonamides (compounds related to sulfanilamide), antibiotics, several iodine-containing compounds used for amoebic dysentery, atabrine and related compounds used for malaria, and the antituberculosis drugs, paraaminosalicylic acid and isonicotinic hydrazide.

THE SULFONAMIDES

Almost 30 years elapsed after Ehrlich's success, before Domagk's discovery in 1935 of the therapeutic value of the sulfanilamide-containing dye, prontosil. This discovery ushered in the active phase of the use of chemotherapy in bacterial disease. Within five years, over 5000 sulfanilamide-like compounds had been synthesized by chemists and tested by microbiologists. The better compounds of this group are especially useful in the treatment of respiratory, intestinal, and urinary infections. During the research efforts on the new sulfonamides, a vast store of useful information was gathered on the absorption, distribution, persistence of high levels in the blood, and excretion of drugs. Slight modifications in the sulfonamide molecule do not interfere with its antibacterial activity but prevent its forming damaging crystals in the kidneys. Some sulfonamides concentrate in the bladder and thus are useful in urinary infections; others are removed from the blood stream by kidney action more slowly and consequently have a sustained action on bloodstream infections.

Sulfonamides — chemical analogues (only slightly different in chemical structure) of the vitamin precursor, para-aminobenzoic acid (PABA) — proved to be successful because they interfered competitively with the microbe's ability to use the PABA. Since para-aminobenzoic acid is built into folic acid by the bacterium, and since most bacterial cell membranes are impermeable to folic acid, bacterial growth is halted by the presence of sulfa drugs. The cells of man are permeable to folic acid, and since that vitamin is present in the diet in sufficient amounts, the sulfonamide inhibitor does not affect human cells or those of other animals. Hence this is a specific inhibitor of bacteria. An intensive campaign was launched by chemists to synthesize similar inhibitors for other vitamins. Although understanding of the chemical reactions in living things has been increased tremendously by these efforts, only a few compounds have been uncovered. When other vitamins were modified, as the chemical structure of para-aminobenzoic acid is modified to yield sulfonamides (Fig. 9.6), the resulting compounds have

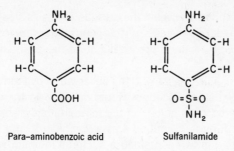

Para-aminobenzoic acid   Sulfanilamide

Fig. 9.6. The chemotherapeutic agent, sulfanilamide, differs from the vitamin, para-aminobenzoic acid, in chemical structure. The carboxyl group (COOH) of the latter is replaced by a sulfonamide group (SO_2NH_2).

shown some promise in cancer therapy but have not been very useful as antimicrobial agents.

The **antibacterial spectrum** is the range of bacterial types against which a drug is active. Sulfonamides have a relatively narrow spectrum when used under practical conditions. The useful spectrum of isonicotinic hydrazide is limited to the bacterium causing tuberculosis. The vitamin niacin, although similar to isonicotinic hydrazide in chemical structure, does not serve as an antidote to the inhibitory action of the drug.

THE ANTIBIOTICS

Antibiotics have been defined as relatively complex chemical substances of microbial origin which display antimicrobial activity. The simple organic acids or alcohols produced by fermentation exert antimicrobial action but are not regarded as antibiotics. Some idea of the impact of antibiotics on our lives can be gained from drug statistics; ten years after their appearance, antibiotics accounted for over half of the total dollar volume of American drug sales. Surgical procedures, which were formerly fraught with danger because of infection, are now routine, and the death rate from infectious disease of bacterial origin has declined dramatically.

In aging cultures, most microorganisms excrete toxic products as a result of normal metabolism. Also, toxic excretions may be produced by a metabolism disbalanced by a modified environment or by autolysis of aged cells. When toxic substances are released into the medium before growth is halted by lack of food, the forms which are resistant to their own inhibitory excretions survive. The loss of complex materials by excretion may then be advantageous to the resistant microorganisms, since the excreted products will inhibit competing organisms.

Antibiotics are produced by microorganisms whose normal habitat is the soil, where competition between the various forms is keen. Molds of the genus *Penicillium,* actinomycetes of the genus *Streptomyces,* and bacteria of the genus *Bacillus* produce practically all of the antibiotics now in commercial use.

The study of penicillin begins with the observation by Fleming in 1928 that a disease-producing staphylococcus would not grow in the region of a petri plate which bordered on a colony of a contaminating green mold. Fleming isolated the green fungus, identified it as a *Penicillium,* and showed that the broth in which it had grown contained a bacterial inhibitor which he named penicillin. After the mold was filtered from the culture broth, the filtrate was not toxic to animals. Though Fleming suggested its use as a therapeutic agent, research languished for almost 12 years until science and technology advanced sufficiently to implement the discovery.

In the commercial production of penicillin a selected high-yielding mutant of *Penicillium chrysogenum* is grown on a liquid medium under vigorous aeration. The medium contains corn-steep liquor (a waste product of whiskey production), sugar and mineral salts, as well as organic chemicals which the mold makes with some difficulty, or at least at a slower rate. When these organic chemicals are present in the medium, more penicillin is produced because the rate is no longer dependent on the slow chemical reactions. Slight modifications in the penicillin molecule may result from different types of additives, the strain of mold, or the degree of aeration. Penicillin G and penicillin V are the most useful variants; the latter is stable in the presence of stomach acid and can be taken by mouth. Other penicillins have been produced that are not destroyed by penicillinase, and are more active against gram-negative bacteria. Generally, penicillin is so easily destroyed by heat, moisture, acid, alkali, and aging that its inhibitory activity is probably useless to the *Penicillium* when it is growing in its natural environment. *Penicillium* molds, because of their slow growth rate, are usually restricted in nature to those environments too dry or too acid for bacterial competitors.

The **therapeutic index** of an antibiotic is computed by dividing the largest dose that can be administered without the patient showing severe toxic reactions by the lowest dose necessary to destroy the infection. If this ratio is large, an excess dosage of the drug insures curative effects and still allows a large margin of safety. Penicillin possesses an extremely favorable therapeutic index, but some patients develop severe allergic-type reactions to it upon repeated exposure.

The action of penicillin is directed against synthesis of the mucopeptide or murein in the cell wall (Fig. 9.7). Its antibacterial spectrum includes the organisms of syphilis, gonorrhea, and most of the thick-walled gram-positive bac-

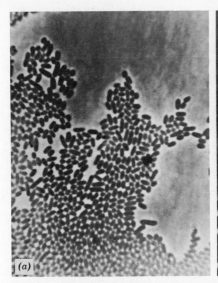

Fig. 9.7. (a) Microscopic appearance of a normal culture of E. coli. (b) The same strain growing in the presence of a small amount of penicillin, which inhibits cell wall formation.

teria, with the exception of the tubercle bacillus. It is inactive, or much less active, against thin-walled gram-negative bacteria (their wall is supported by layers other than the mucopeptide) and completely inactive against animal cells and mycoplasmas (these have no cell walls), and against plant cells (these have cell walls of cellulose).

The enzyme penicillinase, which destroys penicillin, is produced in strains that are naturally resistant, or in resistant mutants that develop from sensitive strains. As with all other drugs a penicillin-resistant organism may restrict the drug from penetrating to the sensitive part of the cell; it may bypass the inhibited reaction, or it may employ a modified process in which the drug is a less effective inhibitor.

The potency of penicillin broths was originally measured in units; one unit of penicillin is the amount which, when placed in 50 milliliters of broth, just inhibits Fleming's strain of *Staphylococcus*. The present unit, the international unit or I. U., is the activity of 0.6 microgram of penicillin G. Although today penicillin is extracted from the mold broth, purified, crystallized, and weighed, the **unit** is still used as the measure of potency; one gram of pure material contains almost two million units.

Following the introduction of penicillin, a systematic search for antibiotic-producing soil microorganisms was undertaken and still continues. Of the hundreds of thousands of strains tested only a small fraction produced active antibacterial substances and, of these, all but a few had to be discarded since the products were too toxic to animals, too unstable, or not as good as others already available.

Five commercially successful broad-spectrum antibiotics produced from actinomycetes of the genus *Streptomyces* warrant our attention. Streptomycin is obtained from *Streptomyces griseus,* and its useful spectrum includes the tubercle bacillus and many gram-negative organisms. Though more stable than penicillin, it is much more toxic. Chemists have been able to decrease its toxicity by attaching hydrogen atoms to the molecule, thus converting it to dihydrostreptomycin. In the treatment of tuberculosis it is often used in conjunction with isonicotinic hydrazide and para-aminosalicylic acid to minimize the opportunity for development of resistant strains. Streptomycin appears to interfere with the action of ribosomes in the bacterial cell and therefore interferes with protein synthesis. In addition to mutants that are streptomycin-resistant, some have been found that are streptomycin-dependent; these will not grow unless streptomycin is present. It is proposed that streptomycin attaches to ribosomes of all types of cells. It prevents protein synthesis in sensitive cells while dependent cells will not synthesize protein unless streptomycin is present. Resistant cells are indifferent. There is also evidence that streptomycin causes a misreading of the genetic code.

Chloramphenicol (also called chloromycetin) is a broad-spectrum antibiotic produced by *Streptomyces venezuelae.* Having a much simpler chemical structure than any of the other major antibiotics, it can be produced inexpensively by chemical synthesis rather than by fermentation. It has a broad spectrum of activity, including gram-positive and gram-negative bacteria, rickettsiae, and even the chlamydiae. Allergic reactions to it are rare, but in a very few susceptible individuals, it has produced a severe blood disorder. Chloromycetin appears to exert its action by inhibiting protein synthesis at a stage involving transfer RNA.

The tetracyclines are much more complex chemical structures consisting, as the name implies, of four attached rings, each made of six carbon atoms along with additional side groups. Chlorotetracycline (also called aureomycin), produced by *Streptomyces aureofaciens,* has a chlorine attached to the tetracycline structure. Oxytetracycline (also called Terramycin), produced by *Streptomyces rimosus,* has an additional oxygen on the tetracycline. Tetracycline itself, which is slightly more stable than the two derivatives, can be produced by chemical modification of aureomycin or by fermentation using a selected strain of *Streptomyces.* It is sold under a number of trade names including Tetracyn and Achromycin. The use of several terms for the same antibiotic is unfortunate. When a new drug is discovered, it is given a name such as penicillin, streptomycin, chloromycetin, aureomycin, or Terramycin.

In the case of the first two drugs, the name became firmly established before the chemical structure was known; the latter three were named according to their chemical structure; chloramphenicol, chlorotetracycline, and oxytetracycline. Later, trade names for the drugs appeared.

All the tetracyclines have low toxicity for animals and broad spectrums of activity, including rickettsiae and the chlamydiae. Organisms resistant to any one of the tetracyclines are also resistant to the others, suggesting an identical mechanism of action. Tetracyclines appear to interfere with the binding of the transfer RNA to the ribosomes, thus preventing the building of the peptide chain.

Three polypeptide-type antibiotics produced by bacteria of the genus *Bacillus* are usually mixtures of several substances and have found limited use in medicine. Relatively toxic when given by injection, they are more commonly used for local application to infected lesions and to the nose and throat. Almost any strain of the genus *Bacillus* will produce these substances, but only three are available commercially. Tyrothricin is produced by *Bacillus brevis;* bacitracin is produced by *Bacillus licheniformis;* and a group known as the polymyxins is produced by *Bacillus polymyxa.* Although high toxicity restricts their use, they have been occasionally successful in systemic treatment where other antibiotics have failed. Most of them function by interfering with the cell membrane of the microorganisms; they usually are lytic for the red blood corpuscles in the animal body.

A number of other antibiotics occupy an important role in chemotherapy because they permit flexibility in treatment of resistant infections. Cycloserine, erythromycin, kanamycin and neomycin are only a few examples. Molds that produce skin infections are inhibited by griseofulvin; deep-seated fungus infections are treated with nystatin, which is one of a series of polyenes. Viruses do not respond to antibiotic treatment. Experience with antibiotics has generally determined the choice for treatment of an infection; for example, streptomycin for tuberculosis; penicillin for syphilis. In other cases, the decision depends on the personal preference of the physician and the effectiveness of promotion by the manufacturer. In stubborn or unusual infections the isolated organism is subjected to sensitivity tests to determine the drug most likely to be successful.

GROWTH STIMULATION

Antibiotics are included in modern feeds for young poultry and livestock. A few grams per ton of feed increases the rate of growth of young animals up to 50%. The mechanism of this stimulation is not known, but it has been explained as (1) removal of certain intestinal bacteria that interfere with absorption of vitamins, and (2) curing of chronic or subclinical infections.

Some microbiologists suggest that such widespread use of antibiotics will increase the incidence of antibiotic-resistant bacteria and could be potentially dangerous.

Physical Methods

Microorganisms may be destroyed, inhibited, or removed without the addition of chemical substances which may leave a residue by (1) heat, (2) cold, (3) desiccation, (4) radiations, (5) ultrasonic waves and (6) filtration.

HEAT

Although a few thermophilic bacteria and blue-green algae will grow at temperatures of 70 C and above, most microorganisms are destroyed by exposure to that temperature for a few minutes. Generations of physicians have boiled syringes, scalpels, and needles for 10 to 15 minutes and successfully sterilized them, apparently destroying even the spores of disease-producing bacteria such as the tetanus bacillus. Of course, the material sterilized in this manner is usually relatively clean with a low contamination load. There is recent evidence that the hepatitis virus occasionally survives this treatment. The addition of alkaline washing soda to retard corrosion also aids in the killing. Where heavy contamination by the more resistant spores of saprophytic bacteria is encountered, boiling for hours fails to achieve sterility. It is standard practice to autoclave such materials at 15 lbs. per square inch of steam pressure (121 C) for 15 to 20 minutes. One of the advantages of steam under pressure is that temperatures above the boiling point of water are attained. A second advantage is the rapidity of heating; in the autoclave any object that is cooler than the steam will have steam condense on it, releasing the heat of vaporization of water; consequently, objects in an autoclave attain the high temperature almost instantly.

Dry heat is much less efficient; heat exchange is slow and a temperature of 160 C, maintained for two hours, is required to obtain the equivalent killing resulting from autoclaving at 121 C for 20 minutes. Organic molecules such as proteins tend to expand when heated and the chains unfold; in the absence of water, the chains have a better chance of refolding in precisely the original pattern upon cooling thereby retaining activity even after considerable exposure to dry heat. The lethal effect of heat is due to protein denaturation; it accelerates the lethal action of disinfectants by its general ability to increase the speed of chemical reactions.

COLD

Microorganisms are often preserved by keeping them in the frozen state; they will remain viable for years if held at the temperature of dry ice (−78 C) or liquid nitrogen (−196 C). During the freezing and thawing process, however, a high die-off of the organisms may occur. This is due partially to the formation of ice crystals which may rupture membranes and to the high concentrations of salts which occur in localized areas of the cell as the water freezes.

Repeated cycles of slow freezing and thawing are used to break up bacterial cells to release enzymes. Rapid freezing minimizes these effects because smaller ice crystals are formed with less damage inside the cell. The incorporation of glycerol or protein into the culture to be frozen, further decreases the lethal effect.

DESICCATION

Many of the bacteria and viruses die rapidly upon drying, and most organisms that are expelled from the mouth by coughing survive only until the droplet of moisture in which they are suspended drys. Exceptions are the waxy-coated tubercle bacilli and the encapsulated organisms that have the limited protection of their water-rich capsules. Of course, spores of bacteria and of fungi live for years in the dry state. Many of the sensitive microbes can be preserved in the dry state if **lyophilized;** this process involves freezing rapidly and then drying in a vacuum, thus removing the water while in the frozen state.

ULTRAVIOLET LIGHT RADIATION

Microorganisms on surfaces exposed to sunlight are destroyed more effectively than can be accounted for either by the heat or the drying which results. The germicidal action of the visible rays of the sun is relatively small compared with the antibacterial action of the ultraviolet range of the spectrum. While much of the ultraviolet light emanating from the sun is screened out by the atmosphere, of the considerable amount which reaches the earth the most highly bactericidal wavelengths are found in the area of 2600 angstrom units (Fig. 9.8). An angstrom unit (Å) is 0.1 nanometer, or 0.0001 micrometer. The germicidal range is from 2100 to 3000 Å. Since the human eye can detect light at about 3900 Å and above, the germicidal rays are invisible to the human eye. The emission spectrum of mercury yields a strong band in the area of 2600 Å, and, consequently, a mercury-vapor lamp is an excellent source of ultraviolet light. Ultraviolet rays, weak in penetrating power, do not pass through ordinary glass. However, they do penetrate a few layers of cells and, therefore, can penetrate a microorganism. They pass through thin layers of pure water, but any cloudiness in the liquid screens out the germicidal rays. The success of ultraviolet light in the treatment of acne probably does not reside in its germicidal power but in its irritating action on the skin cells, which results in greater circulation.

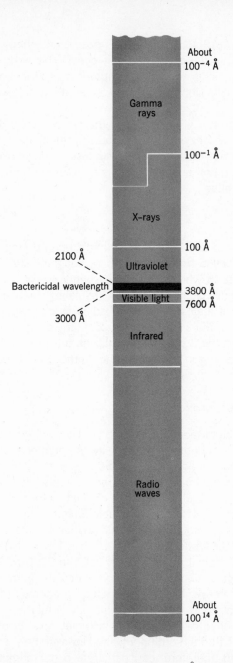

Fig. 9.8. Electromagnetic waves are measured in angstrom units (Å). Gamma rays and X-rays have some antimicrobial action but the wavelengths usually used in microbiology occupy a very narrow region in the ultraviolet zone.

The nucleic acids, found in high concentrations in microorganisms and constituting the essential materials of the DNA, absorb ultraviolet light very strongly. The wavelengths most strongly absorbed are almost identical to those with the highest germicidal activity. Since only absorbed radiations can bring about changes, the death of microorganisms results from a chemical modification in the nucleic acid. When two thymines are next to each other on the same DNA chain they release the complementary adenines on the adjacent chain and unite chemically with each other; such a combination is called a **dimer.** When the affected portion of the DNA codes for some enzyme indispensable to the cell (containing the newly formed dimer of thymine) the ultraviolet light has produced a lethal mutation and the cell is dead. When the defective code produces a different enzyme or when it removes one not necessary for life, the result is an ordinary mutation. These radiations also coagulate proteins and produce hydrogen peroxide, which is in itself a powerful killing agent. Hydrogen peroxide will react with certain organic components in the cell to produce organic peroxides which are strongly germicidal and quite persistent.

Ultraviolet light is used effectively for destroying bacteria in air and on exposed surfaces. Ultraviolet lamps are used in hospitals, in microbiological laboratories, especially where dangerous pathogens are used, and in drug manufacturing where aseptic packaging precautions are required.

When organisms killed by ultraviolet light are immediately exposed to visible light, many of the cells will revive. **Photoreactivation** is the process by which visible light serves as an antidote for ultraviolet; however, this process is effective only in very special situations. The visible light activates an enzyme, which is inactive in the dark, that breaks up the thymine dimer and restores the DNA intact. There are other DNA-repairing enzymes in the cell that act in the dark by cutting out and replacing the damaged portion. Quantitatively, they are not as dramatic as photoreversal. Excessive exposure to visible light will kill bacteria if a dye is present in the medium which will absorb the visible rays. This effect is called **photodynamic action** or **photodynamic sensitization.** Killing probably results from the formation of peroxides in the medium by the interaction of the light, the dye, the water, and dissolved oxygen.

IONIZING RADIATIONS

Ionizing radiations have shorter wavelengths than ultraviolet light. Soft X-rays, hard X-rays, gamma rays, and cosmic rays are the best known of this group. They are often classified relative to their source, such as cathode rays or roentgen rays. Ionizing radiations are those which have sufficient energy to ionize water molecules. The absorption of a single electron is sufficient to cause the death of a bacterium, probably by the release of a shower of ions following the track of the electron through the cell, which in turn damages the DNA. This results in a mutation which may be lethal.

The expense of using ionizing radiations to kill was formerly thought to render any extensive application impractical. However, as byproducts of the development of atomic energy, sources of high-energy radiations are available and are being applied on a limited scale in the food industry.

ULTRASONIC WAVES

High-frequency sound waves are useful as a laboratory tool for breaking up cells, but have as yet no other practical application.

Summary

Microbiology developed as a science only after the microbiologist was able to destroy all organisms in his test tube except those which he chose to maintain. The destruction of unwanted microbes has been one of the most serious challenges to the ingenuity of this science. The development of chemotherapy extends the selective action of chemical inhibitors to the interior of the living animal, where the chemicals used are destructive to the microbe but tolerated by the host. Less dramatic, but no less effective, are the procedures developed for the control of microbes in the environment which permit the sick and the well to live in association. A beginning has been made in the problem of understanding the mechanisms of death.

TEN

Fungi and Actinomycetes

The fungi are microorganisms that are somewhat more complex in morphological structure and are higher in evolutionary development than bacteria. It has been demonstrated that they possess the eucaryotic cell type exhibited by the higher protists. Molds and yeasts are frequently collectively grouped under the term **fungi** which includes, in addition, the slime molds, mildews, rusts, smuts, and puffballs. Diverse forms and transitional stages occur between the molds and yeasts, making group boundaries difficult to draw at times. For example, some of the slime molds are very similar to some protozoa. Transitional types occur between all of the groups, and some exhibit bacterialike forms. Many of the actinomycetes display characteristics of some of the molds and have often been considered as moldlike organisms. In the light of present-day information, actinomycetes are bacteria; they exhibit the procaryotic cell of the lower protists and have a cell diameter of about 1.0 μm. The actinomycetes are presently placed in the class *Schizomycetes* with bacteria, in the order *Actinomycetales*. In spite of the complexity of classification, a study of the classical and typical forms offers a good beginning to the understanding of these protists.

The Molds

GENERAL CHARACTERISTICS AND OCCURRENCE

Although most forms are highly filamentous, the molds are certainly primitive plants since they possess no chlorophyll, roots, stems, leaves, or vascular system. A number of unicellular molds also exist. Asexual and sexual reproduction occur principally by spore formation. The chemical nature of their cell walls is heterogenous, but cellulose and chitin are major constituents. Motility is usually absent in these forms although the reproductive cells are frequently motile, moving by means of multistranded flagella. Saprophytic and parasitic nutritional types are known, but, in general, their nutritional and environmental demands are less stringent than many microbes.

Optimum temperature for their growth ranges between 20 and 30 C, and an acid environment is usually employed in the isolation of molds. The pH of the medium, adjusted to 5.6, will be inhibitory to most bacteria but will permit vigorous mold growth. The growth of molds is limited by their aerobic nature. Molds will not thrive in the absence of oxygen and are, therefore, limited to surface growth. In liquids, they grow as a floating mat or **pellicle;** in solids, they penetrate only as far as oxygen can diffuse. Molds grow more slowly than bacteria and consequently do not participate in the early stages of organic decomposition. When a plant residue decomposes in the soil, the sugars and proteins are consumed rapidly by bacteria. The rapid growth of aerobic bacteria removes oxygen and delays mold development. Only when the readily available foods are gone does the bacterial respiration slow down sufficiently to allow oxygen to diffuse into the material at a rate sufficient to support mold growth on the more refractory portions of the plant material. Of course, if the material is too dry or too acid for bacterial growth, molds will participate in the earlier stage of decomposition.

Molds are widely distributed in nature since they produce tremendous numbers of spores that are readily spread through the air. Although mold spores can survive many years in the dry state, they do not show the resistance to dryness or other adverse conditions that bacterial spores do. No organic substance appears to be safe from mold destruction, since the molds can consume a wide variety of foods. Unusual chemical solutions such as photographic developer frequently support mold growth. Molds require some moisture but can absorb water from moist air. Although substances can be dried sufficiently to prevent mold growth, slight moisture will cause molding of grain (Fig. 10.1), clothing, or leather. Molds thrive in fruit and vegetable tissues that are too acid for bacterial growth.

STRUCTURE

A typical mold is composed of a multitude of fine threads or filaments spreading in all directions. Thus, they appear cottony or woolly. Each filament is a **hypha** (plural, **hyphae**); a mass of hyphae represents a **mycelium.** Hyphae are about 5 to 10 μm in diameter and can vary greatly in length. Some of the hyphae of a mold are embedded in the food material or substrate on which the mold is growing. These **vegetative hyphae** anchor the mycelium and absorb nutrients. Those not submerged are **aerial hyphae**

Fig. 10.1. Aspergillus glaucus, **a common cause of the molding of wet grain, is seen here growing from the tip of a kernel of wheat. The fruiting bodies bearing the conidia grow out of the mycelium anchored to the grain.** Photo courtesy C. M. Christensen.

which function in absorbing oxygen. Water from moist air may condense on the aerial hyphae, thus supporting mold growth on fairly dry medium. Aerial hyphae also bear the fruiting bodies or asexual spores and are referred to as **fertile hyphae.** The reproductive spores exhibit a variety of forms and, therefore, serve as the principal means of identification and classification. Asexual spores may have different pigments which are responsible for color imparted to molds. Spores can also differ in size (small to large), shape, numbers, and arrangements.

In some molds, the hyphae have transverse cross walls, **septa** (singular, **septum**), dividing the hyphae into cells. A pore or opening in the septum permits the cytoplasm to pass from one cell into another. In hyphae where septa are absent, the nuclei are more or less evenly distributed in the cytoplasm. These hyphae are spoken of as being **coenocytic.** Some molds are **dimorphic** since they produce filamentous growth under one set of cultural conditions but show only unicellular yeast-like growth under different conditions (Fig. 10.2).

Several means of reproduction other than the union of two nuclei to form a zygote have been observed in molds. If a bit of hypha containing at least one

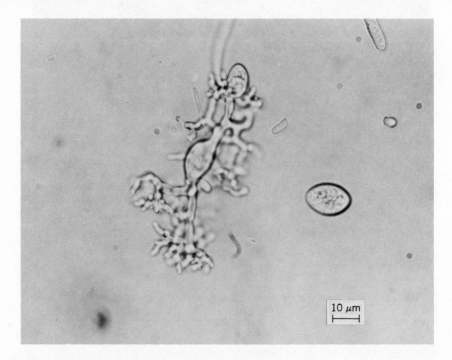

Fig. 10.2. **Light microscope picture of the basidiomycete,** Lenzites saepiaria, **demonstrating dimorphism. There is a normal spore to the right of the mycelial mass. Below and to the right is a typical swollenlike cell.** Photo courtesy J. J. Perry.

nucleus, along with enough cytoplasm to support it, is torn away from the mycelium and planted on a medium, a new culture will be successfully started. This process of reproduction, **fragmentation,** is one of the major ways of establishing laboratory cultures. Asexual **spore formation** is perhaps the most common means of reproduction. Molds produce tremendous numbers of spores, and new cultures arise when they germinate. Asexual spores are either exogenous or endogenous, depending on the way they are formed. Exogenous spores, **conidia,** are produced at the tip or sides of fertile hyphae, **conidiophores** (Fig. 10.3). On the other hand, endogenous spores develop in sac or spore case-like structures called **sporangia** (singular, **sporangium**) on fertile hyphae, **sporangiophores** (Fig. 10.4). With rupture of the sporangium, the **sporangiospores** are released. Motile spores, **zoospores,** are often distinguished from nonmotile spores, **aplanospores,** for classification purposes.

SEXUAL REPRODUCTION

Gametangia is the general term to describe the sex organs of certain molds which are capable of producing the sex cells, the **gametes.** In molds, the male gametangium is the **antheridium;** the female, the **oogonium.** A particular mold species may produce both male and female gametangia on the same mycelium and is, therefore, **monoecious.** Such molds are considered to be **homothallic** since a single pure culture can reproduce sexually. Other species produce only one kind of sex organ on a mycelium and are, therefore, **dioecious.** Such an individual can reproduce sexually only if its gametangia are fertilized by the other sex. These are the **heterothallic** molds.

Other molds produce structures which are morphologically indistinguishable as antheridia or oogonia. When sexual union occurs in these molds, only a strain designated as + exchanges nuclei with a strain designated as −; two + strains or two − strains never recombine. Since both partners in the recombination are morphologically indistinguishable and behave in an identical fashion, they cannot be designated as male or female.

When mold colonies of two different mating strains of the same species are grown close to each other, the hyphae may fuse (**plasmogamy**), and nuclei from one hypha may pass into the other. The resulting mold with two different nuclei in the same cell is known as a **heterocaryon.** The nuclei can divide independently as the mycelium grows so that the heterocaryon will exhibit the hereditary characteristics of both nuclei, being thus a sort of hybrid. This heterocaryotic condition, which is widespread in nature, may be maintained indefinitely and permits the development of a mold possessing desirable properties from both of the parent strains even though actual fusion of the nuclei does not take place.

Sexual reproduction occurs only when nuclei fuse to form a diploid cell,

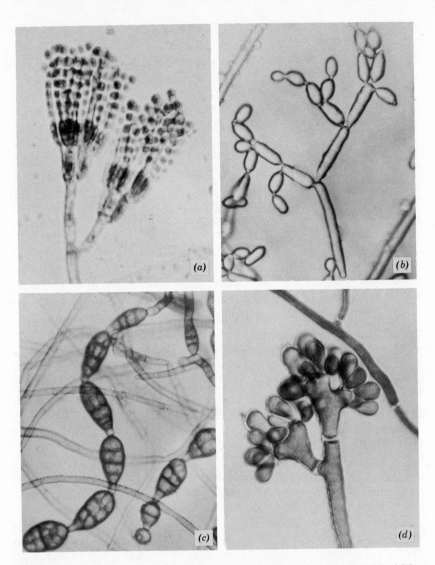

Fig. 10.3. Asexual spores and some mycelium of four frequently encountered molds. (a) Penicillium, **(b)** Chladosporium, **(c)** Alternaria, **(d)** Botrytis. Photos courtesy C. M. Christensen.

the **zygote.** This is followed by meiosis, in which the number of chromosomes is reduced to half, resulting in the haploid condition in the newly formed spores. The spores will have some of the characteristics of each of the parents as a result of the reassortment of the genetic material from the two nuclei that fused before division into the sexual spores. The mycelium that arises from the germination of sexual spores is not to be confused with the heterocaryon, which contains nuclei from both parents and, therefore, **all** the characteristics of both parents. A large variety of structures are involved in the formation of the sexual spores and in the life cycles of molds. Students interested in this aspect are advised to consult a modern textbook on mycology.

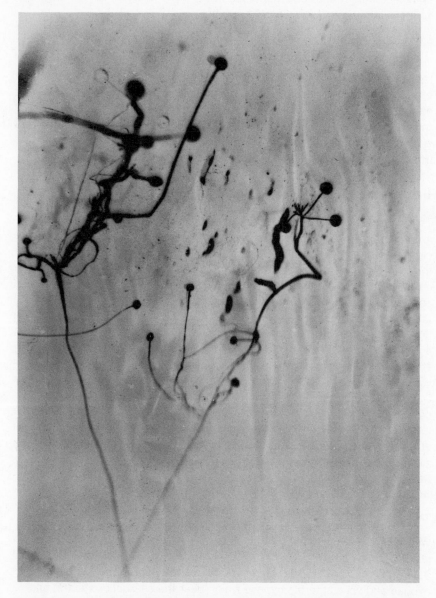

Fig. 10.4. **The phycomycete,** Rhizopus stolonifer, **produces endogenous spores in sporangia on sporangiophores.** Photo courtesy W. LeStourgeon.

Representative Molds

A popular classification of the molds places them in the division *Mycota* of the plant kingdom. Within this division are two subdivisions; one, *Myxomycotina*, containing the true slime molds (class *Myxomycetes*); the other, *Eumycotina*, containing classes of the true fungi.

Slime molds show both protozoal and fungal characteristics within their life cycle. Slime molds are saprophytic and prefer to grow in moist, dark environments. They are frequently found growing in damp soils, on dead leaves, and the bark of dead trees. Some taxonomists consider them to be protozoa due to the absence of cell walls surrounding the amoebalike cells of one stage of the life cycle. The cellular structure that typifies the slime molds is a multinucleated mass forming a free living **plasmodium.** Movement is amoebalike. The plasmodium subsequently sends up stalklike elevations which are quite plantlike in appearance.

THE PHYCOMYCETES

This class of lower fungi includes the more primitive water molds which grow on decaying plant material floating in water. Some of the soil molds are closely related to them. Some taxonomists strongly suggest that members of this group of molds be separated into six distinct classes based principally on the presence and location of flagella on cells or spores. They are here considered as a collective group. As a rule, the *Phycomycetes* have nonseptate mycelium and bear asexual endogenous sporangiospores in sporangia (Fig. 10.4). Motile and nonmotile forms are demonstrated. Both saprophytic and parasitic *Phycomycetes* occur. Many of them are anchored to their substrate by primitive mycelial structures called **rhizoids.** The sexual stage involves the fusion of hyphae from two different mycelia. In some forms, the two hyphae that fuse are morphologically distinct. One hypha passes its nuclei into the other. Hence, the former can be designated as the male, and the latter female. In other *Phycomycetes*, the fusing hyphae are indistinguishable, and the strains must be designated as + or −. The resulting spore is a zygospore. A typical example of this group is the common black bread mold, *Rhizopus stolonifer* (*R. nigricans*).

THE ASCOMYCETES

Well-known representatives of this class are the green and grey molds of the genus *Penicillium* (Fig. 10.5) and the black and brown molds of the genus *Aspergillus* (Fig. 10.6. See page 177). All of the *Ascomycetes* have septate hyphae and special fertile hyphae which bear the asexual exogenous spores called **conidia.** These asexual spores are the major method of reproduction and the spread of the molds. In the penicillia, the conidia appear as green or grey spheres (about five μm in diameter) budded from brushlike tufts on the white or colorless mycelium. The brown or black spores of the aspergilli are budded from a clublike structure at the top of aerial hyphae. Some *Ascomycetes* are simple while others are more complex; for example, the common bread yeast is a simple *Ascomycetes*. This large group of fungi also reveal the formation of sexual **ascospores** in a saclike structure called an **ascus.**

THE BASIDIOMYCETES

The best known *Basidiomycetes* are the mushrooms, which produce a vast mycelial network and, under favorable conditions of moisture and temperature, send up the fruiting body (Fig. 10.7. See page 178). The visible cap of the mushroom is made of densely-packed hyphae. The sexual spores are borne on club-shaped cells that result from the fusion of two nuclei and subsequent meiosis on clublike structures, basidia, found on the gills on the under side

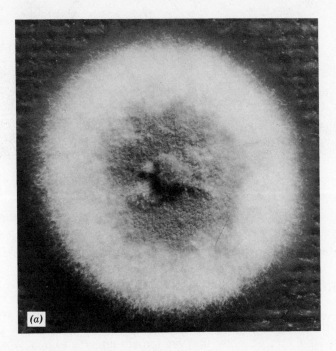

(a)

Fig. 10.5. Colonies of Penicillium **species showing distinctive details; (a)** Penicillium chrysogenum, **which produces much of the commercial penicillin; (b)** Penicillium rimosus, **which has no commercial use.** Photo courtesy Pfizer, Inc.

Fig. 10.5(b)

of the cap. Like other molds, the mushrooms produce tremendous numbers of spores and thrive on decomposing organic matter. The *Basidiomycetes* also include important plant pathogens, the rusts, smuts, and bracket fungi.

THE DEUTEROMYCETES

This class has frequently been referred to as the *Fungi Imperfecti* and represents a diverse group of molds in which no sexual spores have yet been demonstrated. Consequently, they are termed imperfect fungi. When and if sexual spores are discovered in this "form-class," they are moved to the appropriate other class.

Importance of Molds

Molds have become very important to man due to the fact that certain members are able to produce life-saving antibiotics such as penicillin. Gluconic and citric acids are produced commercially by the activity of certain molds. Soya sauce is produced by the decomposition of soy beans by a selected strain of *Aspergillus* with a mixed population of bacteria. Roquefort and Camembert cheeses are ripened and flavored by mold action. In aerobic soils the molds participate in the decomposition of various types of organic matter. The destructive action of molds is observed on posts, railroad ties, and piling as well as in tent fabrics, ropes, paints, papers, plastics, and

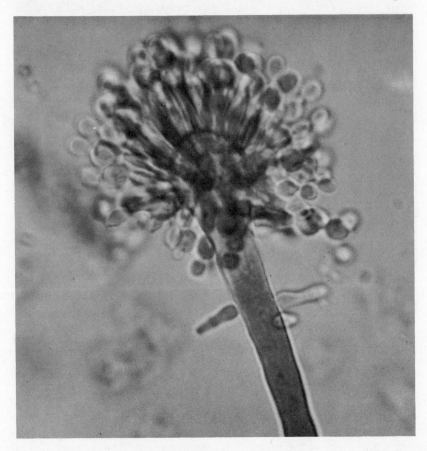

Fig. 10.6. Detail of a more highly magnified fruiting structure of *Aspergillus* **showing the conidia and the club-shaped cell from which they originate.** Photo courtesy C. M. Christensen.

Fig. 10.7. Pleurotus ulmarius. **The fruiting bodies of a basidiomycete growing from a tree trunk. The mushrooms are supported by a vast network of mycelium submerged in the substrate of wood or soil. These aerial structures bear the fungal spores which are disseminated after the mushrooms dry.** Photo courtesy C. M. Christensen.

leather articles. Mold spoilage of foods and feeds constitutes a major part of the economic loss in the food industry. Molds are the most important causative agents of plant diseases. In addition to the rusts and smuts, the mildews, rootrots, and many of the blights are caused by molds. Genetic studies are made using molds.

A few molds produce disease in animals and man. In addition they may cause allergies. Because of their aerobic nature, molds more frequently attack surface tissues. Those pathogens infecting the skin are called **dermatophytes.** Ringworm and athlete's foot are common dermatophytic diseases. Other pathogenic molds are able to infect the lungs and other tissues within the body. Such diseases are referred to as the deep mycotic diseases. The more common diseases caused by molds are discussed in Chapter 16.

Yeasts

Fungi that live primarily as unicellular organisms are called **yeasts.** These organisms are less widespread than molds or bacteria and are found primarily on the surface of plants, soils of orchards, and in the digestive tracts of insects. The diameter of most yeast cells range from four to six micrometers, which is sufficiently large to allow the observation of internal eucaryotic cell structures such as the nucleus, vacuoles, and storage granules under the light microscope, especially after staining. The typical asexual reproduction of yeasts occurs by **budding** (Fig. 10.8). Some of the protoplasm produces a bulge in the cell wall; the bud separates when it approaches the approximate size of the mother cell. A number of yeasts reproduce by binary fission, like bacteria. All yeasts are nonmotile. Some reproduce sexually by forming ascospores. Yeasts are classified in the class *Ascomycetes*.

The budding vegetative cells contain the diploid number of chromosomes. When the cell stops budding, meiosis takes place which reduces the chromosomes to the haploid number in the newly-formed ascospores. In the rather unusual life cycle of the yeast, the ascospores fuse before germinating to yield diploid vegetative cells. In most other biological forms, fusion takes place prior to the formation of sexual spores.

Importance of Yeasts

The baker's, brewer's, and distiller's yeast is an ascomycete called *Saccharomyces cerevisiae*. Anaerobically, the organism converts sugar almost quantitatively to ethyl alcohol and carbon dioxide. Certain wild yeasts, called filmforming yeasts, or torulae, grow on the surface of organic solutions. They oxidize the solutions of sugar, alcohol, or organic acids to form carbon dioxide and water. Yeast spoilage may occur in those foods which contain sugar but are too acid for bacterial growth and too anaerobic for mold growth.

A number of yeasts produce plant diseases and a few cause animal diseases. These will be discussed in Chapter 16.

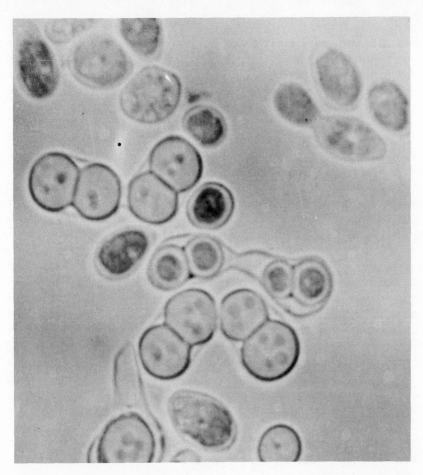

Fig. 10.8. Yeast cells showing the budding method of cell reproduction. Yeasts are much larger than bacteria and show definite compartments within the cell. Note the budding cell in the upper right hand corner and the four ascospores in the center.

The Actinomycetes

The moldlike bacteria of the order *Actinomycetales* and class *Schizomycetes*, commonly called actinomycetes, are gram-positive filamentous forms but differ from the true molds because the hyphae are of bacterial dimensions, about one μm in diameter (Fig. 10.9). The actinomycetes are also more bacterialike because they are procaryons; some produce flagella which are of the bacterial type; some forms show chemoautotrophy; many are antibiotic sensitive; they possess murein and not chitin or cellulose in the cell walls; and some actinomycetes are strict anaerobes.

The typical actinomycete has a branched mycelium and produces such compact colonies that removal of a portion with an inoculating needle is extremely difficult. Unlike yeasts and molds, the actinomycetes do not grow on acid media. They prefer more alkaline conditions for maximum growth. Although they are often studied with the molds (since they form hyphae with true branching), evolutionary hypotheses on their origin tie them closely to the bacteria as previously discussed. The genus *Streptomyces* comprises the most moldlike of the actinomycetes since it produces conidia on aerial hyphae. True endospores are unknown. The surface of the colony has a powdery appearance (Fig. 10.10). The branch cells are aerobic filaments. A few species are thermophilic. They carry out a significant part of the decomposition of organic matter in the soil. The characteristic odor of *Streptomyces* is evident in freshly turned soil. The plant disease, potato scab, is caused by *Streptomyces scabies.* Many actinomycetes produce antibiotic substances. The useful antibiotics, streptomycin, the tetracyclines, chloram-

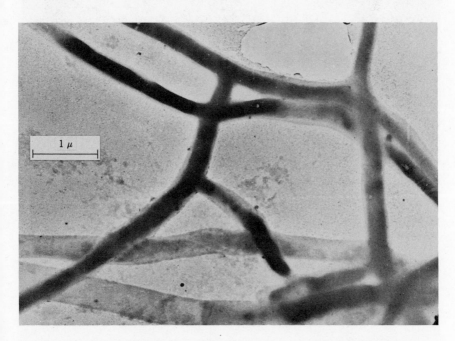

Fig. 10.9. **Electron micrograph of the actinomycete,** Actinomyces. **Several nonseptate hyphae with frequent branching can be observed. The cytoplasm within the hyphae has shrunken away from the cell wall and is of uneven density.** Photo by K. Polevitzky-Zworykin and R. F. Baker, courtesy A.S.M. LS-125.

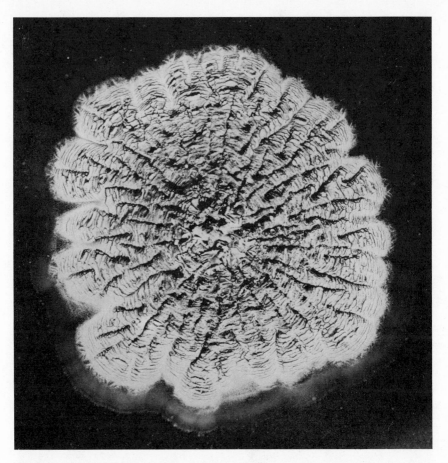

Fig. 10.10. **A colony of the actinomycete,** Streptomyces rimosus, **which produces the oxytetracycline antibiotic, Terramycin.** Photo courtesy Pfizer, Inc.

phenicol, and neomycin, are produced by species of the genus *Streptomyces.*

The anaerobic actinomycetes of the genus *Actinomyces* and the aerobic members of the genus *Nocardia* are less complex in structure. They produce no conidia, but young cultures form a mycelium which later fragments into short rods or cocci. The disease "lumpy jaw" in cattle as well as lesions in other animals and man are caused by actinomycetes of the genus *Actinomyces.*

The genus *Mycobacterium,* which includes the tubercle bacilli, is classified with the actinomycetes. It produces no mycelium and is a transitional form between the moldlike organisms and true bacteria.

Summary

The diverse morphological forms exhibited by the fungi range from unicellular to complex filamentous types. The lower molds are primarily aquatic. The more highly evolved fungi are associated with terrestrial plants or their products. Many of the terrestrial fungi are of great importance to man and his environment; a few are pathogens, many are important in the cycling of nutrients in nature, and some are agents in industrial processes including the production of life-saving antibiotics. The actinomycetes are filamentous *Schizomycetes,* and are considered by some taxonomists to be transitional forms between the bacteria and the molds. They occupy an important position in modern microbiology because they are the major producers of the useful antibiotics. A few are pathogenic to man and animals. Soil fertility is enhanced by their growth because they are the dominant microorganisms involved in the breakdown of organic matter at neutral pH values. This increases soil fertility by releasing utilizable nutrients for plant life.

ELEVEN

The Algae, Protozoa, and Complex Schizomycetes

The extremely diverse groups of microorganisms discussed in this chapter have distinctive characteristics and types, as well as members that appear to be transitional forms between the various groups of microorganisms.

The Algae

Algae, such as the seaweeds, pond scums, and water blooms, are chloro-phyll-containing primitive plants without differentiated roots, stems, and leaves. They are widely distributed throughout nature, being found in fresh and salt waters, shaded moist soils, bark of trees, moist rocks, and even the air. A number of algae are **epiphytes** in that they grow upon other plants but are not parasitic upon them. Frequently, such forms are found growing on buildings, telegraph wires, and similar objects. Moisture for development is obtained from the air. Many algae are found to be associated with bacteria, protozoa, and fungi to form a microbial community called a **plankton.** Over 15,000 species have been described, ranging from the tiny primitive blue-green algae to the very large brown algae such as the giant kelps. Algae are found in the greatest abundance in the upper layers of water bodies. Few thrive at depths greater than 25 feet because of the lack of light; some of the brown algae, however, can grow at about 300 feet. Aquatic forms may be found free floating or attached to rocks or bottom sediments in the shallow or shore line.

All algae, except the blue greens, possess a eucaryotic cell type. Blue-green algae demonstrate the procaryotic cell type and, therefore, are considered by many taxonomists to be more bacterialike and are classified within the plant kingdom phylum with the bacteria. Another popular system of classification entails the grouping of all algae into seven or eight divisions based on pigmentation, reproduction, food storage products, cell wall composition, and flagellar number. Only the major divisions will be presented here (see Table 11.1).

Unicellular, filamentous, and multicellular or colonial forms are exhibited by the algae. They are autotrophic, using absorbed water, carbon dioxide, and minerals from their medium, and are photosynthetic since they absorb and utilize light energy. Some of the blue-green algae are capable of nitrogen fixation, but the other algae require ammonia or nitrate for their nitrogen requirements. A few require vitamins or other growth factors for growth. Many algae, when growing in the dark, are capable of shifting from a photosynthetic to a respiratory type of metabolism, using various organic compounds as energy sources.

A number of algae are able to establish a symbiotic relationship with a mold, usually an ascomycete. Such an association is called a **lichen.** Lichens possess the ability to devour various nutrients, especially minerals, at extremely low concentrations. Each member of the association is capable of existing in the free-living form. Consequently, lichens exist as long as independent growth for each member is unfavorable. In areas of high industrial air pollution, the number of lichens is reduced or eliminated. It is thought that the algae serve as organic nutrients for mold growth; the mold, in turn, supplies water, minerals, and protection against dryness and intense light to its partner. Lichens are slow-growing colonies and distributed primarily in adverse conditions such as surfaces of rocks.

Both asexual and sexual types of reproduction are exhibited in this group of microorganisms. Three types of asexual reproduction are known to occur: fragmentation, where a small piece or fragment of the plant develops into a new organism; asexual spores of various types; and cell division in which a mother cell gives rise to two daughter cells. In sexual reproduction, the process of meiosis occurs, whereby the gametes contain half the normal number of chromosomes originally in the parent cell, the normal number

Table 11.1   Common Names of the Major Divisions of Algae

Division	Common Name
Cyanophycophyta	Blue-green algae
Chlorophycophyta	Green algae
Chrysophycophyta	Golden algae and diatoms
Phaeophycophyta	Brown algae
Rhodophycophyta	Red algae

being restored in the zygote either through conjugation (a special case of modified sexual reproduction) or fertilization. The majority of algae reproduce sexually; the basic life cycles are similar.

With the exception of the blue-greens, chlorophyll is localized in specialized cytoplasmic structures, the **chloroplasts.** Other accessory pigments are also found here. Often their green color is masked by red, brown, yellow, or blue pigments which assist in the photosynthetic process by the absorption of light of varying wavelengths and transferring the energy to chlorophyll.

CLASSIFICATION OF ALGAE

Blue-green algae (*Cyanophycophyta*). These algae are more related in chemistry and cell structure to the bacteria and are, therefore, classified by certain taxonomists in the same phylum (Fig. 11.1a). Chloroplasts are absent. Instead, the photosynthetic pigments are localized in peripheral membranes called **chromatophores.** Mitchondria are absent. There are no flagellated forms. A blue pigment, **phycocyanin,** along with the chlorophyll results in the blue-green color. Reproduction occurs by cell division or by asexual spores. Sexual life cycles have not been observed. In various species, the cells separate after division and are free living. In others, the cells are held together by gelatinous sheaths in colonies or filaments. Some blue-greens are able to use gaseous nitrogen from the air, which enables them to grow in areas where nitrogen compounds are absent from the medium. This ability accounts for their growth on volcanic rock and in similar areas where other plant life fails to develop.

Oscillatoria, Nostoc, and *Gleocapsa* are three common genera of this group of algae. The latter genus has been accepted as a likely candidate for incorporation into the cycling of food and wastes in future space vehicles.

Green algae (*Chlorophycophyta*). These forms are widely distributed in nature especially in fresh waters. They are similar to higher plants in that all have cellulose in the cell walls, starch as the principal food storage product, and chlorophylls and some of the carotenoids localized in the chloroplasts. Although unicellular forms are more common, there are colonial and filamentous species (Fig. 11.1b,c,d). Members of the genera *Chlorella* (Fig. 11.2) and *Scenedesmus* reproduce asexually. *Chlamydomonas,* a unicellular motile algae, demonstrates both asexual and sexual life cycles.

Golden Algae and Diatoms (*Chrysophycophyta*). The prevalence of certain carotenoid pigments confer a golden, golden brown, or yellowish green coloration to these algae. One major fresh-water group possesses only chlorophyll along with golden pigment and are principally unicellular forms. Starch is not found as a reserve food, but oil is stored by several species. The golden brown **diatoms** are included as a genus within this division (Fig.

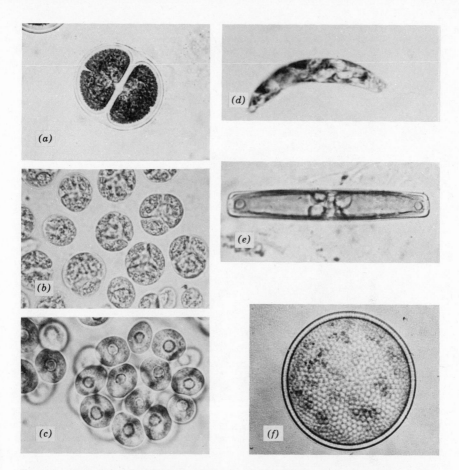

Fig. 11.1. **Light microscope pictures of some representative algae: (a)** Chroococcus **sp., a blue-green; (b)** Tetraciptis **sp., green; (c)** Chlorococcus **sp., green; (d)** Euglena **sp., green; (e)** Coscinodiscus **sp., a diatom; (f)** Pennularia, **a diatom.** Photos courtesy H. Bold.

11.1e,f). They are widely distributed in all types of waters and serve as a food source especially to animal life in the deep sea. Their cell wall is composed of large amounts of pectin and is impregnated and hardened with silica (silicon dioxide). The diatoms are pillbox shaped. During asexual cell division, each daughter cell retains either the cover half or the bottom half of the pillbox and grows a new necessary matching member. Some forms exhibit sexual reproduction. Motile and nonmotile species exist. Most diatoms are unicellular forms, but a few species are filamentous or colonial types. The remains of the silica walls of the diatoms accumulate on the ocean floor, and deposits of fossil diatoms over 3000 feet thick have been discovered during oil well drillings.

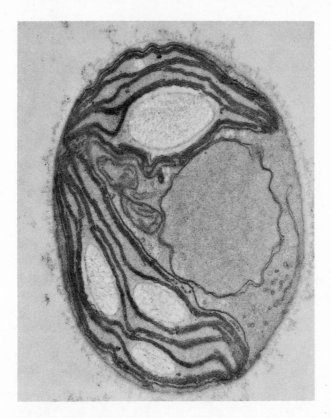

Fig. 11.2. An ultrathin section of the green alga, Chlorella pyrenoidosa, as shown by electron microscopy. Note the internal membranes which are not demonstrated in similar sections of the bacteria or blue-green algae.

Brown algae (*Phaeophycophyta*). These are the common seaweeds found on the surface and depths of the ocean. The possession of an organelle of attachment, the **holdfast,** allows these algae to firmly attach to rocks and even other algae. There are only a few free-floating forms. Microscopic to large plants of 150 feet in length exist. Cellulose, pectin, and a gelatinous material, **algin,** compose the cell wall. Sexual reproduction is characteristic of this group although fragmentation does occur in some species.

Red algae (*Rhodophycophyta*). These nonmotile algae are large, principally marine plants usually found in the subtropical and tropical seas. Only a few members are not multicellular. A red pigment masks the green color of the chlorophylls which are present in smaller amounts. Starch is frequently stored as free cytoplasmic inclusions. The cell walls of some species contain a gelatinous polysaccharide, **agar.**

The algae are an important link in the biological nutrition chain in the sea. They are the builders of organic material used by the small sea animals, which in turn, serve as food for the larger animals. They have been referred to as the "grass of the sea." Some scientists anticipate that, with the development of proper methods of collection and culture, algae may some day be a direct source of human and animal food. One species of brown algae is used as food in Japan. Several green algae have been extensively used in studies of photosynthesis and respiration. Diatomaceous earth (fossils of diatoms) is used as an abrasive in toothpaste and as a filter aid in the filtration of liquids in sugar refineries. It can also be manufactured into an insulating material.

Algin from brown algae is purified and used commercially as food stabilizers in ice cream and marshmallows. **Agar** from the red algae is extracted with hot water for use as a solidifying agent in bacteriological culture media. Photosynthesis by algae promotes fish life by releasing oxygen into the water. In areas of intense sunlight, sewage purification procedures have been developed which involve the oxygenation of the effluent by algal photosynthesis.

The tremendous growth of algae in lakes may ruin the lake for recreational purposes during the height of the water bloom. Increasing amounts of phosphates and nitrates in lakes and rivers due to pollution enrich the waters enhancing algae growth to an extent that these waters fill more readily with the undesirable growth. Even worse is the period following, when the putrefaction of algal cells release decomposition products that are extremely unpleasant in odor.

Algae are the only major group of microorganisms with no pathogenic species for humans, but the decomposition products may produce illness and even death in animals drinking the water during the period of greatest algal putrefaction. Even relatively small amounts of growth of certain species of algae in drinking water may be responsible for obnoxious odors and tastes.

Protozoa

Although these unicellular organisms are classified as animals, primarily due to the absence of a cell wall, the protozoa include some forms that are closely related to the algae (*Chlamydomonas, Euglena*). On the other hand, other protozoa are quite similar to the water molds, and still others are clearly similar to the bacterial spirochetes (Fig. 11.3). It should be emphasized, however, that all protozoa demonstrate the eucaryotic cell type. A few

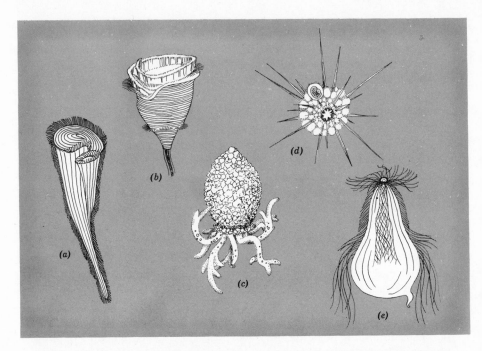

Fig. 11.3. Protozoa exist in a tremendous variety of shapes and cell arrangements. Some representatives are illustrated in this figure. (a) Stewtor polymorphus (heterotricha) (b) Vortecella convallaria sessiline peritrich (peritricha) (c) **DiAlugia urceolata** (lobosa) (d) **Actinophrys sol** (heliozoa) (e) **Trichonympha campanula** (hypermastisina).

species exist in colonial forms. As might be expected of such a diverse group, protozoa are widely distributed in nature wherever there is moisture. In surface waters and in the upper six inches of soil, they live on the small algae and bacteria. In the gut of the termite and in the rumen of cud-chewing animals, they feed on cellulose-digesting bacteria, thus performing an indispensable function in the conversion of cellulose into absorbable nutrients. The intestines of animals and insects contain protozoa existing as commensals. Autotrophic and heterotrophic types of nutrition are represented in this group of organisms. A few species are parasitic. Asexual and sexual reproduction occur among the protozoa. Two or more nuclei, existing in different shapes, may be present in various species. Frequently, the nuclear membrane does not disappear during cell division. Most protozoa possess a chitin-containing outer membrane called the **pellicle.** Many forms have a mouth for the ingestion of food.

Recently the phylum has been divided into four major subphyla or divisions: *Sarcomastigophora, Sporozoa, Cnidospora,* and *Ciliophora.* The *Sarcomastigophora* contain protozoa that possess flagella or pseudopodia (false feet) as locomotory organelles. No spore formation is evident, and division occurs principally through asexual reproduction. The protozoa that move by means of pseudopodia have no fixed shape and, therefore, constantly change in form and size as the cell membrane is stretched into temporary projections by the flowing of the protoplasm. Pellicles are absent. Since the genus *Amoeba* is one of the best known forms of this type, this flowing type of movement is frequently referred to as **amoeboid movement.** The cells engulf food particles, such as bacteria and algae, by flowing around them, thereby enclosing the food in a vacuole, where it is digested. Other contractile vacuoles collect excess water and waste materials inside the cell and move to the cell boundary where they are discharged.

Asexual reproduction involves simple division of the nucleus, followed by cell division. If sexual reproduction occurs, the cells may fuse (or, in some cases, flagellated gametes produced by the cells conjugate) to form a zygote, which in turn, develops to produce daughter amoeboid cells.

One group of marine species develop as flexible amoeba but secrete silica as they grow, forming either an external shell or an internal skeleton. During the millions of years that these forms have been living and dying and leaving skeletons and shells on the earth, they have been changing in structure. Experts in microscopic examination of geologic deposits can determine the age of layers of sedimentary rock by the appearance of these fossil protozoa. This information is especially useful in the identification of the geological formation in which the drill is operating during oil well exploration.

Endamoeba histolytica, another member of this division, is prevalent in the tropical and subtropical areas. It is the causative agent of **amoebic dysentery,** a chronic debilitating disease, which is occasionally fatal. Many healthy carriers are involved in its spread. Diagnosis is made upon finding the organism by microscopic examination of the stool.

The flagellated protozoa of the *Sarcomastigophora* move by whiplike flagella which push or pull the flexible cells through their liquid environment. Flagellates generally have a leaf-shaped structure and are often elongated like a spirochete. Reproduction is almost always asexual; the cells split longitudinally, unlike the transverse fission of bacteria. **Cysts** are formed by enclosure of the cytoplasm in a thick cell wall which protects the organisms during long periods of drying. A group of parasitic flagellates are the **trypanosomes;** most trypanosome diseases involve an insect vector for transmission. *Trypanosoma gambiense* causes African sleeping sickness and is transmitted to man by the tsetse fly.

The divisions *Sporozoa* and *Cnidospora* contain protozoans that produce spores. Only a single nucleus is present, and all members are parasitic. The sporozoans have complex life cycles. Malarial parasitism involves both an animal (man or monkey) and an insect, the *Anopheles* mosquito. The female mosquito injects motile spores into the blood stream of the animal; these attack and destroy red blood corpuscles. The development of the organism is synchronized so that the emergence of each new generation from the red corpuscles causes the recurring and periodic chills and fevers typical of malaria. A few of the parasites mature into sexual forms, which do not develop further in the human body. A feeding mosquito, however, will pick up the zygotes that result from sexual combination, and the protozoa complete their life cycles in the body of the insects. There, the sporozoites eventually form and invade the salivary glands of the mosquito. When the mosquito feeds, the sporozoites may enter the new victim.

The last division, *Ciliophora,* is characterized by members possessing **cilia** as organelles of locomotion. Generally, two types of nuclei are present, and conjugation is a special case of modified sexual reproduction. Most members of this group are free living. The ciliates are exceedingly highly developed. The movement of these single-celled animals is controlled by short hairlike projections, the cilia, which move as if coordinated by a nervous system. In the best known ciliate, the *Paramecium* (Fig. 11.4), food enters the cell through an oral groove and the food vacuole so formed passes lengthwise through the cell. The vacuole finally empties the undigested food into the medium through an anal pore. Other vacuoles carry water and dissolved substances through the cytoplasm.

These organisms possess a **macronucleus,** which controls metabolic activities, and a **micronucleus,** which is involved with reproduction. In asexual reproduction, both nuclei divide prior to longitudinal fission. In sexual reproduction, however, the macronuclei disappear following conjugation of two cells; division of the micronucleus in each cell is followed by a reciprocal nuclear exchange by the conjugating partners; thus, each fertilizes the other. In each species of *Paramecium,* there may be from two to six mating types; conjugation does not take place between members of the same mating type.

The only species of the ciliates known to be parasitic for man is *Balantidium coli,* an intestinal parasite which causes a type of dysentery.

IMPORTANCE OF PROTOZOA

Protozoa are of importance to microbiologists for many reasons. Some members cause diseases of man, such as dysentery, malaria, and sleeping sickness. These are discussed in more detail in Chapter 17. Through their feeding activities, they aid in the control of bacterial populations in soil and water. They are a link in the food cycle in the sea; they consume the smaller algae and bacteria and are in turn consumed by the larger animals. Geologists have found useful fossil protozoa in revealing the age of sedimentary rock.

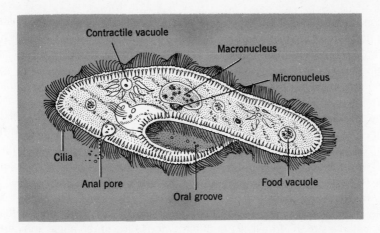

Fig. 11.4. The protozoa, Paramecium, is frequently encountered in pond water where it feeds on bacteria and small algae. Although unicellular, it shows structural detail far more complex than that observed in bacteria.

Complex Schizomycetes

In the classification of the *Schizomycetes,* the orders *Pseudomonadales* and *Eubacteriales* constitute the true bacteria. The other orders contain more complex forms, although they share many of the same characteristics possessed by the true bacteria. These forms are sometimes referred to as the **"higher bacteria."** Some are transitional forms between the true bacteria and the algae, protozoa, and fungi. Others represent evolutionary deviations from the bacteria which have not led to currently existing higher forms of life. All of the higher bacteria possess the procaryotic cell type. Except for the order *Actinomycetales,* the study of these extremely interesting biological entities has been seriously neglected.

The order *Spirochaetales* includes the unicellular, helical or spiral bacteria that are flexible because the cell wall is not rigid. Flagella are absent and, instead, motility is achieved by an **axial filament** which consists of bundles of elastic fibers intertwined between the involutions of the corkscrew-shaped cell. These characteristics differentiate the spirochetes from the true bacteria, spirilla, which possess rigid cell walls and move by flagella. Scientists consider the spirochetes to be transitional forms between the bacteria and some of the protozoa. Spirochetes do not stain readily and are best observed as living specimens under the dark-field microscope or in negatively-stained preparations.

The *Spirochaetales* are divided into two families, *Spirochaetaceae* and

Treponemataceae. The former family consists mostly of free-living forms and their normal habitats are the mouth, genitalia, and the intestinal tract of man and animal. The members are considerably larger and can exhibit long cells, ranging up to 500 μm in length. The genus *Cristaspira* are normally associated with oysters and other bivalve crustaceans. Members of *Spirochaeta* are found in sewage or decomposing vegetable matter.

The smaller, pathogenic species are in the *Treponemataceae.* Of this group, *Treponema pallidum,* which causes syphilis, is the best known (Fig. 11.5). Organisms in the genus *Leptospira* are found in scrapings taken from the base of the teeth and gums of most individuals, even those with seemingly healthy mouths. Yaws, infectious jaundice, and relapsing fever are other human diseases caused by spirochetes.

The order *Myxobacteriales,* or **slime bacteria,** appear to be nonphotosynthetic relatives of the free-living nonfilamentous blue-green algae and certain protozoa. Their cells are nonflagellated, and their mechanism of movement remains a mystery although it has been suggested that the gliding movement results from the secretion of slime. The members of the genus *Myxobacteria* are flexuous gliding bacteria generally found in soil, dung, and decaying vegetable matter. They glide over solid surfaces as single cells or as tongue-shaped cytoplasmic aggregates. Some undergo a developmental cycle culminating in raised aggregates (fruiting structures) consisting of slime-encased masses of "resting cells," or **microcysts,** or of resting cells enclosed in large cysts.

The **sheathed bacteria** of the order *Chlamydobacteriales* grow as individual cells, each with its own cell wall, in chains enclosed in a common gelatinous sheath or **trichome.** In some genera, the sheath is encrusted with iron residues formed as a result of the cells' metabolic activities. These are hetero-

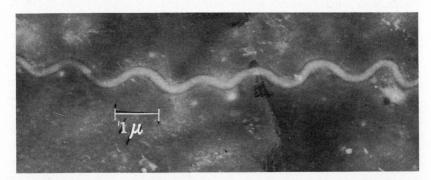

Fig. 11.5. This spirochete, Treponema pallidum, **was observed in an electron micrograph made from exudate of a lesion of a syphilis patient.** Photo by Oskay, courtesy H. E. Morton, W. F. Ford, and A.S.M. LS-254.

trophic water-dwelling species, and the filaments are either free-floating or attached to the bottom or sides of the water body in which they live. The end cells in the sheath may slip out and form new filaments. In some cases, motile conidia may be produced in the sheath. At various intervals, these can be observed swarming away from the parent filament. The sheathed bacteria may become a nuisance in aerated waters where a relatively high iron content is present resulting in clogging of pipes. *Sphaerotilus* is a genus within this group.

Members of the order *Beggiatoales* are filamentous **gliding bacteria** that develop as trichomes of usually nonflagellated bacilli with a continuous wall. They are considered the colorless counterparts of the filamentous bluegreen algae. These organisms live in sulfide springs and other locales where hydrogen sulfide is available. The hydrogen sulfide, oxidized to sulfur, is deposited as granules within the cells. With a depletion of sulfide, the organisms can oxidize the sulfur to sulfate for energy, forming sulfuric acid. A widely occurring species in this order is *Beggiatoa albus,* which multiplies by fragmentation of the ends of the filaments.

The order *Caryophanales* consists of large trichomes of flagellated cells with a common wall. Sheaths are absent. These algalike organisms are saprophytes found principally in organic waste materials.

Members of the order *Hyphomicrobiates* reproduce by buds which form on the ends of hyphaelike structures. Budding is not exhibited. Thus, these organisms are referred to as the **budding** bacteria. The mother cell is usually attached to a solid substrate. The motile cells formed by budding may become detached to start another colony. *Rhodomicrobium* is a photosynthetic genus of the budding bacteria (Fig. 11.6).

An unusual group of microbes, the **stalked** bacteria, is classified under the true bacteria in the family *Caulobacteriaceae,* but shows greater complexity than once believed. One genus that has been widely studied is the iron bacterium, *Gallionella.* These are small kidney-shaped bacilli that are found on the end of a twisted ribbon of iron hydroxide which anchors the cells to a solid surface (Fig. 11.7). When a cell divides, either the ribbon branches or one of the daughter cells breaks free to start a new colony. They are frequently found in water habitats where iron is plentiful, such as iron springs and draining bogs.

Summary

Algae are regarded as the grasses of the sea since they are the primary producers of organic material in aquatic environments. It is estimated that

Fig. 11.6.   **Micrograph of** Rhodomicrobium, **a photosynthetic budding bacterium.** Photo courtesy H. C. Douglas: J. Bacteriol., **58,** 409, 1949.

approximately 70% of the atmospheric oxygen that is being currently produced comes from algae. Many algae demonstrate complex life cycles. Some are macroscopic but these forms demonstrate a relationship to the microscopic forms, thus illustrating a degree of evolutionary development. Microbiologists are concerned with the algae because of the nuisances they create in recreational water areas (especially when these waters are enriched by pollutants), their role in the food cycle of the sea, and the special products obtained from them. Protozoa are scavengers of bacteria and smaller algae. A wide number of diverse forms exist. Although there are no pathogenic algae, the protozoa are involved in a number of human diseases. It has been suggested that protozoa may offer a system whereby the biochemistry, ecology, and evolution of host-parasite relationships could be better investigated. The spirochetes occupy a position between the protozoa and the true bacteria. Because of their somewhat greater cellular complexity, they, together with the slime, stalked, budding, and sheathed forms, are frequently referred to as the "higher bacteria."

Fig. 11.7. The small rod-shaped Gallionella secrete iron hydroxide from one side of the cell to produce the complex twisted ribbon which anchors the cells to a surface.

TWELVE

Viruses, Rickettsiae, and Mycoplasmas

The microbiologist encounters biological entities that are smaller and, in some respects, less complex than the bacteria. As science has become aware of the role of viruses in human disease, the techniques of bacteriology have been modified to deal with the special problems of viral culture and study and many unique and intricate laboratory procedures have been introduced. The new techniques have also been useful in studying rickettsiae and mycoplasmas.

Viruses

A number of transmissible diseases are caused by agents so small that they can pass through the pores of filters which will retain bacteria. These agents, too small to be viewed with the light microscope, are called **viruses,** or filterable viruses. Beijerinck in 1898 confirmed Iwanosky's filtration experiment and reported the first substantial evidence of a virus as the cause of a plant disease. Soon, other reports associated viruses with various diseases. Foot and mouth disease was found to be of virus etiology in 1898, and was the first virus to be isolated from an animal. Variation among members of this group is so wide that there is no simple definition that can describe all viruses. Factors such as filterability and infectivity were early important properties in characterizing viruses. However, certain bacteria were known to possess one or both of these factors.

One of the most striking properties of viruses is their size. The unit of measurement for viral particles is the nanometer (nm), which is 0.001 μm. Viruses range in particle size from about 10 to 300 nm. Although the majority of viruses are much smaller than the usual bacteria (1000 nm), a few bacteria are as small as the largest viruses. Viruses were first described as the agents which passed through a bacterial filter, and filterability was an early criterion of a virus.

The size of virus particles is chiefly determined from electron microscope micrographs. The estimates may not be fully accurate because of distortions caused by the preparation for examination. Measurements by other methods, such as ultrafiltration and ultracentrifugation, are in reasonable agreement.

Electron micrographs of viruses have demonstrated no uniformity in shape or structure, although there appears to be a limited number of morphological types. The shape is characteristic for a particular virus and is therefore useful in identification. Spherical, cuboid, rod-shaped, and tadpole-shaped (Fig. 12.1) virus particles have been reported. Electron microscopy has demonstrated that viruses once thought to be spherical are actually icosahedral (20 triangular faces with 12 corners).

At the present time, a biological entity must possess five collective characteristics before it may be called or classified as a virus. As mentioned previously, all viruses are filterable. However, filterability is no longer used as the sole criterion. (1) A virus must occur as an intracellular, obligate parasite in order to replicate its own kind, the replication taking place only in certain susceptible host cells. (2) Host dependence is furthered by the fact that a virus has no built-in metabolic machinery and (3) also must utilize the ribosomes of its host during replication. (4) Viruses differ from all other forms of life because they possess only one type of nucleic acid, either RNA or DNA, which serves as the genetic material. (5) Finally, a virus does not grow in size.

A virus, chemically speaking, is **nucleoprotein.** An infectious virus particle when existing in the extracellular state is called a **virion.** The genetic material, either DNA or RNA, is surrounded by a protein box or **capsid.** The nucleic acid can be either single- or double-stranded. The nucleic acid and the capsid collectively are called the **nucleocapsid.** The capsid is made from many structural subunits, the **capsomeres.** Certain animal viruses have an additional lipid-containing envelope surrounding the capsid. Some bacterial viruses possess other structures which will be discussed later.

The virus itself is, then, little more than a piece of genetic material with the ability to penetrate a susceptible cell. When only the nucleic acid of a virus is introduced experimentally, complete viruses are produced. Once inside the cell, the virus provides the genetic code which directs the enzymatic apparatus of the host cell to produce more viruses. When this process causes no observable change or damage to the host, the virus infection is termed **la-**

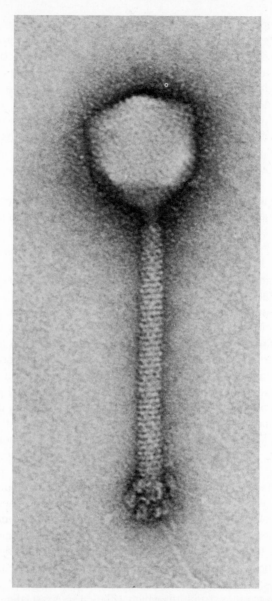

Fig. 12.1. An electron micrograph of a negatively-stained SP50 Bacillus **bacteriophage demonstrating a head (capsid) and tail.** Photo courtesy A. K. Kleinschmidt.

tent. When much of the host cell activity is diverted from the cell's own necessary activities, observable cell damage or destruction results and a diseased condition is recognizable.

Because the replication of a virus is accomplished through the reactions of the host cell, scientists have had difficulty in finding chemical substances that will selectively check virus growth without poisoning or harming the host. Although there may be a few exceptions, the infectious diseases that do not respond to the currently available chemotherapeutic agents may be correctly diagnosed as of viral origin.

Crystalline preparations of certain viruses appear to retain infectious ability indefinitely. This suggests that viruses are inanimate and should not be classified as living entities. Certainly the free virus or virion, as it is observed under the electron microscope and is subjected to chemical analysis, is only the resting stage in the cycle of viral development.

Viruses have two important definitive attributes of life: the ability of **replication** and the ability of **mutation.** Viruses must make use of another living cell for replication, but all other forms of life also reproduce only in a suitable environment. The suitable environment for a virus exists within a host cell. The property of mutation permits living things to survive in a changing environment and participate in organic evolution. In viruses, these changes appear to be analogous to genetic changes arising from mutations in higher forms of life. Recorded viral mutations include alterations in virulence, adaptation to a new host, alteration in tissue affinity, and emergence of antigenic variants. Naturally-occurring viral mutations are exemplified in the almost unlimited ability of the influenza virus to reproduce antigenic variants due to changes in its protein capsid, and, possibly, in the evolution of the vaccinia virus from the variola virus. Viral adaptation to an unnatural laboratory animal, with simultaneous loss or reduction of virulence for the natural host, has been induced in a number of instances. This tendency toward adaptation has been exploited in preparing vaccines for immunization against naturally-occurring viral diseases, such as polio, measles, and mumps.

Most viruses can be preserved in a suitable medium for long periods in a freezer held at the temperature of solid carbon dioxide (−76 C) or of liquid nitrogen (−196 C). Most free viruses are inactivated by elevated temperatures which kill most vegetative bacterial cells. Many can survive long periods in the dry state. Viral strains vary greatly in their resistance to various chemical and physical agents. Agents that cause viral inactivation either disrupt the virus particle or alter the viral protein or nucleic acid in such a way that the virus can no longer infect or replicate.

One early means of classification of the multitude of known viruses was by the limited type of host cell viral replication. Four classes of viruses were established: **bacterial, animal, plant,** and **insect.** Host specificity was soon discovered to be unsuitable as a means of classification because viruses capable of infecting numerous plant and animal species were isolated. However, most viruses exhibit great species specificity.

Early microbiologists observed a special sort of degenerative change in certain laboratory cultures of bacteria. The change could be transmitted by placing a drop of filtrate from an infected culture onto an agar culture of the susceptible bacteria. The filterable agent was identified about 50 years ago as a virus which was parasitic to bacterial cells. Since the colonies appeared "nibbled," the name **bacteriophage,** commonly abbreviated to **phage,** was applied. This term was derived from the Greek word, **phagein,** meaning to eat.

Any particular phage is highly host specific and will attack only one species and perhaps only a few strains of that species. However, probably every known type of bacterium is susceptible to one or more phages. A phage-susceptible cell can mutate to become resistant to one phage type, yet remain susceptible to other phages. Viruses against actinomycetes, yeast, algae, and molds have also been reported. They, too, show host specificity.

Bacteriophages affect susceptible bacteria by causing **lysis** or dissolution during the stage of active bacterial growth. Infection occurs when a few drops of a phage-containing filtrate are added and spread on the surface of an agar plate which has been heavily seeded with a susceptible bacterium. If the filtrate has not been diluted, no growth (or at most a few colonies of resistant bacteria) will appear. If sufficiently diluted, the filtrate addition produces small, clear areas called **plaques,** which are distributed in the background or **lawn** of bacterial growth (Fig. 12.2). Each infectious phage particle infects an individual cell, causing eventual lysis and release of new viral progeny, which in turn infect neighboring bacterial cells and cause lysis resulting in the ultimate clear area. Since each plaque corresponds to an infectious phage particle in the inoculum, a phage count or assay may be made in the same manner that dilution plating is used to count bacteria. Differences among phages can be recognized by variation in their plaque morphology such as size, turbidity, and type of edge. Consequently, the **plaque morphology,** which is characteristic under specific conditions for a given phage, is useful in the identification of phages.

Bacteriophage can be demonstrated also by adding phage-containing filtrate to a broth culture of a susceptible organism. Phage should be introduced when the bacteria are inoculated into the broth or in the early stages of bacterial growth. Clearing of the culture, either partial or complete, will occur as a result of phage activity. Prolonged incubation may produce a recurrence of turbidity, resulting from the growth of mutant cells which were resistant to phage action.

A number of morphological types of phages have been reported when studied by electron microscopy. Some phage particles appear tadpole-shaped because they possess a head and tail structure. Certain phages of

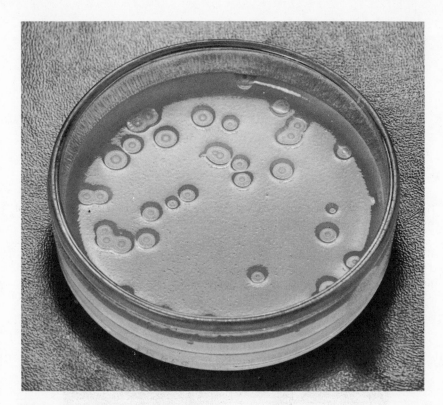

Fig. 12.2. To isolate bacteriophages and to make quantitative studies, plate counts of plaques are prepared. Plaques appear as clear areas of lysis in a lawn of bacteria growing on the agar surface.

this morphological type can also exhibit additional tail organelles (Fig. 12.3). The tail of many phages is a good bit longer than the diameter of the head; in others, the tail may be so short as to escape detection. Other phage particles may be void of any tail structure, and a few phage species exist as rod-shaped structures. Phages are composed of nucleoprotein. The capsid head is composed of protein which houses and protects the genetic material. The tail structure, if present, is also composed of protein and serves in the capacity of attachment to the susceptible bacterial cell. Although the majority of phages possess double-stranded DNA, others contain either single-stranded DNA or RNA.

A series of phages, the T-phages, which are able to attack *E. coli* have been intensively studied. Consequently, these coliphages, possessing double-stranded DNA, have become the model for various aspects of virus replication. Certain phages of this group possess rare ornate structures on the tail (Fig. 12.4). When infectious particles of **virulent** phages are placed into a culture of sensitive bacteria, the phage must come into physical contact with

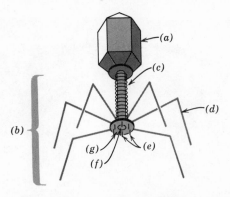

Fig. 12.3. An artist's concept of one of the morphological types of bacteriophages. (a) capsid; (b) tail; (c) sheath; (d) tail fiber; (e) tail spikes; (f) hollow core; (g) tail plate.

the cell. This is accomplished by random collision between the two entities. The course of viral replication involves four major steps: **adsorption, penetration** or **infection, replication,** and **release.**

During the adsorption process, the tail fibers of the coliphage unwind from around the sheath structure of the tail and, along with the tail spikes, aid in

Fig. 12.4. An electron micrograph of the coliphage T-4.

attachment to the cell. The distal tip of the tail, the organelle of adsorption, then becomes attached to a specific **receptor site** on the cell wall (Fig. 12.5). The phage receptor sites are under the control of the bacterial nucleus, and therefore **phage-resistant mutants** can occur by mutation and subsequent modification.

The tail is used more or less as an inoculating needle. The sheath contracts toward the head, exposing a small core which runs the length of the tail structure and opens into the head. The exposed core penetrates the rigid cell wall, and the nucleic acid is injected into the bacterial cytoplasm. The head and tail remain attached outside, and thus play no role in subsequent events in the viral growth cycle. The adsorption and infection steps usually require only a matter of a few minutes. The nucleic acid, serving as a new genetic code, now takes control of the economy of the cell. Although the host continues its energy-producing activities, the cell's biosynthetic machinery is directed to the synthesis of new phage nucleic acid and proteins. The syntheses of these materials occur independently, and thus no infectious phage particles can be detected early in the infection period. This is known as the **eclipse period.** Soon there is an assemblage of their components and mature phages appear within the cytoplasm. During maturation, lysozyme formation is induced by the phage DNA and begins to break down the rigid nature of the cell wall thus becoming osmotically sensitive. Lysis of the cell eventually occurs, releasing a large number of new viral progeny into the environment capable of initiating the same lytic cycle (Fig. 12.6). The period from phage infection to lysis of the cell is referred to as the **latent period.** An average burst releases about 200 phage particles. However, the burst size varies depending upon the phage, host cell, and certain environmental conditions.

Not all phages, however, conform to this model system. For example, some

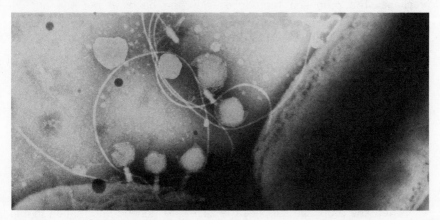

Fig. 12.5. An electron micrograph showing phages adsorbed to a cell of Bacillus subtilis. Photo by H. Reiter: J. Virology, **3,** 581, 1969.

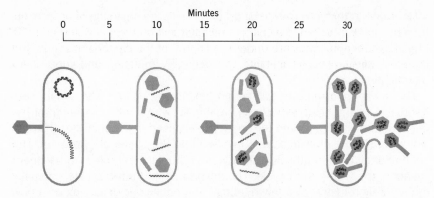

Minutes

| 0 | 5 | 10 | 15 | 20 | 25 | 30 |

Fig. 12.6. **Lytic cycle of a phage.** The phage is adsorbed tail first to the bacterial cell wall and the DNA fibril from the head penetrates into the bacterium. It serves as the genetic code for the synthesis of viral DNA and proteins. Maturation occurs upon the condensation of a DNA molecule and assemblage of the protein parts. The cell eventually bursts, releasing a new crop of phages.

phage receptor sites have been found to be located in the pilus of the cell; others, more rarely, in the flagellum. Some phages do not possess a tail, and their mode of infection presents a mystery to the virologist. Phages that possess either single-stranded RNA or DNA have certain modifications within the replicative process of the nucleic acid.

In addition to virulent phages, which have no alternative but to initiate the lytic cycle upon infection, there also exist **temperate phages.** When the DNA of a temperate phage is injected into the bacterial cell, it may initiate the lytic cycle. More frequently, such nucleic acid associates itself in a more or less symbiotic relationship with the bacterial chromosome and behaves as if it were part of the hereditary apparatus of the bacterium. Such bacterial cells rarely show evidence of the presence of the nucleic acid of the virus, or **prophage.** The prophage reproduces with the bacterial nucleus and is passed on to daughter cells. This process may continue for many generations. Bacterial cells containing a prophage are immune to attack by the same or related phages, and are said to have **lysogenic immunity.** Often, a cell carries more than one prophage. In old cultures or in cultures exposed to radiation, the balance between the prophage and the bacterial nucleus is often disturbed. The prophage then detaches from the chromosome and, like virulent phages, assumes control of the cell economy and initiates the lytic cycle. Lysis of the cell releases a burst of free temperate phage particles which have the ability to attack sensitive cells. Because of this latent potential for lysis, cells containing prophage are called **lysogenic,** and the whole phenomenon is called **lysogeny.** Such loss of individual cells in old cultures is more than balanced by the immunity against phages that lysogeny confers. Often, the prophage supplies the necessary genetic information to the lysogenic cell to exhibit a new characteristic not displayed by the nonlysogenic cell. For example, the poison (toxin) responsible for the bacterial disease, diphtheria, is produced by the diphtheria bacteria only when the prophage is present. Various other phenotypic modifications induced by the presence of a prophage have been reported.

Phage typing has been successfully applied in the differentiation of bacterial strains of the same species. The particular bacterial isolate is tested for its susceptibility to lysis by a number of related and different temperate phages. The culture is not lysed by the phages for which the cell carries a prophage (i.e., has lysogenic immunity). Phage typing is often used in the tracing of the source of a particular infection, such as food poisoning, and for other similar epidemiological studies.

ANIMAL VIRUSES

For years animal viruses could be maintained in the laboratory only by culturing in susceptible animals. Characteristically, animal viruses require specific host species and even certain tissues or cells in the host organism. Fortunately, such specificity applies less to embryonic tissue, and most viruses can be cultured in embryonated chick eggs (Fig. 12.7). The development of tissue culture techniques greatly enhanced the study of animal viruses. Various types of tissues may be grown in petri plates or test tubes by supplying exacting nutrients and environmental factors to the cells.

The majority of animal viruses are nucleoprotein although some, in addition to the protein capsid, have an additional surrounding lipid-containing envelope or shell. Both RNA- and DNA-containing viruses are known, but no single-stranded DNA viruses have been isolated.

Often, the method of replication for the T-even phages is used as a model system for all viruses. Although there are many similarities, there are evident differences between bacterial and animal virus replication. For example, animal viruses do not have organelles for adsorption as most of the bacterial viruses do. The virus appears to get into the cell by the process of pinocytosis, in which the cell membrane folds inward into the cytoplasm forming a vacuole. The cell obtains certain nutrients in this manner, and perhaps the virus gets into the cell by mistake. Although the intact virus particle is ingested, the capsid is soon uncoated, freeing the nucleic acid. The capsid evidently has no role in the replicating viruses, and is just for protection and housing of the nucleic acid. Under the direction of the new genetic code, the cell synthesizes new viral nucleic acids and proteins. The various macromolecules are assembled to form mature viruses. The latent period of animal viruses is considerably longer (hours to several days) compared with the latent period of bacterial viruses (minutes to hours) from the bursting bacterial

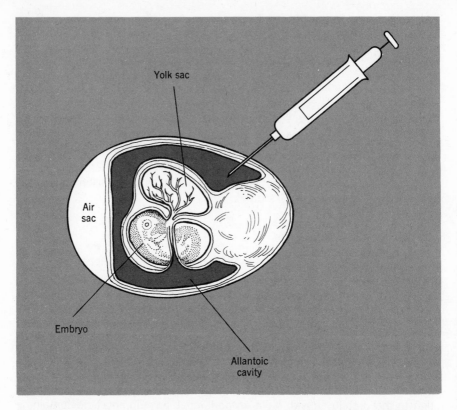

Fig. 12.7. Diagram of a developing chick embryo showing the following sites of virus inoculation: yolk sac, the allantoic cavity, and the embryo itself. If the shell over the air sac is punctured, the outer membrane of the embryo (chorioallantoic membrane) drops away from the shell and can be inoculated with virus.

cell. The animal cell extrudes viral progeny over a period of time before ultimate death and lysis of the cell occurs. The total yield of animal viruses per cell is usually greater than the phage burst size.

Viral infections cause a variety of cellular alterations which depend on characteristics of the virulent virus and the susceptibility of the host cell. In some infections, the cell may show no apparent effect; in others, lysis and death result. The presence of the virus may influence cell metabolism adversely by coding for the formation of chemical substances foreign to the normal healthy cell. This may cause pathology by reducing the production of some normal metabolite. Certain viruses cause cells to fuse to form large multinucleated cells. Others appear to induce an unusual, abnormal increase in cell division, resulting in a swelling of the affected cells or tissues. In some viral diseases, the infected cells show inclusion bodies not present in normal cells. These may appear only in the cytoplasm and are helpful in diagnosis of the disease, as is the case with **negri bodies** in rabies and the **Guarnieri bodies** in smallpox (Fig. 12.8). Or they may be found in the nucleus, as in warts and in herpes simplex (fever blisters). It is thought that the inclusion bodies consist of aggregates of virus particles.

Viral infections may be localized; in warts, only the cells near the entry point of the viruses are affected. The influenza and common cold viruses affect the cells of the respiratory mucous membranes causing a less localized effect. In other viral infections, the invading viruses are disseminated throughout the body through lymph nodes, blood stream, and lymphatics to various body organs. With the reoviruses, both intestinal and respiratory tract tissues are affected by the same virus. Multiplication of the viruses occurs during this spreading effect. In all types of viral diseases, many infections are inapparent or subclinical because (1) the invading virus is not sufficiently virulent; (2) the host has an adequate defense mechanism to suppress virus replication to a level that fails to yield an apparent infection; or (3) the virus

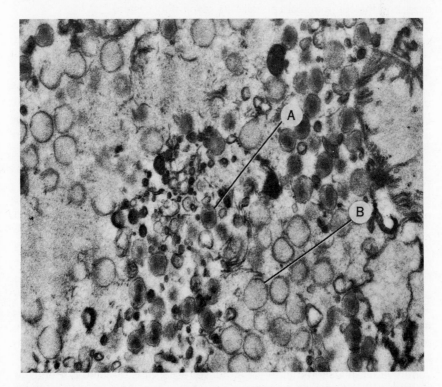

Fig. 12.8. An electron micrograph of a thin section cut through a Guarnieri body reveals the developing vaccinia virus. (a) A mature virus particle showing a dark center. (b) An immature virus particle. Photo courtesy N. McDuffie.

is unable to reach the principal organ or tissue where the major replication for that particular virus occurs.

Animal viruses have also been classified very simply according to the nature of the disease which each produces. However, such a method has proven undesirable since there are viruses which cause the same symptoms of disease, yet are very different from each other; for example, in the type of nucleic acid they possess. Also, there are certain viruses which can produce two or more different types of disease. Attempts have been made to classify animal viruses on the basis of the organ or system they infect, but this has proven undesirable. For example, virus diseases which affect primarily a specific organ or tissue include mumps, which affects the salivary glands; trachoma and the several types of viral conjunctivitis, which affect the eyes; and hepatitis, which affects the liver. The **dermotropic** diseases, such as herpes simplex and warts, localize primarily in the skin and mucous membrane; the **neurotropic** group, such as rabies, poliomyelitis, and the encephalitides, attacks the nervous system. The **pneumotropic** group, including influenza, and the common cold attacks the respiratory system.

At the present time, there is no generally accepted classification scheme of animal or other viruses. Quite often, the various animal viruses are placed into two major divisions based on the type of nucleic acid possessed (Table 12.1).

Most viruses show a high immunizing ability. Second attacks of such viral diseases as measles, smallpox, or mumps are uncommon. The apparent lack of immunity to the common cold is attributed to the dozens of viral strains capable of initiating the disease symptoms. Immunity to one strain does not confer protection against the others.

A phenomenon is now known which involves the interference of viral replication at the cellular level. It appears to be an earlier type of viral interference than afforded by antibody protection. Early reports indicated that the presence of virus in a cell inhibited the replication of other related and even unrelated virus particles. Such protection against viral replication could be conferred on neighboring, yet noninfected cells. The extracellular factor was named **interferon,** and is protein in nature. Interferon appears to be specific for cells of the same species. The virus need not be infectious or active in order to induce interferon since ultraviolet- and heat-treated viruses can induce the production of interferon as well as various nonviral agents. Interferon apparently accounts for recovery from viral infection whereas antibody prevents a subsequent infection. Apparently, there are different types of interferons since they differ in molecular weights and mechanisms of operation in viral replication interference. Active research is directed presently to investigate the possibility of using such substances with antiviral activity in medicine.

Table 12.1 Tentative Classification of Animal Viruses

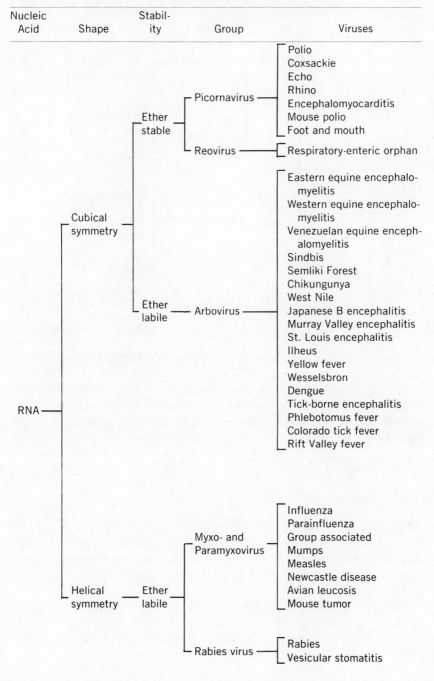

Nucleic Acid	Shape	Stability	Group	Viruses
RNA	Cubical symmetry	Ether stable	Picornavirus	Polio Coxsackie Echo Rhino Encephalomyocarditis Mouse polio Foot and mouth
			Reovirus	Respiratory-enteric orphan
		Ether labile	Arbovirus	Eastern equine encephalomyelitis Western equine encephalomyelitis Venezuelan equine encephalomyelitis Sindbis Semliki Forest Chikungunya West Nile Japanese B encephalitis Murray Valley encephalitis St. Louis encephalitis Ilheus Yellow fever Wesselsbron Dengue Tick-borne encephalitis Phlebotomus fever Colorado tick fever Rift Valley fever
	Helical symmetry	Ether labile	Myxo- and Paramyxovirus	Influenza Parainfluenza Group associated Mumps Measles Newcastle disease Avian leucosis Mouse tumor
			Rabies virus	Rabies Vesicular stomatitis

Table 12.1 *(Continued)*

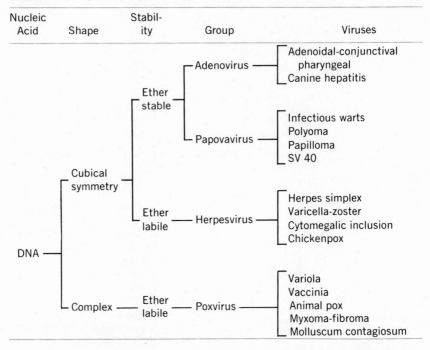

Nucleic Acid	Shape	Stability	Group	Viruses
DNA	Cubical symmetry	Ether stable	Adenovirus	Adenoidal-conjunctival pharyngeal / Canine hepatitis
			Papovavirus	Infectious warts / Polyoma / Papilloma / SV 40
		Ether labile	Herpesvirus	Herpes simplex / Varicella-zoster / Cytomegalic inclusion / Chickenpox
	Complex	Ether labile	Poxvirus	Variola / Vaccinia / Animal pox / Myxoma-fibroma / Molluscum contagiosum

A large number of viral diseases of animals are presently known. Many are primarily human diseases, such as smallpox, chickenpox, measles, mumps, influenza, poliomyelitis, yellow fever, dengue fever, encephalitis, and infectious viral hepatitis. Certain viral diseases are primarily animal but are transmissible to man, such as cowpox and rabies. Certain viral diseases of animals are not transmissible to man, such as distemper, hog cholera, and fowl pox. Certain types of malignant neoplasias (cancer) are definitely known to be of viral origin (Fig. 12.9).

PLANT VIRUSES

Viral infections are widespread throughout the plant kingdom. Only the conifers and lower forms such as the ferns appear not to be susceptible to viral disease. Affected plants do not ordinarily die of the infection, but disease may cause considerable reduction in yield from plants of economic significance.

The tobacco mosaic virus (TMV) in 1935 was the first virus to be crystallized, and its rod-shaped structure has been studied in detail (Fig. 12.10). All viruses of higher plants appear to contain a single strand of RNA.

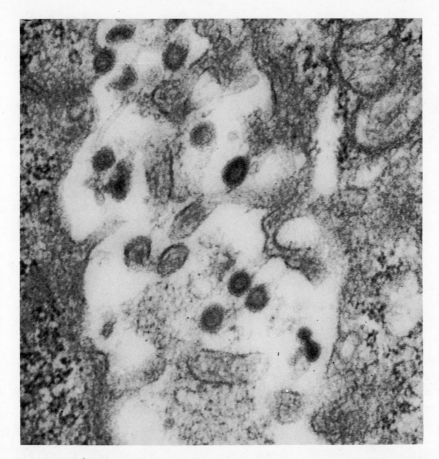

Fig. 12.9. An electron micrograph of material from mouse mammary carcinoma showing viral particles. Photo courtesy N. McDuffie.

The tobacco mosaic virus causes the loss of millions of pounds of tobacco annually. The same virus can cause disease in the tomato plant with a great reduction in yield. The "curly top" disease infecting sugar beets may destroy an entire crop. "Potato blight" and "peach yellow" are similarly destructive.

The manifestations induced in plants by viruses are varied and may simulate mineral deficiencies, fungal diseases, or even genetic mutations. The symptoms may be classified broadly in three major groups: **mosaics, malformations,** and **ring spots** and **necrosis.**

Mosaics. One of the most common plant virus diseases is the mosaic. One type of mosaic disease affects the reddish and bluish pigments of flow-

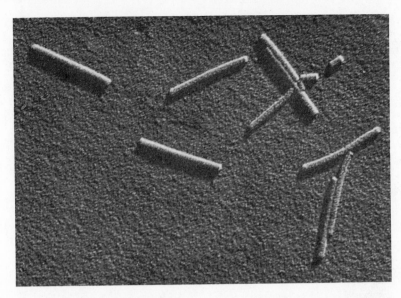

Fig. 12.10. Electron micrograph of shadow-cast tobacco mosaic virus. Photo courtesy Virus Laboratory, University of California, Berkeley.

ers, causing a variegated or mottled effect, the so-called flower breaks, which can be quite attractive. Another type of mosaic virus seems to inhibit the production of chlorophyll in leaf tissues in an irregular pattern over a range from scarcely visible mottling of various shades of green to a brilliant variegation of yellow or even white areas.

Malformations. Most mosaics cause some slight leaf distortion; however, in extreme cases, viruses cause deformity and abnormal growth of part or all of an affected plant. The entire plant may be stunted or dwarfed. Rosette formations or abnormal growths in the form of tumors and galls may appear.

Ring Spots and Necrosis. A number of viruses cause the formation of rings by altering chlorophyll or by killing the tissue. Plants may be infected naturally by various virus-infected insects and other leaf-chewing bugs. Experimentally, plants are infected by rubbing a mixture of an abrasive material and virus on the leaf surface. This causes a rupturing of the cell walls, thus allowing viral particles to penetrate and infect. During replication, viral progeny spread to neighboring cells through protoplasmic bridges between cells. Lesions are formed on the leaf wherever viral destruction has occurred. Such a technique may be used for enumerating the number of viral particles in a given suspension, as each lesion theoretically is initiated by a single virus.

INSECT VIRUSES

Only three orders of insects are known at the present time to be susceptible to viral disease. They are the *Lepidoptera* (moths and butterflies), the *Hymenoptera* (ants, bees, wasps), and the *Diptera* (true flies). Only the larval forms are susceptible to viral disease. The most common insect viruses cause **polyhedral disease,** in which large polyhedral crystals occur in the infected tissues. These crystals contain virus particles and may appear in the host's cell nucleus or in the cytoplasm. The nuclear polyhedral viruses cause the body contents of the larva to liquefy. The skin ruptures easily and liberates the virus-rich fluid. In the cytoplasmic polyhedral diseases, the caterpillar host tends to dry up rather than to liquefy. Thus, the cytoplasmic infections spread more slowly than the nuclear type.

The second largest group are the **granuloses,** in which virus particles are enclosed in a granule or capsule-type structure. Polyhedral structures are not seen. The granulose viruses are restricted to caterpillars of the Order *Lepidoptera.* Some of the hosts of this group are insects of great economic significance. A highly infectious suspension of granulosis virus is readily prepared from only a few infected insects. When the suspension is sprayed over caterpillar-infested fields, the ensuing viral infection produces biological pest control and is just as effective as the use of controversial chemical insecticides.

The third group of insect viruses, the **nonencapsulated,** is similar to the viruses of animals and plants. The virus is disseminated throughout the larval body, producing neither polyhedral bodies nor granules.

The study of insect viruses has developed in laboratories peripheral to the mainstream of microbiology. Control of certain insect pests through induced epidemics of viral disease has shown considerable promise. Since, however, only three orders of insects are known to be affected by viral disease, control of insect populations by viral infection is limited in applicability.

The Rickettsiae

The rickettsiae are microorganisms intermediate in size between the viruses and bacteria (Fig. 12.11). Morphologically, they resemble the bacteria; biologically, because of their obligate parasitism, they are related to the viruses. The generic name *Rickettsia* was given in honor of Howard Taylor Ricketts, who in 1909 first described bacillary bodies in the blood of patients suffering

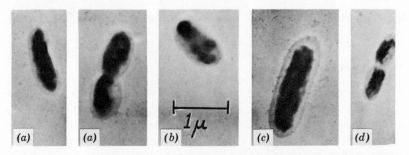

Fig. 12.11. Electron micrographs of the rickettsiae associated with human disease. (a) Epidemic typhus; (b) Endemic typhus; (c) Rocky Mountain spotted fever; (d) American Q fever. Photos by Plotz, Smadel, Anderson and Chambers: J. Exp. Med., LXXVII, 355, 1943, courtesy A.S.M. LS-25.

from Rocky Mountain spotted fever. In 1910, in the course of studies on typhus fever (during which time he contracted the disease and died), he saw similar organisms in patients and in infected body lice. His observations have been confirmed and extended by many later workers.

Rickettsiae principally exist as spherical forms or as short rods and they show considerable variation in size, ranging around the limit of visibility with the light microscope (i.e., to about 0.2 μm). Diplococcoid forms and short chains are seen frequently. With one exception, the Q fever rickettsiae, they are not filterable. Although considered to be gram negative, they stain poorly with ordinary bacterial stains, and special stains are usually employed in their study. Giemsa's stain for blood cells and protozoa have been widely used. Like viruses, they are obligate intracellular parasites, some multiplying in the cytoplasm and others in the nucleus. However, they are found at times in body fluids and excretions. Cultivated only in living cells, they appear to grow most luxuriantly in cells multiplying rapidly. Multiplication occurs until the cell is full and bursts, releasing other rickettsiae to invade noninfected cells. In the laboratory, they are conveniently grown in the yolk sac of embryonated eggs.

Being of the procaryotic cell type, the rickettsiae possess a very simple nucleoid composed of DNA. A cell wall, containing murein, and a membrane, surrounding the granular cytoplasm, is also evident in electron micrographs of rickettsiae. The cell membrane is unusually permeable to macromolecules of the host, and the intracellular parasites need not synthesize them. Unlike viruses, rickettsiae contain certain of the enzymes involved in metabolism, possess both RNA and DNA, divide by binary fission, and are sensitive to a variety of chemotherapeutic agents. Like the vegetative cells of bacteria, they resist heat and bactericidal chemicals to a certain degree, but most exhibit low resistance to drying.

Rickettsiae are widespread throughout nature in a variety of arthropods and mammals. With the exception of the louse, which is killed by the rickett-siae causing typhus fever, they do not harm their natural host. Mammals become infected with rickettsiae through the bites of infected ticks, mites, lice, or fleas. Q fever is an exception to this mode of transmission. Man, in general, acts only as an accidental host. Once transmitted to man, the rick-ettsiae quickly spread by the blood to other parts of the body and tend to localize in endothelial cells of small blood vessels. Their presence causes swelling and an increased cellular division. The diseases they cause in man are characterized by fever, rash, stupor, and terminal shock. A high fatality rate is reported in untreated cases. The groups of rickettsiae pathogenic for man are listed in Table 16.2 and discussed in Chapter 16.

Mycoplasmas

Many years ago, a disease of the pleural membranes and lungs of cattle, sheep, and goats was found to be caused by minute organisms which were subsequently referred to as pleuropneumonia organisms (PPO). Over the years, a number of similar organisms, pleuropneumonialike organisms (PPLO), were isolated from sewage and from the mucous membranes and urogenital tract of man. These unusual cells exhibit extreme variability in their shape (Fig. 12.12) because they possess no cell wall. No special cell structures such as flagella or spores have been detected. Their mode of rep-lication has yet to be definitely established. Such unusual bacteria are now classified in the genus *Mycoplasma*. Some cases of primary atypical pneu-monia in man are of mycoplasma origin.

Mycoplasmas produce very tiny colonies on an agar medium supple-mented with blood serum, and the fried egg-appearing colonies so formed are seldom large enough to see with the naked eye (Fig. 12.13). In diseased animals and in tissue cultures, they grow in the cytoplasm of the infected cell. They have shown resistance to certain chemotherapeutic agents but show no sensitivity to other drugs. Due to the absence of a cell wall, they are much more vulnerable to adverse environmental conditions than vegetative cells of bacteria.

Certain true bacteria give rise to soft, protoplasmic forms exhibiting all the variations in size and shape of the mycoplasmas. The incidence of such pro-toplasmic forms is greater in bacterial cultures grown either in high salt concentrations or in the presence of penicillin which prevents rigid cell wall formation. Termed **L-forms** (after the Lister Institute where intensive studies on these types were conducted), these organisms, however, may frequently revert to the typical bacterial form with a cell wall. Some microbiologists be-lieve that both mycoplasmas and L-forms have a common ancestor.

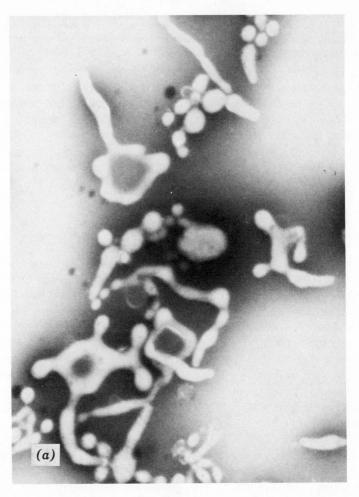

Fig. 12.12. Cells of Mycoplasma pneumoniae. (a) Electron micrograph demonstrating the pleo-morphic characteristics of the cells due to absence of cell walls. (b) Scanning electron micrograph showing cells as they lie in a colony. Photos courtesy E. S. Boatman.

Summary

The viruses are smaller than the bacteria and are obligatively parasitic, but they still retain sufficient individuality to be considered living entities by many microbiologists. They differ greatly in their mode of replication from any other form of life. They usually produce pathologic manifestations in their hosts, and are extremely specific in their association with their host cells. The host-parasite relationship can also exist in a symbiotic relation-ship (i.e., lysogeny). The primary constituent of viruses is nucleic acid,

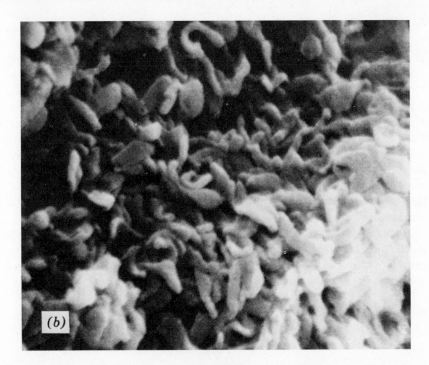

Fig. 12.12(b)

Fig. 12.13. "Fried-egg-like" colonies of a mycoplasma. Photo courtesy J. G. Tully.

Viruses, Rickettsiae, and Mycoplasmas

which serves as the genetic code that directs the cell to replicate more viruses. Since they exhibit a genetic apparatus similar to bacteria and other procaryotic organisms, viruses are used as genetic tools in many genetic research laboratories. Mutation and genetic recombination help to give rise to new virus types. Evolutionists propose that the DNA-containing viruses originated either independently or were derived from degenerative cells. Viruses are of economic importance because they cause a number of human, animal, and plant diseases.

Rickettsiae have a wide variety of enzymes but are still parasitic. There are several diseases of man and animals caused by rickettsiae. Insect vectors are important in their transmission to the susceptible hosts. Elimination of such vectors has significantly reduced the incidence of rickettsial diseases.

Mycoplasmas are unusual procaryotes because they never possess cell walls. Unusual cultural conditions are therefore necessary for these osmotically-sensitive cells. They may be regarded as bacteria that are natural spheroplasts. A few are pathogenic to man; for example, atypical pneumonia.

THIRTEEN

Microorganisms as Disease Producers

Health and disease are complex states which are subject to much variation among individuals. In general, health implies that all parts of the body are functioning in a satisfactory manner. Recognizable departure from normal function is regarded as disease. A condition of malfunction resulting from some constitutional or physiological defect or from a mechanical or chemical injury is restricted to the affected individual. Disease caused by an agent which can be transmitted from an affected individual to one who is not affected is termed **infectious** or **transmissible.** If it is spread by contact, it may be spoken of as **contagious.**

Evidence from skeletal remains indicates that diseases have afflicted man since prehistoric times. However, the concept that disease results from natural causes is relatively modern. Primitive people interpreted the natural as supernatural and regarded disease as an invasion of the body by a devil. The Greek physician Hippocrates (460-370 B.C.) broke with this **demonic** theory and fathered the humoral theory of disease causation which persisted for hundreds of years.

The epidemics of the Middle Ages led some observers to assume that the cause of the disease was a living agent. As the idea of a **contagium vivum,** or a living causal agent for contagious diseases, began to crystallize, the development of the microscope and the discovery of bacteria provided a solid basis for the new theory. Here were organisms not previously known which could be implicated in disease; thus, the **germ theory** was born. The concern of early experimentation with the origin, rather than the significance of microorganisms, delayed development of the germ theory.

The similarity between the fermentation process and the progress of infectious disease had long been noted. When Pasteur established the biological nature of fermentation, the germ theory of disease received strong support. Lister, a Scottish surgeon, reasoned that microorganisms cause the septic conditions which so frequently followed surgery. By operating under a spray of phenol solution and by using sterile dressings for surgical wounds, he dramatically reduced post-operative sepsis. His stature was such that his acceptance of the germ theory greatly influenced other members of the medical profession to accept the theory.

Additional support was provided by the success of Pasteur in controlling pebrine. In studying this disease of silk worms, which threatened the silk industry in France, Pasteur found microscopic bodies in diseased worms that were not present in healthy worms. Reasoning correctly that these bodies caused the disease, he instituted successful control measures directed toward the living agent.

Although many investigators had noted microorganisms in various disease studies, the relationship of the organisms to the disease was not clear. It had been claimed that anthrax could be transmitted by blood, free of the rod-shaped bodies commonly found in material obtained from diseased animals. Robert Koch, a German physician, grew the rod-shaped organisms in pure culture from pieces of spleen of animals suffering from anthrax (Fig. 13.1). He showed that the organisms were able to cause anthrax in susceptible animals and that the disease could be transmitted from animal to animal. Furthermore, the disease was distinct; it could not be initiated by other similar rod-shaped bacteria. Koch's rapid progress in methods of isolation and study of bacteria ushered in the golden age of discovery in bacteriology. Over a short period of years, the bacterial causes of many diseases were established.

Koch's Laws of Disease, or **Koch's Postulates,** established the conditions which must be satisfied for definite proof that a particular organism is the cause, or **etiological agent,** of a given disease.

1. The organism must be found in every case of the disease and should agree in distribution with the lesions observed.

2. The organism must be isolated and maintained in pure culture outside the body of the infected animal.

3. The introduction of the pure culture into a susceptible animal should reproduce identical symptoms of the disease.

4. The experimentally produced disease should agree in characteristics with the naturally-occurring disease and should show the presence of the organism.

Not all disease, however, can be made to fit Koch's Postulates. For example, the organism associated with leprosy, the spirochete *Treponema pal-*

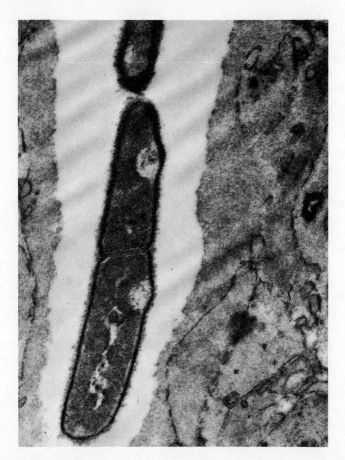

Fig. 13.1. Electron micrograph of a thin slice of the spleen of a mouse infected with anthrax. The pathogen, Bacillus anthracis, is in the intercellular space of the tissue. Photo by Roth, Lewis, and Williams: J. Bacteriol., **80,** 722, 1960.

lidum which causes **syphilis,** rickettsiae, and viruses have not been cultivated on artificial culture media in the laboratory.

The principles embodied in Koch's Laws are not limited to infectious disease but are equally applicable to proof of a cause-and-effect relationship between an organism and a physiological or a biochemical process. For example, they may be adapted to prove the relationship of *Azotobacter* to nitrogen fixation or of yeast to alcoholic fermentation.

Most bacteria are **saprophytic,** obtaining their nutritive needs from non-living sources. The remainder exist in ecological relationships ranging from the simplest sort of loose independent association between two living organisms to an obligate intracellular habitat in which one member of the association obtains its nutritive requirements from the other member. The latter sort of relationship is termed **parasitism,** and the members are termed **parasite** and **host,** respectively. If the members are so well adapted to each other that the association is relatively inconsequential, with neither member showing evidence of harm or special benefit, the association is termed **commensalism.** If, however, the relationship between the two is still in a state of evolution toward a biological balance and the association results in injury or disease to the host, the parasite is said to be **pathogenic.** Almost all pathogenic bacteria are parasitic. One notable exception is *Clostridium botulinum,* an obligate saprophyte that grows in food and produces a poison which is later absorbed from the digestive tract. *Clostridium tetani,* a bacterium widely distributed in the soil, is often introduced into the body through a deep laceration. It will not grow in living, healthy tissue. Proliferating only in dead tissue, it produces a powerful toxin which causes the clinical symptoms known as **tetanus** or **lockjaw.**

It is but a short step from parasitism or commensalism to pathogenicity, in which the organisms cause infection by invasion and growth in the host. Some parasites commonly live as commensals but, under suitable conditions, can invade and cause infection. Such organisms are called **opportunists.** As they are practically always present, they can take advantage of any opportunity to invade.

Although the terms **pathogenicity** and **virulence** are frequently used synonymously, a subtle distinction exists between them. Pathogenicity properly describes the capacity of a group or species of microorganisms to produce disease. Virulence, the relative pathogenicity of a particular isolated culture of an organism, depends primarily on (1) the organism's ability to invade and grow, and (2) its capacity to produce poisons or **toxins.** Almost all gradations and combinations of these two characteristics occur among microorganisms.

Some species of microorganisms which are ordinarily saprophytic may mutate to become virulent. Furthermore, some cultures of ordinarily pathogenic groups are not virulent. Disease production is a complex affair and results when, and only when, the host and the organism meet under conditions which favor disease. If an organism has higher virulence or a host shows lower resistance than usual, an ordinarily inconsequential association between host and organism results in disease. In some cases, a host may be harboring and disseminating the infectious organism, yet show no outward or apparent symptoms of the disease. Such an individual is called a **carrier.**

Lack of balance between host and pathogen is not sufficient to result in disease. An adequate number of pathogenic organisms must be present in an environment permitting contact with the host and subsequent growth. The number of cells of any pathogen necessary to start an infection varies. In some cases, the number is almost always high due to low invasive capacity; in other cases, only a few are necessary due to high invasive power.

Some pathogenic types are fairly exacting as to the site of their entry into the host while others are able to cause an infection from one of several **portals of entry.** However, one avenue will almost certainly be more favorable than others.

RESPIRATORY DISEASES

The portal of entry is most frequently the respiratory tract. Infection follows inhalation of an airborne pathogen through the mouth or nasal passage. Examples are measles, smallpox (Fig. 13.2), and pulmonary tuberculosis.

GASTRO-INTESTINAL DISEASES

Infection follows the ingestion of contaminated food, water, or milk. Examples are food poisoning, dysentery, Q fever, and bovine tuberculosis.

UROGENITAL OR VENEREAL DISEASES

Infection results from sexual contact. Examples are gonorrhea and syphilis.

TRAUMATIC DISEASES

The organisms invade through breaks in the skin or a mucous membrane, or are implanted by bloodsucking insects or arthropods. Examples are staphylococcal infections, tetanus, anthrax, rabies, and yellow fever.

CONGENITAL DISEASES

Infection is transferred through the placenta. It is only necessary that the mother be infected and that the disease be transmissible to the fetus. Examples include syphilis and German measles.

Under special circumstances, some representative of a group may invade through other portals or perhaps through some portal that cannot be clearly recognized.

How Microorganisms Cause Disease

Each case of a disease is an individual matter. Degree or severity is determined by the conditions of resistance of the host and by the virulence of the pathogen at the time of contact. There are, however, two common mechanisms of disease production.

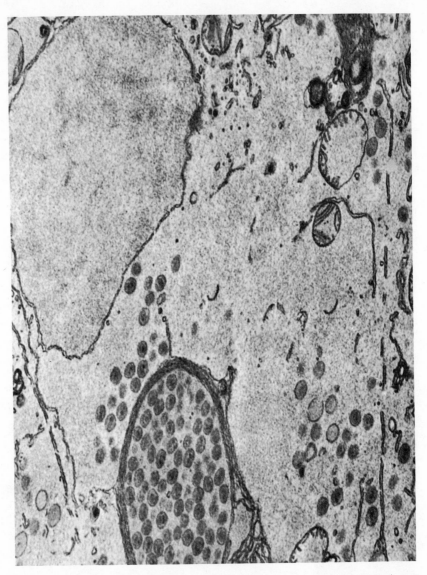

Fig. 13.2. Electron micrograph of vaccinia (smallpox) virus showing localization in tissue cells. Photo courtesy N. McDuffie.

1. Microorganisms may produce disease by **mechanical means.** Heavy growth of the organisms interferes with the functions of certain tissues or vessels.

In some diseases of plants, the water-conducting vessels become plugged by massive bacterial growth or by gums formed as a result of the bacterial injury so that the plant wilts and dies. In whooping cough, bacterial growth along with toxin production interferes with the action of the cilia of the respiratory tract so that they are unable to sweep out accumulations of mucus.

2. Microorganisms may produce disease by **chemical means.** Two types of poisons are most commonly involved: (1) **exotoxins,** or true toxins, which are formed by the organism and diffuse into the surrounding environment while the cells that formed them are still alive and intact; and (2) **endotoxins,** which are within the intact cell, principally the cell wall, and are liberated only after the cell dies and disintegrates. Exotoxins are formed primarily by virulent gram-positive bacteria such as *Corynebacterium diphtheriae* (diphtheria), *Clostridium botulinum* (botulism), and *Clostridium tetani* (tetanus). On the other hand, disease-causing gram-negative bacteria, in general, produce endotoxins. The enteric bacteria, such as *Salmonella*, elucidate disease in this manner. However, there are exceptions to this generalization. In addition to an endotoxin, *Shigella dysenteriae* (bacillary dysentery) produces an extracellular neurotoxin. Exotoxins and endotoxins are distinct in composition, nature of the symptoms caused by each, and other characteristics (Table 13.1).

In addition to their capacity for toxin production, many pathogens produce accessory chemical substances, some of which are enzymes, and aid in lowering the efficiency of the defenses of the host. **Leucocidins,** poisons to the phagocytic white blood cells, are produced by many virulent staphylococci,

Table 13.1 Toxins

	Exotoxins	Endotoxins
Location:	extracellular	intracellular
Symptoms:	muscle spasms, nerve poison	fever, headaches
Potency:	highly toxic	toxic only in large doses
Antitoxin Production:	effective	relatively ineffective
Heat Sensitivity:	destroyed by mild heat (labile)	heat stable
Alterability:	altered by formaldehyde (toxoidable)	unalterable (nontoxoidable)
Chemical Nature:	protein	lipopolysaccharide

streptococci, and pneumococci. **Hemolysins,** which dissolve red blood corpuscles, are made by certain staphylococci and streptococci. The formation of coagulase is associated with the virulence of staphylococci. This enzyme causes the formation of coagulated blood fibrin and, thus, reduces the efficiency of phagocytosis of the fibrin-coated bacteria as well as reducing circulation in the infected area. A number of bacteria, notably the staphylococci and streptococci, produce enzymes called **kinases,** which are able to dissolve fibrin clots formed by the host as a defense mechanism. These are referred to as **staphylokinases** and **streptokinases,** respectively. Another enzyme, **hyaluronidase,** causes the breakdown of hyaluronic acid, a cementing substance which holds body cells together. The spread of the organism through the tissues is, thus, facilitated. The presence of a **capsule** by a bacterium affords it protection against phagocytic activity of the host's tissues and, therefore, is considered to be a factor of virulence for certain bacteria such as the pneumococci.

It is now known that certain disease symptoms are caused by a **state of hypersensitivity.** Although the formation of immune substances is favorable to the host, immune substances are sometimes responsible for some of the pathology in rheumatic fever and tuberculosis. This phenomenon is discussed in detail in Chapter 14.

How the Host Defends against Disease

Three defense mechanisms may protect an individual when exposed to an infectious organism. They include both nonspecific factors, which are operative against any invading organism, and specific factors, which are operative against a particular organism.

The first defense is the **body covering.** In general, pathogenic organisms must enter the body tissues before harmful effects can be produced. Very few are able to pass through the intact skin. Likewise, the mucous membrane is an effective barrier against most pathogens. There are other protective factors in addition to these mechanical barriers: secretions of the skin, such as lysozyme; tears, which tend to wash organisms from the eyes; hairs of the nostrils, which act collectively as a filter; the acidity of the gastric juices; and mucus, which traps invading microbes and in which they are ejected or excreted.

The second line of defense of the body is a **cellular response.** Phagocytes are protective cells which destroy microorganisms and other foreign substances by ingesting and digesting them (Fig. 13.3). There are two general types of phagocytes: **free forms,** which include the white blood cells (**leucocytes**) and wandering **macrophages;** and the **fixed forms,** which are found in the lymph nodes, spleen, liver, bone marrow, and connective tissue.

Microorganisms and Man

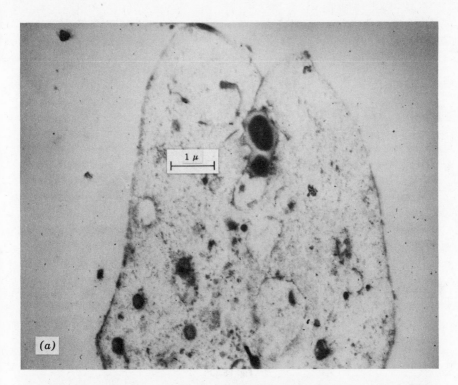

Fig. 13.3. **Electron microscopic study of the phagocytosis of** Staphylococcus. **(a) A micrograph of a thin section of a portion of a neutrophil showing the nearly complete entrapment of two staphylococcal cells. The pseudopods of the neutrophil have completely surrounded the bacterial cells and have started to coalesce. (b) A thin section of a portion of a neutrophil showing a staphylococcal cell completely entrapped.** Photos by J. R. Goodman, courtesy A.S.M. LS 355 and LS 356: J. Bacteriol., **71,** 547, 1956; **72,** 736, 1956.

The phagocytes are sometimes called "scavenger cells" because of their capacity to remove foreign material from the body. The **inflammatory response** is initiated by those organisms that escape phagocytosis. Inflammation is characterized by the dilation of the blood vessels; increased vascular permeability, allowing additional plasma to localize at the site of the invaders; and the passage of blood cells, especially **polymorphonuclear leucocytes,** through the capillary walls into the affected tissues. The latter cells dominate the inflammatory exudate. Unfortunately, conditions are not always favorable for phagocytic activity. Since some pathogens escape these protective cells, they must be dealt with by other means if the invaders are not to cause disease.

The first two defenses are nonspecific, operating against any organism or foreign material. **Natural resistance** is based on a genetic or hereditary basis, rather than as a consequence of previous contact with the infectious organism. One individual may possess a natural resistance to a disease to which

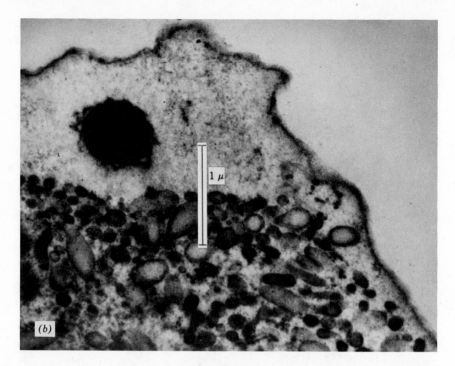

Fig. 13.3(b)

others are susceptible. In any extensive outbreak of an infectious disease, a few persons do not become infected, although they have been freely exposed. Some species do not develop a disease to which other species are susceptible. Certain diseases occur spontaneously only in man, or in one or very few animal species. For example, man does not develop hog cholera or distemper, while rats are not subject to diphtheria of human beings. Or, resistance may be racial, in which one race of a species is more resistant than another race of the same species. The wild mouse, for example, is quite resistant to pneumococcal infection, although the white mouse is highly susceptible. Age and sex are sometimes factors affecting natural resistance.

The third defensive mechanism, **immunity,** consists of the properties of the host which confer specific resistance. Immunity implies relative or absolute resistance. **Acquired immunity** is the development of resistance by susceptible individuals. There are two types: **active immunity** and **passive immunity.** Active immunity develops as a result of the activity of the body cells of the individual who becomes immune; he is active in producing his own immune state. Such immunity is achieved by having a recognizable or subclinical infection. Passive immunity is obtained by the transfer of an immune substance which was developed by another. The body cells of a passively immunized person do not participate in the production of the immune state. The body becomes temporarily immune through the introduction of antibodies which were produced actively by another. Further discussion of these two types of immunity is presented in Chapter 14.

Immunity can be induced **artificially** in several ways. This is done by the use of **vaccines** or other agents of biological origin.

1. Living, **attenuated** organisms of reduced virulence may be inoculated. Various methods of attenuating pathogenic microorganisms have been described, and several practical immunizing procedures are extensively used. The best known example involves the use of vaccinia virus for immunization against smallpox.

2. Dead organisms may be inoculated. Millions of persons have been immunized against typhoid fever by the use of a suspension of killed typhoid cells. The Salk vaccine for prophylaxis, or protection, against polio consists of killed virus.

3. For those diseases in which the clinical manifestations are caused by an exotoxin, a **toxoid** may be inoculated. The toxoid is a toxin so treated that it loses its poisonous qualities but retains its immunizing or prophylactic power. Toxoids are inoculated as a prophylactic against tetanus and diphtheria.

Passive immunity also may be acquired artificially. Artificial immunization, by the passive transfer of immune substances, may be prophylactic or it may be therapeutic. Prophylactically, tetanus antitoxin can be used to protect a previously nonimmunized person (not sensitive to horse serum) against tetanus following injury. Large doses of specific antitoxins are employed therapeutically in the treatment of diphtheria and tetanus.

Various biological agents that are used for immunization purposes are presented in chapter 14.

Summary

A pathogenic organism can cause disease only when it is able to overcome certain defense mechanisms of the animal host so that it can establish itself in tissue favorable for its growth in sufficient numbers. There are several mechanisms by which bacteria are capable of causing disease, including the important ability to produce a biologic poison, or toxin. A wide variety of pathologic conditions in the host can be exhibited due to these disease-producing organisms. The animal body possesses both nonspecific and specific factors that facilitate in defense mechanisms. The tremendous advances that have been made in the control of infectious diseases resulted from (1) the resolution of problems on the nature and spreading of the pathogens and (2) the artificial enhancement of the immune response.

FOURTEEN

Antigens and Antibodies

An understanding of the phenomenon of immunity requires a knowledge of antigen-antibody reactions. The science which deals in these reactions is called **immunology.** An important part of immunology, **serology,** deals with the chemical changes in the blood serum that result when the body is exposed to certain foreign chemicals.

Some individuals suffering from infectious diseases are treated by an injection of the serum (the liquid part of the blood after clotting) from a recovered case of the same disease. Such treatment often affords temporary protection against the specific disease in otherwise susceptible individuals. A protective factor in the serum of the recovered case is thus transferred from one person to another. This factor cannot be demonstrated in the serum of one who has not had the disease, because it is produced in response to the infection. Such a factor is termed **antibody.** The substance which stimulates the body to produce antibodies is designated **antigen** (antibody stimulator).

For many observations of immune reactions there are no readily apparent explanations. Furthermore, workers in this field are not always in full agreement about all of the recorded phenomena. This discussion is restricted to fundamentals about which there is little or no difference in interpretation.

Antigens

A complete antigen must possess two fundamental properties:

1. It must stimulate the body to produce antibodies.
2. It must react with the antibody in some demonstrable way.

Most complete antigens are protein molecules containing aromatic amino acids, and are large in molecular weight and size. However, it has been demonstrated that other macromolecules, such as pure polysaccharides, polynucleotides, and lipids, may serve as complete antigens.

However, certain other materials, incapable of stimulating antibody formation by themselves can, in association with a protein or other carrier, stimulate antibody formation and are the antigenic determinants. These determinants are referred to as **incomplete antigens** or **haptens** and they are able to react with antibodies which were produced by the determinant-protein complex.

However, before an antigen can stimulate the production of antibodies, it must be soluble in the body fluids, must reach certain tissues in an unaltered form, and must be, in general, foreign to the body tissues. Protein taken by mouth loses its specific foreign-protein characteristics when digested in the alimentary tract. It reaches the tissues of the body as amino acids or other altered digested products of protein. Consequently, it no longer meets the requirements for antigenic behavior. Although most foreign native proteins are antigenic, a few are not, simply because they are insoluble in the unaltered state. Silk is used as a surgical suture since this protein is insoluble and, therefore, does not provoke antibody formation. In general, the greater the taxonomic separation between the antigen and the immunized animal, the better the stimulation of antibodies.

MICROBIAL ANTIGENS

Antigens of microbial origin which diffuse outside of the cell are called **extracellular.** Antigens of this nature include the exotoxins and extracellular enzymes. The bacterial cell, which consists of a complex of many different proteins, polysaccharides, and other macromolecules contains a complex of different antigens, which are carried as a unit so long as the cell is intact. Toxic antigens that are firmly attached to the cell wall are called **endotoxins.** The total of this multiplicity of antigens is termed the **antigenic mosaic.**

HUMAN BLOOD GROUPS

Although the antigen must generally be foreign to the species to be immunized, blood cell proteins which are antigenic for some members of a given species may occur in other members of the same species. Such antigens are called **isoantigens,** and the corresponding antibodies, **isoantibodies.**

Blood groups of man are determined by two isoantigens of the red blood corpuscles, which are designated A and B. The naturally occurring homologous isoantibodies for these antigens are designated a and b, respectively.

The distribution of the isoantigens and isoantibodies among the four major blood groups is shown in the following simplified tabulation:

Group	Isoantigen on Cell	Isoantibody in Serum
O	none	ab
A	A	b
B	B	a
AB	AB	none

The discovery of these groups made possible the present safe and extensive use of blood transfusions. Blood from a donor must not have antigens homologous to the antibodies of the recipient. The isoantibody level of **plasma** or of **gamma globulin,** is sufficiently low to allow injection of these substances without regard to blood type. Dilution below reacting level occurs in the blood of the recipient.

Other antigens occur in blood corpuscles; therefore, compatibility tests of donor and recipient bloods are made to avoid transfusion accidents. One of these additional antigens is the heritable Rh factor, which occurs in about 85% of Caucasians. Actually the Rh factor is not a single antigen but a multiple group of proteins. Mating between an Rh positive male and an Rh negative female can result in a fetus with Rh positive blood cells. Repeated pregnancies of this type may result in **erythroblastosis fetalis;** fetal blood accidentally enters the maternal circulation at the time of delivery and Rh positive cells of the developing child stimulate the mother to produce homologous immune antibodies. In a later pregnancy these maternal anti-Rh antibodies will enter the fetal circulation and if the concentration is high enough, will cause destruction of fetal blood corpuscles (Fig. 14.1).

Antibodies

An antigen introduced into the body causes certain cells to develop the capacity to react with the antigen in a manner different from their behavior before exposure to the antigen. This altered reactivity results in the production of specific antibodies. For a long time, antibodies could be described only in terms of their reactions. However, it is now possible to separate the antibodies from other blood serum proteins; the major amount being the gamma globulin fraction. Antibodies are also found in the first milk of lactating animals, in tears, and bound to certain tissue cells. The antibody molecule is composed of four chains which are firmly associated through chemical bonding. Each chain is made up of amino acids. Antibodies with different specificities are therefore possible due to the differences that may occur with respect to the kinds, numbers, and sequences of amino acids making up the antibody.

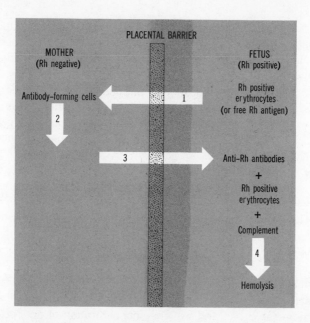

PLACENTAL BARRIER

MOTHER
(Rh negative)

FETUS
(Rh positive)

Antibody-forming cells ←———— 1 ———— Rh positive
erythrocytes
(or free Rh antigen)

2 ↓

3 ————————→ Anti-Rh antibodies
+
Rh positive
erythrocytes
+
Complement

4 ↓

Hemolysis

Fig. 14.1. The circulation of this Rh negative mother and her Rh positive child are separated by the placenta. Should some of the child's Rh positive red corpuscles get into the mother's circulation (1), they would start the formation of antibodies against the Rh protein (2). These anti-Rh antibodies may diffuse back into the child (3), and react with the child's red blood corpuscles causing them to lyse (4). Such a child may be stillborn or may be suffering from lack of red corpuscles. Usually step 1 occurs at the time of delivery; therefore step 4 is rarely reached during a first pregnancy, but rather in a subsequent pregnancy. Of course, step 2 (followed by steps 3 and 4) could result if the Rh-negative woman received a blood transfusion properly typed as to the four major blood groups, but improperly typed as to Rh.

Many proposals have been offered to explain the cellular responses to antigens that lead to antibody formation. Leucocytes now appear to be deeply involved in the events leading up to antibody synthesis. It has been proposed that the pus cells or **neutrophilic granulocytes** which are active in the phagocytosis of foreign materials initiate the cellular response. If the neutrophil is unable to destroy the phagocytized substance, it itself is destroyed, releasing the foreign and cellular materials into the body fluids. **Monocytes** and/or **macrophages** then arrive to engulf the materials. A small amount of messenger RNA is released from macrophages and passed to the lymphocytes, many of which contain the repository of immunological memory. With the presence of the antigen-induced messenger RNA, the mitotic control is deactivated, and the lymphocytes, which have the potential of developing into plasma cells, undergo rapid division. The actual antibody synthesis is done by the **plasma cells** in the deep tissues.

Antigenic response to constituents of one's own tissues should theoretically never occur; that is, the animal body can distinguish chemical differences between "self" and "nonself." It would appear that "recognition of self" is a matter of **tolerance.** One explanation suggests that the clones of cells capable of producing antibodies are killed or mitotically inhibited by the constant presence of a high concentration of the "self" antigens. This is an important concept in desensitization in allergies. Recent evidence suggests that various pathological conditions, such as lupus erythematosis, are manifestations of immunological reactivity between autogenous antigen and autoantibody so induced. Such manifestations are termed **autoimmune diseases.**

Antibody formation cannot ordinarily be detected for several days or weeks after an injection of antigen. Successive doses of antigens raise the antibody content to a higher level than could be achieved by a single injection (Fig. 14.2). Often **adjuvants** are used in association with antigen injections, because these materials enhance and prolong the antibody response through their ability to maintain the antigen in the tissues for a longer period of time.

If no further contact with the antigen occurs, antibodies begin to decrease or disappear. However, a fresh injection of the same antigen will usually result in a more rapid rise and a higher concentration of antibody than occurred with the initial injection. Such "booster" doses of immunizing antigens are given from time to time after the initial immunization to maintain protective antibody levels. Since the body cells seem to react as if in

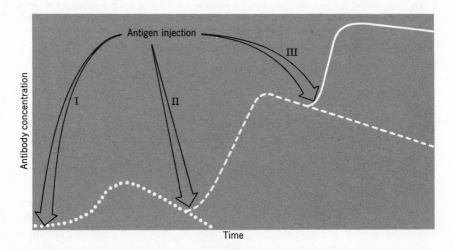

Fig. 14.2. Antibody formation occurs more quickly and to a higher level with succeeding injections of antigen. This also explains why a person, once immune, may not succumb to a later infection, even though his antibody titer has fallen almost to zero.

memory of their prior experience in the production of antibodies, this phenomenon has been termed the **anamnestic reaction.**

Experimental data supports the concept that one pure antigen stimulates the production of a single type of antibody, not of several. This highly specific relationship between antigen and antibody is known as the **unitarian hypothesis.**

Complement

A nine-component system of serum proteins, **complement,** may also take part in antigen-antibody reactions. This reaction complements certain reactions between cell or particulate antigens and their antibodies. Complement occurs in the blood serum of all mammals. Unlike antibody, it does not increase in quantity with immunization. It is heat sensitive and may be inactivated by being kept at 56 C for 30 minutes. Although not specific for individual antigens, it is an essential participant in several types of immune reactions. Once the cellular antigen has been sensitized (reacted with its homologous antibody), complement is able to bring about the lysis of the cellular antigen.

Types of Antigen-Antibody Reactions

The reaction between an antigen and the homologous antibody it stimulates is highly specific. The type and nature of the reaction is determined by the nature of the antigen and the environment in which the two reacting members are placed.

TOXIN-ANTITOXIN REACTION

If the antigen is an exotoxin, the reaction with homologous antibody (**antitoxin**) neutralizes the toxicity. The reaction renders the exotoxin harmless and this is demonstrated by inoculating the reaction mixture into a susceptible animal.

PRECIPITATION REACTION

If a soluble antigen is brought into contact with its homologous antibody, a precipitate will form. The antibody in such a reaction is specifically referred to as a **precipitin.**

If the antigen is particulate (that is, whole cells, cell wall fragments, flagella), the reaction with homologous antibody causes the antigens to adhere to one another in clumps. This is called **agglutination,** and the antibody in this case is referred to as an **agglutinin** (Fig. 14.3).

Agglutination and precipitation (alterations in surface characteristics of protein molecules) are fundamental to many immune reactions. Basically, each results in a decrease in the dispersion of the antigen. The undispersed antigen may be less effective in production of pathologic effects; furthermore, the living body is capable of more efficient disposal of the antigen when it is assembled into the larger units resulting from agglutination or precipitation.

LYTIC REACTIONS

If the antigen consists of cells which are subject to disintegration, a **lytic reaction** may result from combination of cellular antigen, homologous antibody, and complement. The lytic reaction results in a rupture or lysis of the cell; hence the antibody is referred to as a **lysin** (Fig. 14.4). The specific lysin for red blood corpuscles is **hemolysin.** The cell-hemolysin complex is lysed in the presence of complement. Generally, gram-negative bacterial cells also undergo a similar breakdown in the presence of homologous antibody and complement. Clearly the antigen and antibody react prior to lysis, since the cells frequently agglutinate before they are broken down.

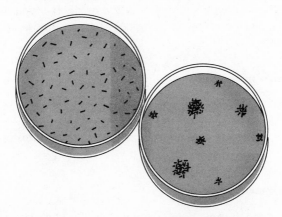

Fig. 14.3. The Widal test indicates the presence of typhoid antibodies in a patient's serum. The diagram to the left represents a dispersed culture of Salmonella *typhi*. To the right, the agglutination of cells after mixing with serum indicates the presence of typhoid antibodies.

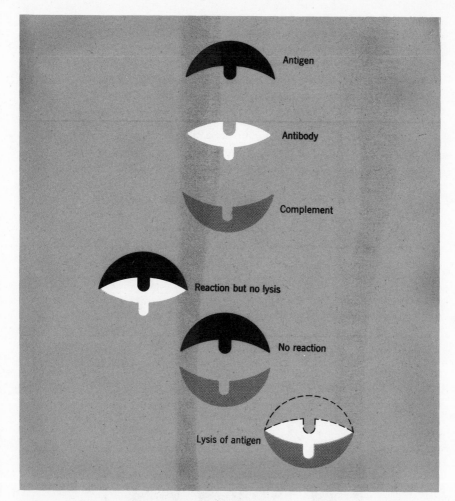

Fig. 14.4. The lytic reaction. Certain bacterial cells and red blood corpuscles lyse when the specific antibody against them is present; however, this reaction will occur only in the presence of complement. Antibody alone or complement alone gives no lysis.

COMPLEMENT-FIXATION REACTION

Antigen-antibody reactions not readily demonstrated by methods previously discussed may sometimes be observed in a special reaction involving complement. Fundamentally the same as the lytic reaction, **complement fixation** is particularly useful when the complement is bound or fixed by the antigen-antibody reaction, but the cellular antigen does not lyse. However, the addition of an indicator-lytic system produces a visible reaction. The most commonly used indicator system consists of a suspension of sheep red blood corpuscles and the anti-sheep corpuscle hemolysin (Fig. 14.5).

This reaction measures the ability of white blood cells to phagocytize bacteria. Although phagocytosis will occur to a certain extent even in the absence of specific antibodies, the efficiency of the activity is enhanced if antibodies for the foreign bacterial antigen are present. A heavy suspension of bacteria is mixed with freshly drawn blood; drops of the mixture are placed on microslides and stained, allowing actual counts of the number of bacteria engulfed by the white blood cells. If the antibody, **opsonin,** is present, a large percentage of the bacterial population will appear inside the white blood cells. Opsonin also markedly increases the destruction of such phagocytized bacteria.

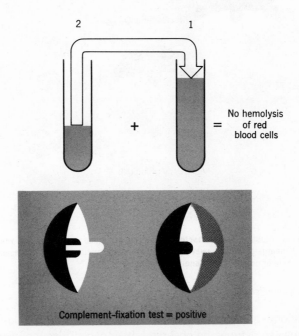

Fig. 14.5a. The Wasserman test is a complement fixation test. To tube 1 is added complement and syphilis antigen and the patient's serum. Since syphilis antibody is present in the patient's serum, antigen and antibody unite, "fixing" the complement as indicated on the right. After a few minutes tube 2, which contains sheep red blood corpuscles and anti-sheep serum, is poured into tube 1. Since all the complement is fixed in the syphilis reaction, there is none left over for reacting with the sheep corpuscle-sheep antibody combination; the sheep red blood corpuscles do not lyse.

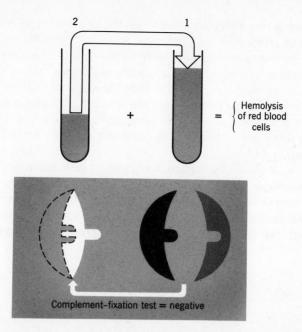

+ = { Hemolysis of red blood cells

Complement-fixation test = negative

Fig. 14.5b. As in 14.5a, but no syphilis antibody is present in this patient's serum. Antibody is not fixed as in 14.5a, so it is left over to react with the sheep corpuscles, and these lyse. The appearance of the tube changes from a turbid blood suspension to a clear red solution.

SPECIAL IMMUNE REACTIONS

A number of the tests used for diagnostic purposes involve special manifestations of immune reactions. For example, antibodies against the spirochete, *Treponema pallidum,* causing syphilis, can be detected by the **immobilization test.** Living cultures of the spirochete remain motile in normal blood serum, but serum from an individual infected with the disease organism prevents active motility. An immune reaction, the **Quellung reaction,** which results in the swelling of the capsule of *Diplococcus pneumoniae* in the presence of specific capsular antiserum, was long used to determine the specific antiserum required in the treatment of pneumococcal pneumonia. This reaction is also used to type the specific strain of pneumococcus so that the prevalent or common types present in a particular population will be known by the epidemiologists. Certain animal viruses readily adsorb to the surface of red blood corpuscles which then agglutinate. These **virus-hemagglutination** reactions permit rapid assay of virus containing fluids.

In many cases, the ability of a particular bacterium to cause disease (virulence) must be determined by its injection into a suitable laboratory animal. However, the **Elek plate** determines the virulence of an isolated diphtheria bacillus without the use of animals. A strip of absorbent paper impregnated with diphtheria antitoxin is placed on the surface of a suitable culture medium. Then a suspected toxin-producing diphtheria culture is streaked perpendicular to the paper strip and then incubated for several days. The antitoxin from the paper diffuses into the medium. If the bacillus produces an exotoxin, it, too, diffuses into the surrounding medium and a precipitate appears as white lines wherever the toxin and antitoxin meet in optimum proportions (Fig. 14.6).

The diagnosis of various bacterial diseases is now accomplished by the **fluorescent antibody** technique. A fluorescent dye or fluorochrome is coupled with antibody gamma globulin against a surface antigen of a particular bacterial cell. Adsorption of the fluorescent antibody complex to the cell surface occurs when they are mixed. By observing the preparation under a fluores-

Fig. 14.6. An Elek plate showing three toxicogenic strains of C. diphtheriae. The virulence of these organisms is evidenced by the white precipitate or "whishers" formed in the medium. The second streaked culture from the top is avirulent due to the absence of a precipitate.

cent microscope, fluorescent outlines of the cells may be seen. Several viral diseases may be diagnosed by employing certain modifications to this technique.

Immune reactions may help to identify an unknown component. If the antigen is known, an unknown antibody can be identified. On the other hand, if the antibody is known, an unknown antigen can be identified. For example, the serum from a patient suffering from a febrile disease may be used in a series of agglutination tests with a series of bacteria which conceivably could be responsible for the symptoms. Agglutination occurring with one and not with the others may determine both diagnosis and therapy.

Antigen-antibody reactions are not limited to diagnosis. Precipitation tests are routinely used to identify unknown proteins from such materials as blood and meat. For example, a blood stain suspected to be human blood can be identified by using the stain as the antigen which is mixed with serum from a rabbit previously injected with human blood. If precipitation occurs with the anti-human antiserum, the stain is identified as being human blood. The same reaction has legal use in the enforcement of food laws. Horse meat, sold as beef, can be similarly identified by the precipitation test. A beginning is being made in matching tissues for tissue transplants.

Types of Immunity

As introduced in Chapter 13, there are two major types of immunity, differentiated on the basis of the source of antibody and the degree of duration of protection. **Active immunity** involves the formation of antibodies by the individual's body (self made) as a result of the presence of living, killed, or attenuated microorganisms which serve as antigens. Active immunity is usually complete and is of long duration, often lasting a lifetime. It may be achieved in one of two ways. **Natural** active immunity occurs when an individual becomes infected with a disease-causing organism and recovers. Many individuals become immune as a result of a subclinical case which never exhibits the observable symptoms of the disease. **Artificial** active immunity, on the other hand, occurs upon the injection of a vaccine (killed or attenuated organisms) or a detoxified product (toxoid) which stimulates antibody formation. Although protection is afforded for many years by this means, "booster" shots are frequently needed to reinforce protection.

Passive immunity involves the receiving of preformed antibodies. That is, antibodies are produced in one individual by active immunity and then transferred to another. Although passive immunity provides immediate protection, in contrast to active immunity, the protection is of short duration — several weeks to six months. **Natural** passive immunity involves the protection that the new borne receives from an immune mother through placental transfer and the mother's first milk. **Artificial** passive immunity involves the receipt of preformed antibodies in a serum obtained from an immune animal or human. This type of immunity is sometimes used when a nonimmune individual has been exposed to an infectious disease or during epidemics. Various biological agents used to obtain artificial active and passive immunity are given in Table 14.1.

Hypersensitivity and Homografts

Unfortunately, not all of the antigen-antibody reactions which occur in the body are beneficial. Under certain conditions, the reactions can cause severe tissue damage in the host. The person experiencing such damage is considered to be **hypersensitive** to the antigen. Although hypersensitivity reactions vary considerably in the nature of their manifestations, all follow a more or less general pattern. The reactions occur after the individual has first been sensitized by exposure to the antigen. A lapse of time is required for antibody formation to take place and for the removal of the antigen from the body. Subsequent to this, exposure to extremely small or to extremely large amounts of the same antigen may induce adverse antigen-antibody reactions to occur in the tissues.

There are two types of hypersensitivities: **immediate** and **delayed.** The types are principally distinguished on the basis of the time between second exposure to the antigen and the appearance of the hypersensitivity response. In the immediate type, the symptoms are quick to appear, usually in minutes. Certain allergies and serum sickness are examples. Anaphylactic shock may result if dosage of the antigen is high. The production of circulating antibodies against the sensitizing antigen is responsible for this phenomenon. Delayed hypersensitivity is characterized by slower evolving symptoms and is demonstrated in the skin sensitivity (tuberculin test) of tuberculosis, poison ivy, and poison oak. The mechanism of this type of hypersensitivity is not yet understood.

Atopy (allergy) is a special type of hypersensitivity in which an **allergen** (antigen) is capable of inducing in an individual a susceptibility to the sensitizing agent. Allergens may be inhaled (pollens, dusts), ingested (certain foods, oral drugs), injected (penicillin), or contracted on the skin (poison ivy, certain fabrics). The induced antibodies to these antigens are called **allergins** and have been chemically identified in a certain fraction in the gamma globulin. Allergic responses include hay fever, asthma, intestinal manifestations, hives, and dermatitis.

In the majority of allergies, the sensitizing antigen stimulates the formation of antibodies which become fixed to certain cells such as blood platelets. Upon subsequent exposure to the antigen, such cells respond as they

Table 14.1   Some Biological Agents for Human Use

VACCINES

Bacterial and Rickettsial Agents

  Killed

    typhoid-paratyphoid
    pertussis (whooping cough)
    cholera
    plague

  Attenuated

    typhus fevers
    Rocky Mountain spotted fever

Viral Agents

  Killed

    rabies
    influenza
    mumps
    measles
    polio (Salk)

  Attenuated

    smallpox
    polio (Sabin)
    measles
    mumps
    German measles

TOXOIDS
(for immunization)

diphtheria
tetanus
staphylococcus

ANTITOXINS
(for prophylaxis and treatment)

diphtheria
tetanus
botulinum
gas gangrene

---

do in inflammation and release histamine and serotonin. Swelling of mucous membranes and skin result. Drugs are available that counteract the histamine effect. Allergists frequently employ desensitization to give relief to patients. Small doses of the specific antigen are injected over a time span so that the antigen eventually ties up all of its corresponding antibody. Few, if any, symptoms are demonstrable. Thus, insufficient antibody is present to react when the desensitized individual is exposed to amounts of the antigen that would normally evoke his allergic symptoms. Desensitization is only temporary, but it often gives relief for several weeks.

Antigen-antibody reactions are also troublesome in **homografts** (organ transplants). The transplant tissues are incompatible due to the foreign proteins of the donor's tissues with respect to the recipient. Antibodies are formed, which then in turn react, resulting in the death of some of the cells and subsequent rejection of the tissue.

Medical research has been active in relieving the effects of the undesirable antigen-antibody reactions. Certain substances, such as cortisones, are injected into the recipient in an effort to suppress the formation of these rejecting antibodies. Cortisone is thought to stabilize membranes, thus preventing loss of antibody from the antibody-producing cells. However, this prevents synthesis of all types of antibodies, including those useful in the protection against disease. Antiserum against human lymphocytes, when injected into the homograft recipient, will also interrupt normal antibody production.

Summary

The ability to form protective antibodies is an evolutionary adaptation which assists the animal body in recovering from infectious diseases. Immunity to further infection by the same agent usually results. Formerly unsolvable medical problems, such as the diagnosis of infectious diseases and the immunization of whole populations against various crippling infectious diseases, are more easily handled as a result of microbiological research into the nature of antigens and antibodies. Unfortunately, not all antigen-antibody reactions are beneficial. Thus, specific activity has created problems such as those encountered in allergies, serum reactions, blood transfusions, and tissue and organ transplants.

FIFTEEN

Infectious Diseases of Man—The Bacterial Diseases

With the possible exception of the algae, infectious disease-producing micro-organisms may be found among all of the major microbial groups. Indeed, even the algae, though not infectious agents, are capable of adding allergens or lethal decomposition substances to drinking water during periods of algal decay and in this sense represent potentially harmful forms. Disease production is a dynamic phenomenon dependent on the high resistance of the host as opposed to the virulent factors manifested by the parasitic pathogen. Each case of an infectious disease is actually very different from every other case. Infectious diseases caused by one kind of microorganism may differ widely in symptomatology, average severity, modes of transmission, and portals of entry.

Vast numbers of different kinds of organisms produce disease, and the disease symptoms caused by microorganisms, even of the same species, may vary widely. This complexity suggests that a rational and systematic consideration of human infectious diseases is impossible. Certainly, a consideration of all of the infectious diseases of man, even in slight detail, is beyond the scope of this text. In Table 15.1, a list of a number of the more prevalent microbial diseases, their causative agents, and the number of reported cases and deaths for the years 1960 and 1967 (for contrast purposes) is given. Table 15.2 lists only deaths due to infectious diseases which the physician does not need to report as cases. These are the classic examples and in actual practice the microbiologist, working closely with the physician, is often called upon to help diagnose and suggest a method for the cure of atypical cases which bear little resemblance to these classic textbook examples.

Table 15.1 Specified Notifiable Diseases

Disease	Causative Agent	Reported Cases 1960	Reported Cases 1967	Deaths 1960	Deaths 1967
BACTERIAL					
Anthrax	*Bacillus anthracis*	23	2	0	0
Botulism	*Clostridium botulinum*	12	5	12	2
Brucellosis (un-dulantfever)	*Brucella melitensis, B. abortus, B. suis*	751	265	6	4
Diphtheria	*Corynebacterium diphtheriae*	918	219	69	32
Gonorrhea	*Neisseria gonorrhoeae*	258,933	404,836	38	11
Leptospirosis	*Leptospira canicola, L. icterohemorrhagiae*	41,666	38,909	844	938
Meningococcal infections	*Neisseria meningitidis, Hemophilus influenzae*	2,259	2,161	644	635
Pertussis (whooping cough)	*Bordetella pertusis*	14,809	9,718	118	37
Rheumatic fever, acute	*Streptococcus pyogenes*	9,022	3,985	734	376
Salmonellosis (ex-cluding typhoid fever)	*Salmonella* spp.	6,929	18,120	82	63
Shigellosis (bacillary dysentery)	*Shigella* spp.	12,487	13,474	153	62
Streptococcal sore throat and scarlet fever	*Streptococcus pyogenes*	315,173	453,351	108	47
Syphilis	*Treponema pallidum*	122,003	102,581	2,945	2,381
Tetanus	*Clostridium tetani*	368	263	231	144
Tuberculosis (newly reported cases)	*Mycobacterium tuberculosis*	55,494	45,647	10,866	6,901
Tularemia	*Pasteurella tularensis*	390	184	3	3
Typhoid fever	*Salmonella typhi*	816	396	21	12
PROTOZOAL					
Amebiasis	*Entamoeba histolytica*	3,424	3,157	88	65
Malaria	*Plasmodium vivax, P. falciparum, P. malariae, P. ovale*	72	2,022	4	4

Table 15.1 (*Continued*)

Disease	Causative Agent	Reported Cases 1960	Reported Cases 1967	Deaths 1960	Deaths 1967
RICKETTSIAL					
Typhus fever (endemic, murine), flea-borne	*Rickettsia mooseri*	68	52	0	0
Rocky Mountain spotted fever, tickborne	*Rickettsia rickettsii*	204	305	11	28
VIRAL					
Hepatitis, infectious	not classified	41,666	38,909	844	938
Measles (rubeola)	Paramyxovirus	441,703	62,705	380	81
Mumps	Paramyxovirus	–	–	42	37
Rabies in man	Rhabdovirus	2	2	2	2
Rubella (German measles)	Paramyxovirus	–	46,888	–	–
Rubella (congenital syndrome	Paramyxovirus	–	10	–	–
Polio, total	Picornavirus	113	41	1	0

Infectious agents may be disseminated from a host in various types of material, often referred to as **primary vehicles.** Saliva, sputum, pus, feces, and urine are examples. Transmission to a susceptible individual takes place directly, as in kissing, or indirectly by **secondary vehicles** or **fomites-**articles such as linens, handkerchiefs, or food and eating utensils which have become contaminated with the infectious material. The infectious agent must enter the host by way of a suitable portal of entry (for example, mouth, skin) before there is any possibility of the pathogen initiating growth within its new environment.

In general, there are four customary modes of transmission of the pathogen to a susceptible host: through secretions of the respiratory tract or other tissues, food and drink, direct contact, and inoculation.

The respiratory diseases tend to appear in epidemic form, large numbers of susceptible persons succumbing in a short period of time. The number of cases usually rises during the fall and winter months since people tend to congregate in crowded, poorly ventilated areas. Although transmission may be direct, as in kissing, discharges of secretions from the respiratory tract of infected individuals are more frequently the primary vehicles for the transmission of these respiratory tract diseases such as diphtheria, whooping

Table 15.2 Non-Notifiable Diseases

Disease	Causative Agent	Deaths 1960	1967
BACTERIAL			
Erysipelas	*Streptococcus pyogenes*	22	14
Pneumonia, as primary cause of death	*Diplococcus pneumoniae, Klebsiella pneumoniae, Mycoplasma pneumoniae*	58,931	55,417
Pneumonia of newborn	same as above	3,544	2,219
Actinomycosis	*Actinomyces isarelii*	28	28
FUNGAL			
Coccidiodomycosis	*Coccidioides immitis*	55	49
Blastomycosis	*Blastomyces dermatitidis*	15	17
Histoplasmosis	*Histoplasma capsulatum*	80	67
Moniliasis	*Candida albicans*	108	94
VIRAL			
Chicken pox	Herpesvirus	126	118
Influenza	Myxovirus	7,872	1,475

cough, pulmonary tuberculosis, influenza, and measles. Forceful exhalation, which accompanies a cough or sneeze, sprays into the air a cloud of saliva together with the microorganisms present in the mouth, nose, and/or respiratory tract. This means of transmission is commonly referred to as **droplet infection.** The microorganisms remain airborne on minute flakes of protein for considerable periods of time and may be readily inhaled into the nasal or buccal portal of entry of other susceptible individuals. Indirect transmission may also occur when secondary vehicles such as eating utensils, linens, foods, and so forth have become contaminated with the infectious material (Fig. 15.1). Also frequently included with this group are the diseases in which material from skin and other tissue lesions serve as the primary vehicle. Tuberculosis, chickenpox, and smallpox are frequently transmitted in this fashion.

A number of diseases are transmitted as a result of intestinal or urinary discharges, serving as primary vehicles, contaminating water or food and entering the mouth of susceptible individuals (Fig. 15.2). Meat, eggs, or milk of infected animals may also serve as primary vehicles. Direct transmission would involve the consumption of improperly cooked food from an infected animal. Examples of this type are brucellosis, bovine tuberculosis, tularemia,

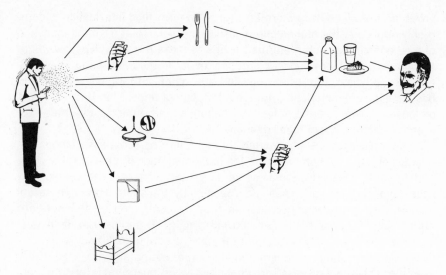

Fig. 15.1. A schematic drawing illustrating modes of transmission of respiratory diseases where infected saliva from a host serves as the primary vehicle.

and gastroenteritis. Indirect transmission is achieved by the infectious material contaminating certain secondary vehicles such as water, food, eating utensils, bed linens, and flies. Common **waterborne** diseases are cholera, polio, typhoid fever, paratyphoid fever, bacillary dysentery (shigellosis), and amoebic dysentery. A number of noninfectious diseases may result from the ingestion of food containing the pathogen or food in which the pathogen has

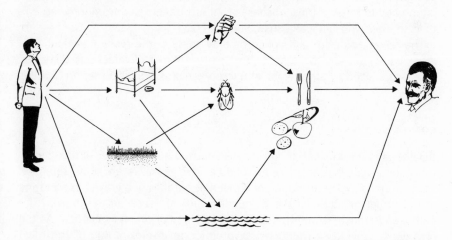

Fig. 15.2. Modes of transmission of intestinal diseases. Intestinal or urinary discharges from the infected host serve as primary vehicles.

grown and produced a preformed toxin. The term **food intoxication** is commonly applied to the noninfectious diseases which result from the ingestion of food containing such preformed toxins as those produced by *Clostridium botulinum* and *Staphylococcus aureus*. Foods, on the other hand, may contain the pathogen and, when ingested, the pathogen multiplies within the host causing characteristic symptoms that do not appear for 18 to 24 hours or longer. Gastroenteritis, infectious hepatitis, and amoebic dysentery are examples of this type of **food infection.**

Contact diseases are transmitted through close personal association. Discharges of infected membranes of the urogenital tract, discharges of lesions, and, in some cases, urine may serve as primary vehicles. The contact diseases syphilis and gonorrhea may be directly transmitted through sexual intercourse. Ophthalmia neonatorum may be contracted by the newborn during birth if the mother is infected with gonorrhea during birth. Transmission in some cases may be indirect through the contamination of secondary vehicles. Others may be transmitted via skin or membrane breakage. Other examples of contact diseases are chancroid, yaws, puerperal fever, impetigo, erysipelas, boils, abcesses, and lymphogranuloma venereum.

Various diseases may be transmitted by inoculation if the primary vehicles are saliva of an infected mammal, blood or tissues of infected animals, or even soil. Transmission may be accomplished through the bite of an infected mammal, blood-sucking arthropods, or through contamination of deep lacerations with soil that contains pathogens. Diseases classified within this group are rabies, tetanus, gas gangrene, bubonic plague, yellow fever, encephalitis, epidemic typhus fever, endemic typhus, Rocky Mountain spotted fever, malaria, African sleeping sickness, and Kala-azar. **Zoonoses** is often used to refer to those diseases that are transmitted from infected animals to man, who is usually an accidental host.

The following discussion concerns only some of the more prevalent bacterial types, chosen to present prominent features of infectious diseases. Diseases caused by other types of microorganisms are discussed in Chapter 16.

CORYNEBACTERIUM

Diphtheria is the most common disease caused by a member of this genus. Diphtheria, usually considered a childhood disease, is an acute infection characterized by the formation of a patch or patches of a grayish membrane in the throat or other localized mucous surfaces such as the tonsils. The pathogen is most frequently transmitted by way of droplet infection. As the organisms grow, they produce a byproduct, an exotoxin, one of the most powerful toxins known. The toxin causes damage and degeneration of the epithelial cells lining the mucous surfaces, which in turn results in the development of an inflammatory exudate consisting of blood cells, fibrin, dead tissue cells, diphtheria bacilli, and other bacteria, all of which are coagulated to form a **pseudomembrane.** In untreated cases, the membrane may clog respiratory passages and cause suffocation unless suitable steps are taken. The toxin may be absorbed locally resulting in a toxemia causing damage to cardiac muscles, liver, kidney, adrenals, and cranial nerves.

The etiological agent, *Corynebacterium diphtheriae,* is an aerobic, nonmotile, nonsporeforming, weakly gram-positive bacillus, which may show considerable variation in its cellular morphological characteristics. The organism is most readily cultivated on a coagulated blood serum medium. Potassium tellurite is frequently added to a plating medium which will inhibit many of the contaminants present in primary cultures from carriers or cases. The colony types of the different diphtheria bacilli on this medium are black or gray and are often distinguishable from other tellurite-reducing microorganisms that may not be inhibited. With alkaline methylene blue stain, the bacilli may, in some cases, stain evenly while others may show an uneven or barring effect. With some strains of the diphtheria bacilli, pink-staining metachromatic granules are seen distributed within the cytoplasm. Laboratory diagnosis not only requires the isolation of the diphtheria bacillus, but its toxicogenicity must also be demonstrated. This is because other bacteria are often morphologically indistinguishable from the diphtheria bacilli but do not produce toxin. Such isolates are often referred to as **diphtheroids.**

Although once considered a dreaded childhood disease, diphtheria now is of secondary importance. An active case may be cured by antitoxin from the blood serum of animals previously immunized with diphtheria toxoid. Recovery is followed by a lasting immunity. Susceptible individuals may be immunized by inoculation with diphtheria toxoid. As a result of widespread immunization of children, this disease has become more and more rare. The immune state may be detected by the **Schick test** in which a small amount of toxin is injected into the skin. If adequate antitoxin is present in the individual, the toxin is neutralized, and no demonstrable reaction occurs. However, if the body is deficient in antitoxin, a small area of inflammation develops at the site of the inoculation of the toxin. Since the artificially-produced immunity of children is not usually reinforced by repeated exposure to the diphtheria bacillus or by booster shots, young adults frequently exhibit a positive Schick test, suggesting that diphtheria is now a greater potential danger to young adults than to children.

Toxicogenic diphtheria bacilli are known to be in the lysogenized state, i.e., the cell is carrying a prophage (See Chapter 12). Evidence indicates that the prophage possesses the genetic code necessary for toxin production. Therefore, nonlysogenic cells are nontoxicogenic and are thus considered to be diphtheroids.

BORDETELLA

Whooping cough, a disease widespread throughout the world, is predominately a disease of young children. Its causative agent, *Bordetella pertussis,* is a small, aerobic, nonmotile, gram-negative, ovoid bacillus and can be cultivated on specialized culture medium such as a glycerol-potato extract agar enriched with blood. Typical colonies appear smooth and possess a pearllike luster. The virulent bacilli are often encapsulated.

A heavy growth of the bacteria forms on the mucous membrane of the larynx, trachea, and bronchi and interferes with the action of the cilia as they attempt to sweep out the newly-formed mucus. Irritation from the accumulation of mucus stimulates spasmodic coughing which ends with an inspiratory crowing sound or whoop. A large number of the bacilli are present in the upper respiratory tract during this catarrhal (cold) stage of the disease and, consequently, the infected individual is highly contagious. Secretions of the mouth and nose serve as primary vehicles to infect other individuals, either directly, through droplets expelled into the air, or indirectly, via contaminated handkerchiefs, towels, and similar objects. The cough is frequently relieved by the expulsion of the mucous plug. Death may result from a throat spasm which cuts off the air supply. Recovery is uneventful unless complications of bronchial pneumonia, bronchiectasis, or otitis media occur. The exact mechanism of disease production is unknown, although evidence suggests an endotoxin as well as a certain amount of tissue sensitivity.

Man is the only natural host of this disease. The incident of pertussis is highest among children under seven years of age, and the mortality is highest among infants under six months. Immunity to the disease with rare reoccurrence is developed upon recovery. Prophylactic immunization of infants, two to six months old, greatly reduces the severity in incidence. The killed bacterial vaccine is usually combined with the diphtheria and tetanus toxoids for immunization. The use of either convalescent serum or gamma globulin helps prevent or modify the disease.

Laboratory diagnosis is rarely necessary in a disease with such pronounced clinical manifestations; however, confirmation of diagnosis may be made by isolation of the bacillus from secretions using the cough-plate method or from a nasopharyngeal swab, the latter yielding a higher proportion of positive cultures. Final identification of the suspected isolate requires serological testing. The fluorescent antibody technique has been used with much success in rapid identification of the bacilli in nasopharyngeal smears.

NEISSERIA

Meningococcal meningitis, caused by *Neisseria meningitidis,* is an acute bacterial infection characterized by sudden onset of fever, intense headache, nausea, signs of inflammation of the meninges (membrane covering the brain and spinal cord), and often a rash. Delirium and coma often appear early, and occasionally an acute overwhelming blood invasion develops so rapidly that the patient dies within five to six hours after the onset of the symptoms.

The infectious bacteria can usually be cultivated from the blood, spinal fluid, and nasopharynx. They grow best on culture media enriched with serum or heated blood (chocolate blood agar) and incubated in an atmosphere of about 10% carbon dioxide. The cells are small, gram-negative, coffee-bean-shaped diplococci. Virulent strains from infected spinal fluid and young cultures are usually encapsulated.

Although epidemics of meningococcal meningitis occur, isolated cases may also develop. Infection usually occurs when a susceptible individual is in close contact with a carrier. Droplet infection is the primary mode of transmission, the upper respiratory tract serving as the portal of entry. During nonepidemic periods, the carrier rate for the organism may be as high as 30%, and during epidemics, it may rise as high as 70 to 80%. Before the advent of modern chemotherapy, the mortality rate was 40 to 50%. The use of sulfadiazine, penicillin, and the tetracyclines has reduced the mortality rate to below 5%.

As suggested by the low ratio of cases to carriers, natural susceptibility is slight. Although the disease may occur at any age, statistics suggest that younger persons are most susceptible. The duration of immunity following recovery is not known, and no generally accepted methods for inducing artificial immunity have been established.

The capsule, which confers antiphagocytic properties, is in part a virulence factor for the organisms. An endotoxin is also known to play a role in the pathogenicity of the meningococcus. The virulence of suspected cultures of the meningococcus is frequently determined by the intraperitoneal inoculation into young white mice. It has been shown that as few as ten meningococcal cells can cause a fatal case in experimental animals.

Gonorrhea, the most common of the venereal diseases, is caused by *Neisseria gonorrhoeae,* a gram-negative diplococcus, similar morphologically to the meningococcus (Fig. 15.3). Since the disease is specific for man, he serves as the only known natural reservoir. The disease usually involves the invasion of the mucous tissues of the genital tract. Although an individual becomes infected through intimate contact, such as sexual intercourse, young girls occasionally have been reported to become infected from fomites which contain pus from active cases.

Laboratory diagnosis employs the demonstration of the organisms within polymorphonuclear leucocytes in gram-stained smears from exudates. Cultures of the organisms are grown on media enriched with serum or heated blood. Optimum growth occurs at 36 C in an atmosphere of 5 to 10% carbon dioxide. Like all *Neisseria,* gonococcus colonies yield a positive oxidase test.

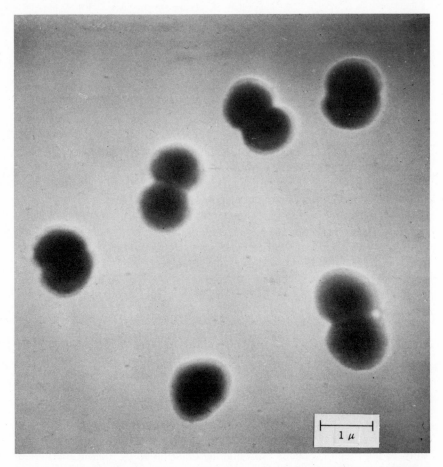

Fig. 15.3.   Electron micrograph of Neisseria gonorrhoeae. **Six diplococci showing various stages of division. This strain was isolated from a male with acute anterior urethritis.** Photo by H. E. Morton, courtesy A.S.M. LS-23.

The development of a dark purple color when a few drops of an oxidase-detecting reagent are placed on suspected colonies serves as a useful screening or presumptive test. Since other types of organisms may give similar results, further laboratory confirmation is required for definite identification of the gonococcus.

Ophthalmia neonatorum, an inflammation of the eye of newborn infants, is contracted during birth by gonorrheal infection from the mother and may result in blindness. Many states require that dilute solutions of silver nitrate or penicillin be applied to the eyes of the newborn child as a prophylactic treatment for this condition.

No apparent immunity develops from an active case of gonorrhea nor can immunity be induced by artificial means at the present time since no vaccine is available for prophylaxis. Treatment has been successful with the use of various chemotherapeutic agents and the incidence of gonorrhea dropped markedly following World War II. However, a high incidence of resistant strains has developed in recent years and a resurgence of the disease has occurred. Gonorrhea is as prevalent today as it was before chemotherapy.

MYCOBACTERIUM

Tuberculosis is a chronic disabling disease which has been one of the greatest killers of man through recorded time. Today, more people die as a result of peptic ulcers or hypertension than tuberculosis. Although any tissue of the body is susceptible to infection from *Mycobacterium tuberculosis*, pulmonary infection is the most common. This respiratory disease is transmitted via respiratory droplets either directly from a case to a susceptible individual or through various secondary vehicles contaminated with the infectious material. It has been estimated that an active case of tuberculosis will infect 15 other susceptible individuals before he is detected and confined for treatment. The waxy nature of the organisms' cell wall confers a relatively high resistance to dryness. Therefore, dust serves as an important secondary vehicle. The prohibition of spitting in public places helps control tuberculosis. Characteristic symptoms include pleurisy, chest pain, coughing, fever, weakness, and weight loss. Characteristic lesions (**tubercles**) form in the lung. The tubercle results from a proliferation and aggregation of certain body cells which respond to stimulation by the presence of the tubercle bacillus. Its products tend to surround and wall off the infection. The bacilli continue to multiply within the nodule, and in some cases, the tubercles break open into the bronchi, thus liberating virulent bacilli. In other cases the walling off may be successful, and the infection may persist for years without further development.

The tubercle bacilli belong to the order *Actinomycetales.* The three principle species are *M. tuberculosis, M. bovis,* and *M. avium,* thus designating their hosts as man, cattle, and birds. Both the bovine and human species are virulent for man, but over 90% of tuberculosis cases are caused by *M. tuberculosis.* The bovine strain usually causes tuberculosis of bones or the lymphatic system and gains entrance to man through ingestion of infected milk or beef. It is now known that mycobacteria other than *M. tuberculosis* can cause pulmonary disease similar to tuberculosis. It is now the practice to study any mycobacteria isolated from human material rather than to discard them as saprophytes. In recent years these "atypical" mycobacteria have appeared with greater frequency. There is no evidence that these organisms are transmitted person-to-person.

The isolation and identification of the tubercle bacilli from such body discharges as sputum, spinal fluid, or urine constitutes diagnosis. In many cases the organism can be demonstrated in stained smears of such discharges. The cells are difficult to stain with the usual dyes, so it is necessary to use an intensifier, such as heat, to drive the primary stain into the cells. But once stained, the high waxy content of the organisms' cell walls renders them resistant to destaining with an acid-alcohol solution. This **acid-fast** characteristic is very useful in the presumptive laboratory diagnosis of tuberculosis. Since the search for acid-fast organisms is an extremely tedious process, a newly-improved fluorochrome staining procedure is used successfully in the rapid screening for the presence of tubercle bacilli. Where this test is positive, confirmation is made using the acid-fast stain. Specimen concentrates are inoculated onto an egg-glycerol-potato medium for cultivation of the organism (Fig. 15.4).

The same laboratory methods are frequently used to serve as a guide to treatment, an evaluation of response to therapy, and an indication of the relative danger of the patient as a source of infection to others.

In response to tuberculosis infection, the body develops a hypersensitivity to bacillary cell substances, making possible the **tuberculin test,** which is of additional diagnostic value. The test involves the intracutaneous injection of small amounts of tuberculin, a protein derivative from cultured tubercle bacilli. An inflammatory allergic reaction at the site of the injection is a positive sign of existing or past infection of adults. A positive test in a child strongly suggests an active case. X-ray examination reveals lung lesions and is used in conjunction with the tuberculin test for diagnosis. However, isolation of the tubercle bacilli is required for a final and conclusive diagnosis.

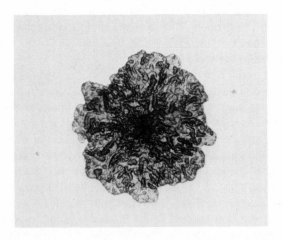

Fig. 15.4. Colony of Mycobacterium tuberculosis. Photo courtesy Communicable Disease Center, Atlanta, Ga.

No completely satisfactory immunizing method has yet been developed, although some success has been reported in Europe with vaccination programs using an attenuated, avirulent strain of *M. tuberculosis*. The organism, B.C.G. (the bacillus of Calmette and Guerin), is used on tuberculin-negative members of special groups in which high rates of infection and serious disease may be expected.

Treatment for tuberculosis involves complete rest, occasional surgery to remove or collapse an infected lung, and chemotherapeutics. Since the bacterial cells localize inside of the impermeable tubercles, they are difficult to reach with chemotherapeutic agents. A decline in the number of cases has generally resulted over the years because of a persistent program of detection and isolation, sanitation measures such as milk pasteurization, and public education about the disease. Early diagnosis and treatment have dramatically reduced the number of deaths resulting from tuberculosis; however, thousands of cases still occur each year, and the death rate among younger people is still high.

STAPHYLOCOCCUS

The staphylococci, normal inhibitants of the skin and mucous membranes, are particularly likely to pass through these barriers if they are broken (Fig. 15.5). Under normal conditions, the invasive characteristics of the staphylococci are balanced by the body defense mechanisms so that invasion is not initiated. However, if the balance is upset by lowered body resistance or a particularly virulent strain, any of a number of disease conditions of varying severity may develop. Most common are **boils, abcesses, carbuncles,** and infections which accompany accidental or surgical wounds. The staphylococci are the most common cause of **osteomyelitis** and are frequently associated with cases of **impetigo.** Staphylococcal **septicemia** (growth in the blood stream) may be particularly dangerous. A fulminating (becoming suddenly severe) form is characterized by profound toxemia and death within a few hours.

Staphylococcal food poisoning is produced by a potent exotoxin which attacks the digestive system, thus called an **enterotoxin.** Virulent staphylococcal cells can be inoculated into food from the skin or mucous membranes of food handlers infected with a toxin-producing strain. Such foods as salads, custards, cream pastries, milk products, and sometimes meat, if allowed to incubate in a warm place for several hours, will become toxic due to the growth of the organisms and subsequent production of the enterotoxin. If the infected food is ingested, symptoms usually begin to develop in about two hours due to the presence of preformed toxin. The enterotoxin causes nausea, cramps, diarrhea, and vomiting. Symptoms persist for several hours, but the illness is rarely fatal.

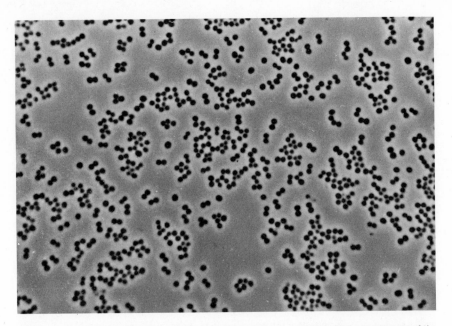

Fig. 15.5. **Phase contrast microscope picture of** Staphylococcus aureus **which causes many of the staphylococci infections in man.** Photo courtesy A. N. Chatterjee: J. Bacteriol., **100,** 846, 1969.

Staphylococcal infection is usually diagnosed by the examination of gram-stained smears of exudates (staphylococci are gram positive) and by isolation of the organism through bacteriological cultivation on the nonselective medium, blood agar, or on one of several available selective media. The ability to coagulate blood plasma differentiates the pathogenic from nonpathogenic forms, **coagulase positive** being the virulent type. *Staphylococcus aureus* is the most predominant disease-producing staphylococcus. Shortly after its introduction, penicillin was an important antibiotic against most staphylococcal strains. However, a number of problems have developed in recent years as a result of the development of highly virulent penicillin-resistant strains for which the antibiotic has no therapeutic value, and epidemics caused by drug-resistant staphylococci are feared, especially in children's hospitals.

STREPTOCOCCUS

A number of streptococcal diseases in humans and other animals are known to be caused by various members of the genus *Streptotoccus*. Streptococci are fairly easily identified in the clinical laboratory (Fig. 15.6). The cells are gram positive. Upon initial isolation, stained cells are ovoid and occur pri-

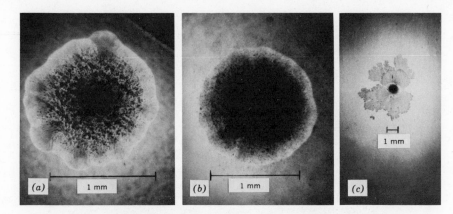

Fig. 15.6. Colony variation of streptococcal colonies. (a) A mucoid or "M" colony in which there are a few notches at the periphery in which "M" to "S" variation is taking place. The colony is moist and glossy; (b) a smooth or "S" colony having a dull and slightly granular surface; (c) an old "M" colony with prominent rough or "R" outgrowths. Photos by A. Pomales-Lebron and P. Morales-Otero, courtesy A.S.M. LS-297: Proc. Soc. Exp. Biol. and Med., **70,** 612, 1949.

marily in a diploid grouping. Stains from exudates demonstrate the more typical chained arrangement. Many of the streptococci possess capsules, which are of a polymer of hyaluronic acid. Many isolates from the mouth, nose, and upper respiratory tract are capable of growing on blood agar demonstrating recognizable changes. The **viridans** group, also called **alpha hemolytic,** causes a greenish halo around colonial growth due to an incomplete breakdown of the red blood corpuscles. Other streptococci produce a soluble enzyme, **beta hemolysin,** which destroys red blood corpuscles (Fig. 15.7); therefore, the colonies on blood agar are surrounded by clear halos. Such organisms are **beta hemolytic.** A third or **gamma** group are nonhemolytic.

The beta-hemolytic streptococci are classified into serologic groups by testing for specific carbohydrates within their cell walls. The grouping (Group A, B, C, etc.) is based upon a precipitation test using extracts from pure cultures and group-specific antisera prepared in rabbits. *Streptococcus pyogenes,* the streptococcus of Group A, is the etiologic agent for approximately 95% of the human streptococcal diseases (Fig. 15.8. See pages 263-265.). A special protein or "M" substance located in the cell wall is associated with the virulence of the Group A streptococci. In addition to the formation of hemolysins, some of the Group A streptococci can produce streptokinase, erythrogenic toxin, and hyaluronidase.

Streptococcal diseases usually can be differentiated clinically according to their portal of entry, tissue localization, and the presence or absence of a skin rash.

If entrance is gained through the respiratory tract, **streptococcal sore**

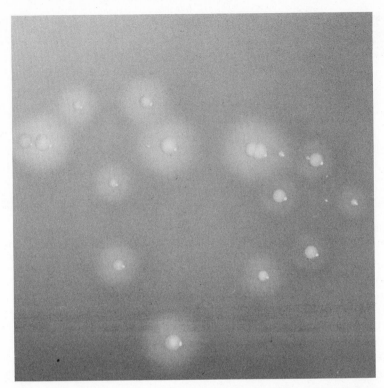

Fig. 15.7. Colonies of Streptococcus pyogenes **growing on blood agar and demonstrating beta** hemolysis.

throat, scarlet fever, bronchial pneumonia, laryngitis, meningitis, sinusitis, and other diseases may develop. Once *Streptococcus pyogenes* has gained entrance to the body, secondary complications, such as **rheumatic fever** or **nephritis** (kidney infection) may develop. We will now deal with the two primary respiratory diseases and with rheumatic fever, which often follows these conditions.

Scarlet fever is streptococcal sore throat in which the infectious agent produces an **erythrogenic** toxin in the patient who has little antitoxic immunity. Symptoms include fever, sore throat, exudative tonsillitis or pharyngitis, and rash caused by the toxin. The rash usually appears on the neck, chest, both of the axilla, elbows, and inner thighs. Diagnosis may be confirmed by the **Schultz-Charlton** test in which a small amount of the antitoxin is injected into the skin of the patient. Fading of the rash at the site of the injection is a positive sign of scarlet fever. Long-lasting immunity against the erythrogenic toxin usually develops within a week of the onset. Both active and passive immunization are used but have little apparent value.

Streptococcal sore throat is sometimes designated scarlet fever without a rash. Absence of the rash suggests that the organism is not a good toxin producer or that the patient has a high degree of antitoxin immunity.

Rheumatic fever is an occasional sequel of upper respiratory tract infection by *Streptococcus pyogenes*. Symptoms usually occur two to three weeks after the streptococcal infection has been recognized. A chronic disease of the heart valves results and tends to recur with each subsequent hemolytic streptococcal infection. A primary cause of death among children aged six to ten years, rheumatic fever is an autoimmune response following the recovery from streptococcal infection. The streptococcal cells and the heart valves have an antigenic component in common; therefore, the antibodies react with the heart valves as well as the streptococci. No practical preventive measures are known except prophylaxis against Group A streptococcal infection.

The same hemolytic streptococci which can cause disease as they enter

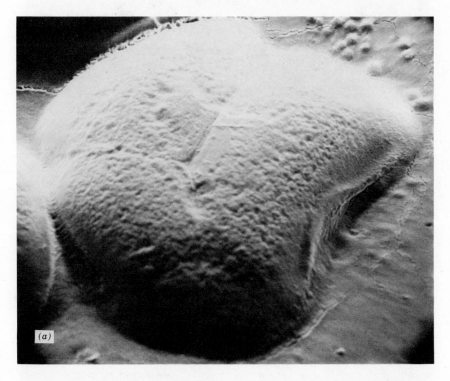

Fig. 15.8. **Electron scanning micrographs of** Streptococcus pyogenes. **(a) Colony magnified 105 ×. (b) and (c) Portion of the same colony magnified 1260× and 6300× respectively, showing the numerous ovoid cells composing the colonial growth.** Photos courtesy I. H. Roth and E. Springer: to be published.

Fig. 15.8(b)

the respiratory route can also cause different disease symptoms if they enter the susceptible host through abrasions on the skin or mucous membranes. A few of the more common conditions which result from direct contact are discussed in the following.

Puerperal fever, or puerperal septicemia, is an acute infection which formerly caused a large proportion of the maternal deaths at childbirth. Most often it is caused by hemolytic streptococci which reach the uterus via contaminated hands or instruments. A number of other bacterial agents including nonhemolytic streptococci, anaerobic streptococci, *Staphylococcus aureus, Escherichia coli,* and *Clostridium perfringens* can cause clinically similar symptoms. With the advent of antibiotic therapy, the mortality rate has fallen more than 80% in the United States in the past 25 years.

Impetigo contagiosum is a nonfatal but often disfiguring disease caused by streptococci in conjunction with staphylococci. Characterized by vesicular lesions occurring on the face or hands, it is transmitted by direct contact with an infected person or secondary vehicle.

Erysipelas is an acute inflammation of the skin which is caused by hemolytic streptococci entering a wound, fissure, or abrasion. The lesion extends peripherally from the site of infection as a red thickening of the skin. The face and legs are most often involved. One attack seems to predispose to subsequent attacks, but reoccurrence can be prevented through chemotherapy.

DIPLOCOCCUS

Lobar pneumonia, the most important and prevalent pneumococcal infection of man, is caused principally by *Diplococcus pneumoniae,* although it is occasionally produced by other pathogens such as *Streptococcus pyogenes, Staphylococcus aureus,* and *Klebsiella pneumoniae.* Acute lobar pneumonia is more common in young adults while a higher morbidity of bronchial pneumonia occurs in infants, young children, and the aged.

The pneumococcus is a gram-positive, small, slightly elongated lancet-shaped coccus which commonly exists in pairs or short chains. It possesses

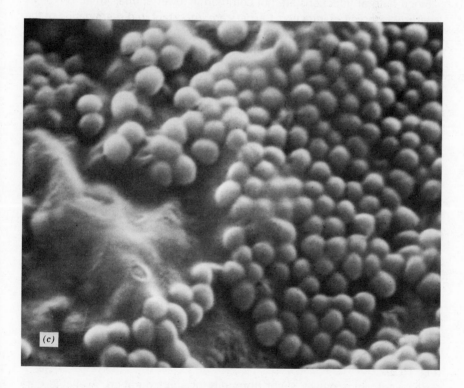

Fig. 15.8(c)

a capsule when isolated from animal exudate or grown on special culture medium such as blood agar. Certain types have been found to be involved in the majority of cases of pneumonia.

Pneumonia is characterized by a sudden onset of chills followed by high fever, chest pain, and a productive fever. The lobes of the lung accumulate fluid and exudate in the air spaces, rendering them nonfunctional. "Rusty" sputum, discoloration due to blood, is frequent in progressive cases. The pneumococci may be distributed to other parts of the body giving rise to localized foci of infection as well as pneumococcemia. Pleurisy and empyema are frequent complications and sequelae of pneumonia. The mortality rate for untreated cases is about 25%. The clinical symptoms associated with pneumonia are believed to be involved with some type of toxin, but the ability of the pneumococcus to invade and initiate the disease is dependent upon the presence of its capsule. Although not a primary toxin, the capsule confers invasive capacity to the cell since it offers impedence to phagocytosis and allows proliferation of the cells in the tissue.

Since natural infections are not common in lower animals, man serves as a natural reservoir of infection. Carriers are important in dissemination of the organism. It has been reported that 40 to 70% of the normal adult population carry one or more types of pneumococci in the upper respiratory tract. Epidemics, however, are rare, and the morbidity rate is low due to the resistance or defensive barriers of man to the organism. The physiological state of the host is quite important in determining the susceptibility to infection. Predisposing factors, such as infection with another disease and the subsequent lowering of natural defenses, exposure to severe cold, alcoholism, and fatigue often lead to possible pneumococcal infection. Therefore, it is possible for a carrier whose resistance is greatly reduced to succumb to the disease. The disease organisms are transmitted chiefly through secretions and discharges from the upper respiratory tract via droplet infections.

Blood agar is the medium of choice for the isolation of the pneumococcus from the various specimens or primary vehicles such as sputum and pleural exudate. The nutritionally-fastidious organism develops small, moist colonies which are surrounded by a zone of alpha hemolysis. Since pneumococci colonies cannot be distinguished from certain alpha-hemolytic streptococci, certain key differential tests must be employed. Pneumococci are positive for inulin fermentation and are bile and optochin soluble; streptococci are negative. The chemical nature of pneumococcus capsules is polysaccharide, of which there are approximately 90 different types. The capsule confers type specificity and permits differentiation by use of the **Quellung** reaction (Fig. 15.9).

Recovery yields only a slight immunity of short duration. Available data

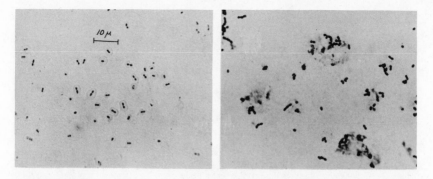

Fig. 15.9. The capsule swelling or Quellung reaction in Ciplococcus pneumoniae. The left picture shows the capsule swelling that occurs when a homologous antiserum is added. Photos by R. Austrian, courtesy A.S.M. LS-324.

indicates that active immunization employing the specific capsular material reduces the cases of infection to the specific pneumococcus type and the carrier rate. However, immunoprophylactic measures are considered impractical due to the multiple types of capsular materials. However, polyvalent vaccines (mixture of the most prevalent and dangerous types) have been used especially during epidemics of respiratory diseases such as influenza.

Measles, whooping cough, and influenza are often complicated by a secondary infection of the lungs in the form of a bronchial pneumonia. The secondary pneumonia, though not always fatal, is particularly dangerous because the patient is usually much weakened by the previous disease. No single microorganism is responsible for cases of all secondary pneumonia, but the hemolytic streptococci, staphylococci, pneumococci, and *Hemophilus influenzae* (Fig. 15.10) are frequently designated as the infectious agents.

SALMONELLA

Members of this genus are potential pathogens which can cause **enteric fevers, septicemia,** and **gastroenteritis.** Endotoxins formed by these bacteria are responsible for the clinical manifestations shown by the infected individuals.

Typhoid fever is an acute infectious disease caused by *Salmonella typhi.* It is characterized by a gradual rise in fever and systemic invasion through the lymphatic system to the blood stream yielding a condition called a **bacteremia** (bacteria in the blood). Diagnosis can be made in the laboratory early in the infection by culturing the blood. Diagnosis in the later stages of

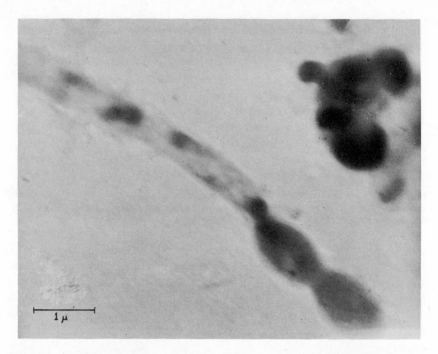

Fig. 15.10.   An electron micrograph of Hemophilus influenzae showing coccobacilli as well as filamentous forms in old cultures. Photo by H. E. Morton and R. G. Picard, courtesy A.S.M. LS-99.

the disease is accomplished by the isolation and identification of the organism obtained from stool specimens. During the second and third week after onset of infection, an increase in the amount or titer of specific agglutinins in the patient's blood serum for S. *typhi* is of diagnostic significance. Typhoid fever is one of a number of diseases which spread as a result of fecal contamination of water and food. The infection by this organism is often initiated by careless food handlers who are carriers of the disease organisms. The mouth serves as the portal of entry in all cases. Prophylactic immunization and improved sanitation have dramatically reduced incidence and death rate.

Paratyphoid fever is very similar to typhoid fever, but generally speaking, less severe. S. *paratyphi*, S. *schottmuelleri*, and S. *hirschfeldii* are commonly associated with the infection in man.

Gastroenteritis is a food infection characterized by an acute onset with diarrhea and is caused by *Salmonella* species, many of which thrive and are localized in the intestinal tracts of animals other than man. The most common species are S. *enteritidis*, S. *oranienburg*, and S. *typhimurium*.

Symptoms usually appear six to 24 hours after the ingestion of food contaminated with a large number of these organisms. They are usually not invasive, but the symptoms result from endotoxins associated with the ingested cells and their growth in the lumen of the intestine.

Shigellosis or **bacillary dysentery** is caused by one or several species of *Shigella*. Symptoms of the disease are intestinal inflammatory diarrhea and watery stools which contain blood, mucus, and pus. Transmission is via contaminated water. The incubation period is about four days. Due to the fact that the organisms do not invade the blood stream, prophylactic use of vaccines is not effective. *Shigella* and *Salmonella* are both gram-negative bacilli which cannot be readily differentiated on the basis of cell morphology or through colonial morphology. *Salmonella* are motile; *Shigella,* nonmotile. A number of differential and/or selective culture media are available for their isolation in the laboratory. They are identified by various fermentation reactions and are confirmed by serological tests.

In the United States this infection is mild in adults and causes relatively few deaths since it can be managed readily in adults with appropriate drugs. This disease is, however, quite severe in infants. In the Orient, a more pathological strain is found that causes large numbers of deaths.

Asiatic cholera remains endemic in India and Southwestern Asia, yet no cases have occurred in the United States for over 50 years. Mild cases resemble gastroenteritis, but the symptoms of severe cases include vomiting and profuse diarrhea with "rice-water" stools. Extreme dehydration and loss of minerals may lead to death. The infectious agent, *Vibrio cholerae,* is a motile, slightly curved, gram-negative rod (Fig. 15.11). Active immunity of short duration can be produced with a vaccine of killed cholera cells. Cholera is a waterborne disease, so sanitation is the best preventive measure. Most strains of *Vibrio cholerae* do not exist in carriers.

A number of potentially harmful bacteria exist as highly resistant forms in the soil because of the spores they form. They can gain entry into the body in contaminated soil during wounding. Such a disease is **tetanus** or **lockjaw**

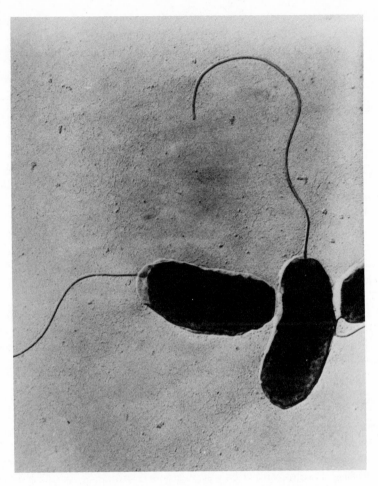

Fig. 15.11. An electron micrograph of Vibrio cholerae. Photo courtesy V. Shaw and C. E. Lankford.

which is an acute disease with a high fatality rate. It is caused by an obligate, anaerobic sporeformer, *Clostridium tetani*. The disease is characterized by painful contractions of muscles of the jaw, neck, and trunk. The anaerobic condition existing in deep wounds are optimum for the development of the organism, which elaborates during growth a highly potent neurotropic exotoxin. Active immunization with tetanus toxoid produces good protection, and a single reinforcing booster on the day of the injury essentially assures that no effects will result from tetanus infection. Passive immunization with tetanus antitoxin is sometimes used for those who have not been immunized with toxoid and show no hypersensitivity to horse serum. Another disease spread by wound inoculation, again primarily through soil contamination, is **gas gangrene,** which is caused primarily by *C. perfringens.* This disease is characterized by a spreading destruction of muscular tissue together with the development of hydrogen gas in the affected tissue. One component of the exotoxin of these pathogens is thought to be lecithinase which breaks down lecithin, an important constituent of cell membranes, thus permitting invasion and destruction of cells. Death can result from general toxemia as exotoxin spreads over the body from the site of infection. Treatment is best effected by surgical removal of the infected tissues and by keeping the wound open in order to limit the growth of this pathogenic obligate anaerobe. Gas gangrene antitoxin helps prevent the development of the symptoms. *C. perfringens* is also the etiological agent for a mild form of food poisoning in man. However, the principal food-poisoning anaerobe is *C. botulinum,* which is discussed in Chapter 19.

TREPONEMA

Syphilis is a venereal disease of man which is caused by the spirochete, *Treponema pallidum.* This disease is transferred from one infected person to another through contiguous personal association, in this case, sexual intercourse. The normal course of the disease involves three stages, the first of which is characterized by the development of a **chancre,** an ulcer with a raised margin. Exudate is usually discharged in which the spirochetes are present in great numbers. During the second stage, the chancre disappears, but the spirochetes have spread throughout the body, and subcutaneous lesions appear on the skin, particularly on the back and in the mouth. Involvement of the heart and central nervous system often characterizes the third stage because of the localization of the spirochetes in these tissues. **Congenital syphilis** can result from placental transfer of the organisms from an infected mother during the early months of pregnancy.

Diagnosis of the primary stage is made by either dark-field examination or a *Treponema* immobilization test using material from lesions. Complement fixation and precipitation tests are often used for serologic diagnosis of later stages. False-positive serologic tests are not unusual and can be excluded by special tests which involve immobilization of the spirochetes by the serum of diseased patients. Although syphilis can be effectively treated with penicillin, erythromycin, or the tetracycline antibiotics, resistant strains are known to occur and, therefore, this disease continues to be a major public health problem.

Yaws is a nonvenereal, direct-contact disease caused by the spirochete *T. pertenue,* which is similar to the syphilis spirochete. Ulcerated lesions of the skin and bone destruction are characteristic of the infection, but the disease is rarely fatal. Diagnosis is usually made by dark-field examination of lesion exudates.

Leptospirosis is a systemic infection which can be caused by a number of spirochetes of the genus *Leptospira,* including *L. icterohemorrhagiae, L. canicola, L. autumnalis,* and *L. pomona.* Transmission to man is usually accomplished by direct contact with water that has been contaminated by urine from infected animals or by direct contact with the infected animals. Cattle, hogs, pigs, rats, skunks, raccoons, and opossums may serve as reservoirs and sources of infection. Symptoms of the disease include fever, headache, and chills. Kidney damage and jaundice may develop in severe cases. Diagnosis is usually confirmed either by isolation of the leptospira from the blood or through serologic means.

ACTINOMYCETES

Actinomycosis is a chronic suppurative process which can be localized in the jaw, thorax, or abdomen. Infection characteristically begins in the back lower jaw or the tonsils and can also involve tissues of the face and neck. Subcutaneous nodules form over the infected area which becomes hard and swollen. Later, the infected area softens and pus drains from multiple openings in the skin. The disease usually progresses slowly with severe injury to the affected parts. Bones as well as soft tissues can be destroyed. Death can result from a secondary invasion by streptococci or other pathogens which invade the injured tissues.

The infectious agent, *Actinomyces bovis,* is a gram-positive bacillary and branching anaerobic form. The organisms tend to grow as a tangled mass of branched filaments or hyphae which are organized into yellowish granular bodies called sulfur granules. The presence of the yellow granules in discharges from draining sinuses is the primary diagnostic characteristic.

Though primarily a disease of cattle, human actinomycosis continues to occur sporadically all over the world. Its frequency is highest in males fifteen to thirty-five years of age, but both sexes and all age groups are susceptible.

Characteristically, a number of infectious diseases of man are transmitted from host to host by means of certain members of the animal phylum *Arthropoda.* The arthropod agent of transmission is designated as a **vector,** and the natural host for the pathogen is designated as the **reservoir.** In the *Arthropoda,* two classes are involved, the *Insecta,* which includes the fleas, lice, flies, mosquitoes; and the *Arachnida,* which contains the ticks and lice. Typically these diseases occur seasonally in correspondence with the life cycle of the vector.

Tularemia is a disease of wild and domesticated animals. Human tularemia is frequently acquired through the direct contact with tissues of infected rabbits. It may occasionally be transmitted to man by the bite of ticks or flies. Symptoms of the disease include chills, fever, and swollen lymph nodes in the area of the original infection. The etiological agent is *Pasteurella tularensis,* a small, gram-negative, nonmotile bacillus. In diagnosing the disease, specific antibodies are detected in the patient's serum or the causative agent is isolated. The mortality rate for untreated cases is about five per cent.

Bubonic plague is an especially severe infectious disease in man which is transmitted from the natural reservoir, the house rat, by the bite of an infected rat flea. Characteristic of the disease is the development of painfully swollen lymph nodes called **buboes.** The acutely inflamed nodes develop in the area which drains the site of the primary infection. Under other conditions the organism may infect the lungs and **pneumonic plague** results which is transmitted like other respiratory diseases. The etiological agent, *P. pestis,* is a small, gram-negative coccobacillus which is nonmotile. Diagnosis is made by demonstration of the bacteria in fluid from buboes. The disease in wild rodents in the western part of the United States is called **sylvatic plague.** The infected wild rodents are the source of sporadic cases in man and serves as a potential reservoir of future epidemics. Although a high mortality rate of 25 to 50% is characteristic in untreated cases, chemotherapeutic agents have sharply reduced fatalities. Pneumonic plague was the black death of the fourteenth century, so-called because of the severe cyanosis which occurs shortly prior to death.

Role of the Microbiologist in Diagnosis of Infectious Disease

An important role is played by the microbiologist in the diagnosis of an infectious disease. The aims in bacteriological diagnosis are to isolate the infectious agent, to identify it in the shortest possible time, and to determine which available chemotherapeutic agent is most promising in the treatment of the patient. Various general procedures must be utilized to achieve these aims.

The collection of the type of specimen appropriate for the isolation of a suspected pathogen must be accomplished in such a manner as to obtain little or no extraneous contamination. Respiratory pathogens are sought in specimens such as sputum, throat and nasal swabs or washings, or pleural exudates. Whenever a patient is suspected of having a bacteremia or septicemia, blood is taken for culture. Sediments of spinal fluid and of urine are examined for pathogens causing meningitis and urinary tract infections, respectively. Stool specimens are examined for bacteria whenever an enteric disease is involved. Pus and exudates would be collected for wound and burn infections. Specimens are promptly transported to the laboratory for microscopic examination and culture.

Direct stained smears (e.g., gram or acid fast) of the pathologic specimen often give valuable information as to the type of pathogen or to the type of predominant microorganisms present. Such information guides the microbiologist, in many cases, to the culture procedure to be used for the isolation of the pathogen.

In order to grow and isolate the pathogen, the specimen is inoculated to the appropriate selective media when it contains bacteria other than the particular pathogen. Selective media, in most cases, are differential in that growth of the pathogen may produce some characteristic reaction or appearance. Selective enrichment media are often inoculated concurrently in case the pathogen is present in small numbers, and it then does not escape detection due to overgrowth by contaminants. Nonselective enrichment media also may be inoculated with a specimen in which the pathogen is likely to be present in pure culture (e.g., blood or spinal fluid) and to detect pathogens other than that which has been anticipated.

A direct antibiotic sensitivity test is frequently performed by streaking a small part of the specimen onto a suitable culture medium and planting appropriate antibiotic test discs on the inoculated surface. Such early information about the antibiotic sensitivity pattern of the pathogen, if present in sufficient numbers to be recognized and differentiated from contaminants, ensures that optimum chemotherapeutic measures would be instituted.

The fluorescent antibody technique is applicable to the more rapid diagnosis of various diseases. Bacteria treated with specific antibody conjugated with an appropriate fluorescent dye will fluoresce in a microscopic dark-field preparation illuminated with ultraviolet light, whereas tissue cells, debris, and unrelated bacteria do not fluoresce under these conditions (Fig. 15.12). This fact has been applied successfully to rapid, direct detection and identification of certain pathogens in pathologic materials, but in many cases, this technique cannot substitute for pure culture isolation and indentification.

Suspected pathogens growing on selective and differential media or on nonselective enriched media are detected by various characteristics such as colonial characteristics, reactions with a dye, pH indicator, or other chemicals. The purity of the isolate must be established. Often, a variety of differential test media are inoculated with the pure culture for its identification. Confirmatory stains are also made.

Colonies or a pure culture of the pathogen are frequently subjected to serologic tests (agglutination, Quellung, etc.) with antisera specific for the pathogen or for certain of its toxins. These serologic tests and the results

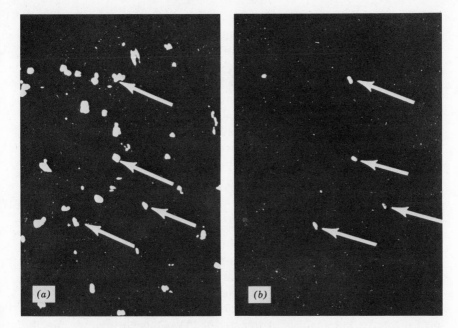

Fig. 15.12. Fluorescent antibody technique. (a) The photograph shows a mixture of Malleomyces pseudomallei and Pseudomonas aeruginosa viewed with visible light and a dark-field condenser. The smear has been treated with fluorescein-labeled antibody of M. pseudomallei. (b) The same field is observed with the fluorescent microscope fitted with a dark-field condenser. Only the cells of M. pseudomallei fluoresce. Other nonspecific cells and materials do not. Photos by B. M. Thomason, M. D. Moody, and M. Goldman, courtesy A.S.M. LS-361: J. Bacteriol., **72,** 362, 1956.

obtained from key differential biochemical and physiological tests provide conclusive identification.

Often, when a pure culture of the pathogen is isolated, another antibiotic sensitivity test is performed to ensure proper chemotherapy for the patient. Due to the wide number of antibiotics and other drugs that are available for this purpose, only a limited number can be evaluated. In general, a select representative group of known value are used for the pathogens usually located in the upper respiratory tract, blood, pus, stool, or urine.

Final identification of certain pathogens may require the determination of virulence for the isolate. Inoculation of suitable experimental animals with the organism or filtrates which may contain a toxin is an **in vivo** method. A specific exotoxin can be identified if it kills an unprotected animal but does not kill an animal protected by a dose of antitoxin to the specific toxin. With certain exotoxin-producing pathogens, such as diphtheria exotoxin and staphylococcal enterotoxin, a specific toxin can be demonstrated and identified **in vitro** by the formation of a precipitate.

Laboratory diagnosis and identification of viral diseases entail different procedures. Often, filtrates of specimens are inoculated into appropriate tissue cultures for isolation of the virus. In certain diseases, the inoculated tissue is examined for the presence of "diagnostic" inclusion bodies, killed cells, or for the presence of specific viral antigens. Many viral diseases are diagnosed by determining whether or not there is an increase in antibody titer during the course of the disease; the complement-fixation test is used in conjunction with paired antisera. Certain viral agents, such as the influenza virus, can be identified through their ability to cause hemagglutination of red blood corpuscles of various fowls. The fluorescent antibody technique has been applied to a limited number of viral diseases.

Summary

Infectious agents may leave the host in various types of materials and be transmitted to susceptible individuals by different routes. The frequency of infection in a given population is influenced by various factors. (1) The incidence of a particular disease is a reflection of the immunity of the overall population to the disease-causing organism. (2) A more highly populated area increases the chances of transfer of the pathogen from an infected individual or carrier to a healthy susceptible host. (3) The mode of transmission for a particular pathogen through a given population is more efficient than others.

Microbiology developed from the stimulus that followed the success of Koch and Pasteur in developing procedures for studying the bacteria involved in infectious diseases. Modifications of their methodology were effective in surmounting the problems encountered with diverse microorganisms. Epidemics have been brought under control, and functional methods for prophylaxis and therapy for most infectious diseases are available. Continued success in these endeavors rests on understanding the microbiology of each disease.

SIXTEEN

Infectious Diseases of Man—Other Microbes

In the previous chapter, infectious diseases caused by bacteria were discussed. This chapter will deal with infectious diseases caused by other microorganisms such as fungi, viruses, rickettsiae, and protozoa. Only the more prevalent or common diseases will be discussed. These pathogenic microbes are transmitted to susceptible hosts via various ways. Consequently, the diseases will be discussed in the light of modes of transmission.

Respiratory Tract Diseases

VIRAL AGENTS

Smallpox, or **variola major,** is a severe infectious disease characterized by a sudden onset of fever, chills, headache, backache, and prostration of three to four days duration. These constitutional symptoms are followed by an eruption of pustular lesions which develop first along the hairline and later on the face, forearms, wrists, and hands. Usually within a week the pustules become enlarged and filled with fluid containing large numbers of the poxviruses. Scales or crust from the pustules disappear after two to three weeks, leaving the characteristic craterlike scars of smallpox. A mortality rate of 30% is characteristic. Although few recognized cases have been reported in the United States in the past twenty years, the disease still remains endemic in South America, Africa, and Asia. **Alastrium,** or **variola minor,** is a milder form of the disease with a mortality rate of less than one per cent.

Smallpox is transmitted directly from person to person by droplet infection or by secondary vehicles contaminated with nasal and buccal secretions. It also may be transmitted from the infectious pustules. Immune persons may harbor the virus and serve for short periods as carriers.

Most cases of smallpox can be diagnosed strictly on clinical grounds. Since mild cases are often difficult to diagnose, laboratory confirmation is sometimes required. The elementary bodies of the virus can be demonstrated in stained smears of skin scraping. The virus can be cultivated from pustular fluid on the chorioallantoic membranes of embryonated chicken eggs, or human cutaneous lesions may be used for antigens in a complement-fixation test with rabbit antiviral serum. The virus may also be cultivated in various tissue cultures.

Recovery from an attack of smallpox usually renders lifelong immunity. Artificial immunization with material from cowpox lesions has been particularly successful. Smallpox vaccination now involves the infection of the susceptible individual on the arm or leg with material obtained from calf lymph infected with the vaccinia virus. Some vaccines are prepared by growing the vaccinia virus in chick embryos. Infants should be vaccinated between the third and twelfth months and again when they enter school. Revaccination is recommended for all persons facing unusual exposure, such as travel to areas where the disease is still uncontrolled.

Chickenpox, or **varicella,** is a mild infection primarily of children, and is characterized by typical skin and mucous membrane eruptions. Cutaneous lesions usually appear on the back and then extend to the face and other parts of the body. Susceptibility is universal among those previously not infected, but in most metropolitan areas, 70% of the inhabitants have had the disease by the time they have reached 15 years of age. Recovery from one attack confers immunity of long duration. In the rare cases where the disease is in adults, the symptoms are more severe. However, adults in contact with chickenpox sometimes contract **shingles;** the skin lesions are restricted to the waist, but the virus appears identical to the chickenpox virus.

Measles, or **rubeola,** an acute disease, is characterized by fever, inflammation of the respiratory passages and the eyes, coughing, skin eruption, and small white patches (Koplik's spots) on the mucous membranes of the mouth. The disease is endemic throughout the world. Death from uncomplicated measles is rare, but the disease often leads to such serious complications as pneumonia, otitis media, and mastoiditis. The disease is transmitted principally by droplet infection.

Since it is a common childhood disease, few persons pass the age of 23 without an attack. Permanent immunity is usually gained after recovery. Offspring of mothers who have had the disease are usually immune for the first several months. The administration of gamma globulin within three days of exposure can prevent measles or may be used to reduce the severity of the disease. In 1962, an attenuated rubeola virus grown in tissue culture was introduced and has served as a suitable vaccine. Complications that might arise may be treated with appropriate chemotherapeutic agents.

German measles, or **rubella,** a mild, febrile viral disease which is characterized by rash of variable character and swelling of the lymph glands below the ear and at the nape of the neck. As the patient usually recovers within a week or less, the disease is sometimes called the "three-day measles."

One attack of this disease results in permanent immunity. Approximately 10% of living infants born to women who develop rubella during the first three or four months of pregnancy have serious congenital defects, such as cataracts, heart disease, and deafness. Passive immunization with immune serum globulin is advocated for adult female contacts who are within the first four months of pregnancy and have no history of rubella. Recently, a suitable vaccine, using an attenuated rubella virus for prophylactic immunization, has been made available. However, the use of this vaccine may present further problems among women of child-bearing age because the virus from recently vaccinated children can be transmitted to such women with possible unfortunate effects on the fetus. Deliberate exposure of healthy female children to the disease before puberty has been recommended by some authorities in the past.

Mumps, or **epidemic parotitis,** is an acute viral infection of sudden onset which is characterized by fever, swelling and tenderness of the parotid glands or, less frequently, the sublingual or submaxillary glands. The involvement of ovaries and testicles is frequent in those few persons past puberty who contract the disease; it may result in sterility. Bloodstream transfer of the virus can result in pancreatitis and meningoencephalitis. Infection of susceptible individuals is by droplet infection. Epidemics among school children and young individuals in similar crowded conditions are not infrequent.

An attack of mumps is usually followed by permanent immunity. An effective vaccine using an attenuated virus was made available to the public in 1966. Convalescent serum, given from seven to 10 days following exposure, usually, protects children from infection. Death from mumps is exceedingly rare. The clinical disease appears less frequently than the other common communicable diseases of childhood.

Influenza is an acute highly infectious disease which is characterized by an abrupt onset with fever, chills, coryza (an acute inflammation of nasal mucous membranes), headache, muscle pains, sore throat, malaise, and prostration. Pneumonia is a frequent secondary complication. Deaths are most frequent among the debilitated, aged, women in late pregnancy, and infants in which the acute illness is neglected.

The disease is ordinarily recognized by its symptoms and confirmed by laboratory examination of the virus from throat washings or by demonstration of significant rise in antibody content of the blood serum of the patient.

Three immunological types of influenza virus (A, B, and C) are recognized. The disease occurs frequently in epidemic proportions and shows a cyclic tendency in which influenza A appears at shorter intervals than influenza B. Influenza C has appeared only in localized outbreaks. The 1957–1958 pandemic of "Asian Flu" was caused by a variant strain (A_2) of influenza A. This pandemic was clinically mild but extremely widespread. Highly fatal pandemics occurred in 1889 and in 1918. The more recent epidemic of "Hong Kong Flu" was caused by a reappearance of the A_2 strain.

The production of active immunity with a multivalent influenza vaccine has the disadvantage of comparatively short duration. Attempts to passively immunize man against all influenzal virus types have not been successful.

The **common cold** is an acute catarrhal infection of the upper respiratory tract characterized by coryza, eye watering, irritated nasopharynx, and malaise which lasts from two to seven days. Although probably never fatal, it temporarily disables more people than any other viral infectious disease and is potentially serious because it lowers body resistance and is frequently complicated by sinusitis, otitis media, laryngitis, tracheitis, and bronchitis. Staphylococci, pneumococci, streptococci, and *Hemophilus influenzae* are frequently involved in the various complications and sequelae of colds. Many people have from one to six colds a year. Recovery from the disease is followed by limited and transient immunity. No satisfactory cold vaccine is available since none could possibly include all of the strains of viruses causing the disease. There are 53 immunological types of the **rhinovirus,** a small RNA virus that causes the preponderance of upper respiratory disease. A DNA virus, the **adenovirus,** also produces infection in the adenoids, tonsils, and eyes as well as a virus pneumonia.

Psittacosis or **ornithosis** is primarily a disease of parrots, parakeets, and related birds but can be transmitted to man. The disease organism is passed directly between their vertebrate hosts. It also infects pigeons and turkeys and is given the more general name ornithosis. It is an acute, generalized infection in man characterized by chills, fever, loss of appetite, sore throat, and spotty pneumonic consolidation. Many mild cases exist, but it can be a very serious disease in older persons. However, antibiotic therapy has greatly reduced the mortality rate. Laboratory diagnosis entails the isolation of the organism from the specimen using either mice or embryonated eggs.

Although once considered to be a virus, the obligate parasite is now classified as *Chlamydia* due to its relatively large size, the presence of both DNA and RNA in the same cell, and sensitivity to the tetracycline antibiotics.

FUNGAL AGENTS

The fungi that characteristically enter the body through the respiratory tract and invade the lungs or other internal parts of the body cause diseases referred to as **deep** or **systemic mycoses.** Infections are common but clinical cases are relatively rare. Since they often produce symptoms similar to other respiratory infections, accurate diagnosis is necessary for suitable treatment. We will now discuss a few of the more common systemic mycoses.

Cryptococosis may involve only the lungs or skin but is often fatal when the infections spread through the blood stream to the brain and meninges. The disease is slowly progressive. The patient may die of respiratory failure after weeks or months of illness.

The causative agent is *Cryptococcus neoformans* (Fig. 16.1). It reproduces by budding but does not form a mycelium. Small masses of the organisms enclosed in gelatinous capsules develop in the infected meninges. They are often large enough to be seen with the naked eye and are similar to the nodules of tuberculosis.

Coccidioidomycosis, a noncontagious disease, is caused by *Coccidioides immitis,* a soil saprophyte (Fig. 16.2). Although subclinical and mild cases ending in full recovery are more frequent than fatal ones, this disease is one of the most dangerous of the fungus infections. Mild cases may involve skin lesions only (desert sore), but the lung infection with symptoms similar to those in tuberculosis is more characteristic of this disease. In fatal cases, the organisms are spread by the blood stream causing a generalized infection of the internal organs. Most fatalities are in the Negro population, the reasons for which are not understood.

A large portion of the cases reported in this country have originated in the San Joaquin Valley of California. The condition is common particularly among vineyard workers. Dried spores of the fungus are inhaled or introduced through skin abrasions. The organisms may be readily identified in the pus of lesions as cyst-like, spherical bodies which have a thick, doubly-contoured capsule.

Histoplasmosis is primarily an infection of the pulmonary system. Lesions in the lungs similar to tubercular lesions are common. Ulcerations of the tongue, pharynx, larynx, and the mucosa of the nose may occur. The spleen, liver, and lymph nodes are generally enlarged.

In the United States, the majority of cases occur in the Mississippi Valley region. Seventy per cent of the people in that area are sensitive to skin tests for the disease, but clinical cases are of only moderate frequency. The organism is found in the soil and is especially frequent in areas fertilized by the droppings of chickens, birds, and bats as well as other animals. Transmission of the disease is by the inhalation of spores. Provisional diagnosis is made by recognizing the organism within the phagocytic cells of blood smears from patients exhibiting a positive skin test. The infectious agent, *Histoplasma capsulatum,* is a small, oval, yeastlike fungus and appears intracellularly. In culture at room temperature, it is a typically moldlike filamentous fungus, and its isolation from tissues serves for definite diagnosis (Fig.

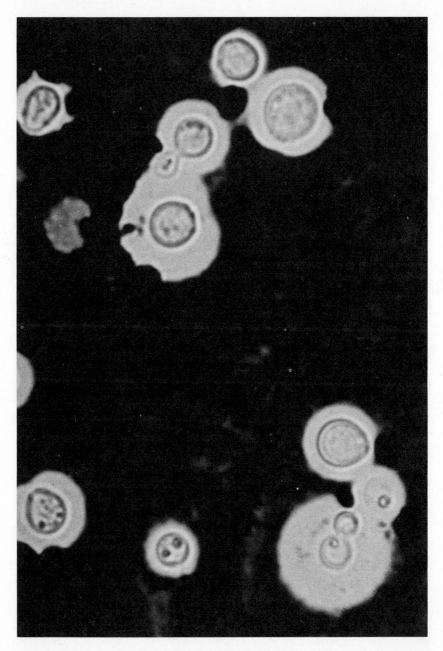

Fig. 16.1. Cryptococcus neoformans in India ink preparation. Wide capsule surrounds mature and budding cells. Highly-refractile cell wall appears as a bright ring. Photo courtesy E. M. Macdonald.

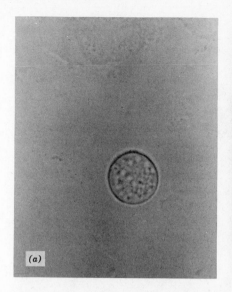 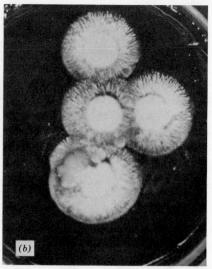

Fig. 16.2. Coccidioides immitis. **(a) Spherule in exudate. Halo is its refractile wall. (b) Colonies on blood agar after one week. Centers are cottony; margins, prickly.** Photo courtesy E. M. Macdonald.

16.3). Old cultures contain large, round to pyriform, thick-walled spores, which are characteristic and are of diagnostic importance.

North American blastomycosis, which is largely confined to the United States and Canada, is caused by *Blastomyces dermatitidis* (Fig. 16.4). The fungus is dimorphic. The yeast phase is exhibited in infected tissue and when grown at 37 C; the mold or mycelial phase when cultivated at room temperature.

The infection appears to start in the lungs but can spread to various visceral organs of the body. Often, crusty ulcerated lesions erupt on the skin. The thick-walled cells are not encapsulated, and their presence in sputum, pus, or tissue preparations aid in diagnosis of this disease.

Foodborne and Waterborne Diseases

VIRAL AGENTS

Poliomyelitis, or **infantile paralysis,** is an acute illness characterized by fever, malaise, headache, and stiffness of the neck and back. In severe cases, damage to motor nerves results in paralysis of voluntary muscles due to the infectious neurotropic virus. Death from respiratory failure may ensue if the respiratory muscles are involved. Many infections of the **nonparalytic** type

Fig. 16.3. Histoplasma capsulatum. (a) **Colonies after two weeks at room temperature. (b) Photomicrograph of a preparation from a colony showing hyphae, microconidia, and macroconidia.** Photo courtesy E. M. Macdonald.

are mild with vague nervous symptoms. Most common are abortive cases in which no nervous system symptoms appear. The disease is transmitted primarily by secondary vehicles contaminated with infectious pharyngeal secretions or feces.

The virus causing poliomyelitis can now be grown in various types of tissue culture which are frequently used to isolate the virus from throat secretions or feces (Fig. 16.5). An increase in complement-fixing or neutralizing antibodies denotes recent infection and is of diagnostic significance. Three different immunological types of the virus have been identified.

Although it is a disease of all ages, children from 1 to 16 years of age are more frequently infected than adults. In many areas of the United States, however, the proportion of cases among older children and young adults is greater than it used to be. In areas where artificial immunization has been widely employed, the occurrence of paralytic cases has progressively decreased as has the total poliomyelitis death rate. In **bulbar polio,** which involves damage to the respiratory center of the brain, the fatality rate is somewhat higher.

Abortive infection gives lasting protection against subsequent infection with the same type of virus. Vaccination with the three strains of formalin-inactivated viruses grown in monkey kidney tissue (Salk vaccine) was made available to the public in 1955 and was highly successful in immunization against this disease. However, the development of an oral vaccine using living, but attenuated viruses, was prepared by Sabin and introduced in 1962. It has been so effective that its use has almost eliminated polio in the United States.

Infectious hepatitis is probably conveyed by ingestion of food or water contaminated by infected feces or urine. Symptoms of the illness include fever and gastrointestinal distress. Liver function is impaired and, in the more severe cases, a yellow discoloration of the skin (jaundice) appears, which accounts for the synonyms **epidemic jaundice** or **catarrhal jaundice** of this disease.

PROTOZOAL AGENTS

The onset of **amoebic dysentery** is usually less abrupt and the symptoms less severe than those of **bacillary dysentery.** Infection is caused by the pathogenic amoeba *Entamoeba histolytica*, which is swallowed in the encysted form with contaminated food or drink. Upon reaching the small intestine, the

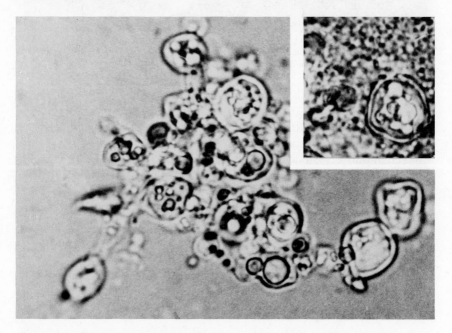

Fig. 16.4. Blastomyces dermatitidis **yeast phase in culture at 37 C is similar to the form seen in tissue (inset).** Photo courtesy E. M. Macdonald.

Fig. 16.5. A purified preparation of polio virus as seen in a shadowed electron micrograph. Photo courtesy Virus Laboratory, University of California, Berkeley.

cyst walls of the organism break down, and four to eight active amoeba are formed from each cyst. They pass into the large intestine and penetrate the intestinal epithelium. Occasionally, the amoeba penetrate small mesenteric blood or lymph vessels and are spread to the liver, lungs, and brain.

Infection with *E. histolytica* is widespread in backward areas. But even in cities with good sanitary facilities, it has been estimated that up to five per cent of the people are carriers. If an infected individual is not properly treated, he may remain a carrier of the amoeba for an indefinite period. Diagnosis is by the identification of the amoeba or its cyst in the feces or of the amoeba in smears from lesions.

Contact Diseases

VIRAL AGENTS

Lymphogranuloma venereum is a venereal disease caused by an organism that is immunologically similar to the etiological agent of psittacosis and is, therefore, not considered a virus as was once believed. The disease is characterized by the development of suppurating lesions, multiple lesions, and enlargement of the genitalia. Laboratory diagnosis entails the demonstration of elementary bodies in stained smears of exudative material. Diagnosis may be aided by the demonstration of complement-fixing antibodies against the lymphogranuloma venereum organism or by a skin test. The disease is worldwide in distribution but occurs most frequently in tropical and subtropical climates. A relatively high frequency of this disease occurs in the indigent population in the southern United States.

FUNGAL AGENTS

Several genera of molds, the **dermatophytes,** are able to infect only the epidermis, hair, and nails resulting in lesions called **dermatomycoses.** Such diseases are often referred to as ringworms or *tineas* (worms) because in premicrobiological times it was thought that these were caused by worms rather than fungi. The skin lesions are circular and have spreading, raised edges. There appears to be a healing over in one portion of the skin while continuing to advance in others. The dermatophytes do not exhibit dimorphism; hyphae, macroconidia, and microconidia are formed and are used in their identification. Transmission is from man to man. Certain animals serve as reservoirs of infection for man. Only two of the more common ringworms will be discussed.

Ringworms of the scalp (*Tinea capitis*) is primarily a disease of children. *Microsporum audouini* is the principal fungus that causes scalp ringworm. Certain fungi, *M. gypseum* and *M. canis* are considered to be primarily of animal origin. Often children become infected from diseased cats and dogs. Certain *Trichophyton* species are also causative agents. Combs, hair brushes, and other articles contaminated with the fungi may serve as secondary vehicles. The infected area of the hair will fluoresce under ultraviolet light, but final diagnosis is accomplished by culturing and identifying the mold.

Athlete's foot (ringworm of the feet) is caused principally by *Epidermophyton floccosum* or certain *Trichophyton* species. The lesion usually begins between the toes, spreading to the soles. If treatment is not initiated, the infection may become chronic. Many infections respond to ordinary skin antiseptics, but those of long standing may require chemotherapy with antifungal antibiotics such as griseofulvin. The highest frequency of this disease is during warm weather when sweating is most pronounced. Infections are common in public swimming pools, showers, and gymnasiums. Disinfectant foot baths should be made available in these public places to reduce the possibility of spreading the disease.

Inoculation Diseases

Rabies is a particularly severe, acute encephalitis caused by a virus. The disease in man is characterized by headache, fever, malaise, and numerous psychological changes including alternating periods of stimulation and depression. Spasms of the throat muscles may occur as the victim attempts to drink. Death may result from paralysis of respiratory muscles. The disease is usually transmitted to man by the bite of a rabid animal or by the entrance of saliva from an infected animal into a scratch or break in the skin. Although dogs are the usual carriers, a large group of wild and domesticated animals, including the fox, coyote, wolf, cat, skunk, raccoon, and bat, also serve as reservoirs and sources of infection. The incubation period for man varies from several days to several months, but it is usually from two to six weeks. Vaccines are available and are prepared from virus grown in rabbit brain or duck embryo (the latter is more widely used) and then inactivated. After possible rabies infection, active immunization is achieved by injecting either vaccine subcutaneously for fourteen successive days. In cases in which the bite is severe, twenty-one successive days of injection are made. Usually, booster shots are recommended shortly thereafter. Diagnosis of the disease in the suspected animal carrier is confirmed by the demonstration of cytoplasmic inclusion bodies, **negri bodies,** in the nerve cells of the brain.

Arthropod-Transmitted Diseases

VIRAL AGENTS

Yellow fever is an acute infectious disease of short duration and is transmitted from one infected human to another by the bite of one species of the *Aedes* genus of mosquitoes. It is primarily a disease of the tropics. Symptoms of the disease include fever, backache, nausea, vomiting, and jaundice. A viremia occurs when the fever is at its highest. The major reservoir of the virus is man. However, monkeys and certain marsupials may also harbor the infective agent.

Laboratory diagnosis is carried out by demonstration of the minute virus in blood by animal inoculation or of a rise in neutralizing antibody titer during the course of the disease.

Although no specific treatment is available for yellow fever, prevention of the disease by mosquito control or by active immunization with attenuated viruses have essentially eliminated it from many areas. It is recommended that yellow fever vaccination be obtained by individuals traveling in endemic areas. The fatality rate is about 30% in infected humans. Outbreaks now rarely occur in the Americas.

For a review of the nature of rickettsiae and the general characteristics of rickettsial diseases, the reader should review that section in Chapter 12. The important rickettsial diseases of man are listed in Table 16.1.

Epidemic typhus fever is a rickettsial disease transmitted by lice which become infected by feeding on the blood of infected humans. Man is the reservoir for the pathogen, and he usually becomes infected by rubbing feces from the lice into the wound caused by the vector's bite. Symptoms of the disease include generalized pains, fever, chills, headache, and body pain. Fatality may be as high as 40% in untreated cases.

The causative agent is *Rickettsia prowazekii* var. *prowazekii*. Confirmation of diagnosis may be made by the **Weil-Felix** serologic reaction in which heterologous antigen, *Proteus* OX-19, gives a positive test for antibodies against the rickettsial pathogen.

Endemic typhus or **murine typhus,** is a milder disease similar in symptoms to louse-borne typhus. It differs from epidemic typhus in that rats are the reservoir, and the rat flea is the vector. The mechanism of transmission is similar in that feces from the infected flea contaminate the bite wound after the flea takes a blood meal. The etiologic agent is *Rickettsia typhi* or *R. mooseri.* The Weil-Felix reaction can be used for preliminary diagnosis, but serologic tests which involve the use of rickettsial suspensions for antigens are necessary to differentiate the two types of typhus. The fatality rate from murine typhus rarely exceeds two per cent.

Rickettsialpox is a relatively mild disease caused by *Rickettsia akari.* The house mouse serves as the reservoir, and the rodent mite transmits the infection to man. Symptoms of the disease include chills and fever, muscular pain, and a generalized rash with no characteristic distribution. Diagnosis is confirmed by a complement-fixation test. Cases appear annually among inhabitants of older apartment houses in New York City, and infrequent cases appear in other crowded cities of the eastern part of the United States. Fatality is less than one per cent in untreated cases.

One of the more severe rickettsial diseases now occurs throughout most of the United States during the spring and summer months. **Rocky Mountain spotted fever** is most prevalent in the western states as a disease of adult males and in the Middle Atlantic states as a disease of children. The infectious pathogen, *Rickettsia rickettsii,* is maintained in reservoirs which include not only mammals, such as rabbits, field mice, and dogs, but also ticks which serve as the vectors of the disease. Once a tick has become infected, it continues to harbor the pathogen for life, and, more important, it may pass the rickettsiae from one generation of the tick to another.

Transmission to man is by the bite of an infected tick or by contamination of the skin with infected tick feces or crushed tissues. Symptoms of the disease include sudden fever, headache, and a generalized rash which spreads

Table 16.1 Groups of Disease-Producing Rickettsiae

Group	Type of Disease	Causal Rickettsia	Natural Hosts Arthropod	Mammal	Mode of Transmission to Man
I Typhus	Epidemic	R. prowazekii	Body and Head louse	Man	Infected louse feces into broken skin
	Endemic	R. mooseri	Rat flea	Small Rodents	Infected flea feces into broken skin
II Spotted Fevers	Rocky Mountain	R. rickettsii	Ticks	Small wild rodents; dogs	Tick bite
	Mediterranean (boutonneuse fever)	R. conorii	Dog tick	None known	Tick bite
	South African Tick bite	R. conorii var. pijperi	Ticks	None known	Tick bite
	Rickettsial pox	R. akari	Mite	House mice	Mite bite
III Scrub Typhus		R. tsutsugamuchi	Mite	Small rodents	Mite bite
IV Q Fever		R. burnetii	Ticks	Cattle, sheep, other animals	Contact, inhalation, tick bite
V French Fever		R. quintana	Body louse	Man	Louse feces into broken skin

to most of the body parts including palms and soles. Diagnosis may be confirmed by the Weil-Felix reaction with *Proteus* OX-19 and OX-2 and by complement-fixation tests. The untreated case fatality is about 20%.

Although **malaria** causes thousands of deaths annually, it has, for all practical purposes, been eliminated as a scourge to man in the United States and Europe through control of the insect vector, the *Anopheles* mosquito. The disease is characterized by the development of cycles of chills followed by fever and sweating which occur every 48 hours for the pathogen *Plasmodium vivax,* and every 72 hours for the protozoal parasite *P. malariae.* A third form of the disease caused by *P. falciparum* is characterized by more frequent paroxysms occurring over a period of less than two days. The time interval between cycles corresponds to the developmental period of the malarial parasite in the erythrocytes of man, since there is a synchronous release of parasites every 24 or 48 hours with the breakdown of the blood corpuscles.

Although higher apes have been suggested as possible reservoirs for *P. malariae,* man is the major sustaining host for all forms. The female *Anopheles* mosquito becomes the vector as she ingests a blood meal from an infected human. The parasites undergo a complex series of changes within the arthropod vector and finally settle in her salivary glands from which they are effectively transmitted to another human host with the next blood meal.

Laboratory diagnosis depends on demonstration of the protozoal pathogens by microscopic examination of blood smears. Mortality rates in untreated cases may be as high as 10%. A number of specific antimalarial drugs, including quinine, chloroquin diphosphate, amodiaquin dihydrochloride, and primaquin diphosphate have provided effective treatment. Fatality rates among treated cases rarely exceed 0.5%.

African sleeping sickness is caused by the flagellated protozoan pathogens, *Trypanosoma gambiense* and *T. rhodensiense.* A number of wild African game animals, unaffected by the disease, serve as natural reservoirs for the flagellates. The tsetse fly is the vector for transmission of the infective agent to man. Metabolic products of the parasites have a paralyzing effect on the central nervous system resulting in symptoms varying from tremors and delusions to deep "sleep." Laboratory diagnosis involves the demonstration of the parasites in blood or spinal fluid smears. An organic arsenic-containing drug, tryparsamide, may be used as a specific for treatment.

Kala-azar, a chronic disease caused by the flagellated protozoan, *Leishmania donovani,* is characterized by irregular fever, dysentery, a dusky skin hue, and enlargement of the liver and spleen. Dogs are frequently infected and may serve as natural reservoirs. Sandflies of the genus *Phlebotomus* serve as vectors of transmission. Diagnosis is dependent on demonstration of the parasites in biopsies of skin, spleen, or liver, by blood culture, and by a group of serologic reactions including a highly specific complement-fixation test. Untreated cases usually result in death from secondary infections. A number of antimony compounds have been used as specifics for treatment.

Summary

Innovations in the procedures used for bacterial diseases have been successfully applied to diseases caused by other microorganisms. Virus diseases have been better understood due to the invention of the electron microscope and the development of tissue culture techniques. These diseases do not respond to current antibiotic therapy, but their control by immunological procedures has been very gratifying. Immunological procedures have been very useful in diagnosing fungal diseases; however, vaccines are almost useless against these pathogens. The control of insects is the most effective procedure in reducing incidence of some of the protozoal and rickettsial diseases.

SEVENTEEN

Microbiology of Water

Potential sources of water for human consumption are **atmospheric water,** such as rain or snow; **surface water,** in lakes, rivers, or streams; and **ground water,** from wells or springs.

ATMOSPHERIC WATER

The bacteria, viruses, molds, yeasts, algae, and protozoa present in rain or snow reflect the microbial population in the air at the time the meteoric water formed. They are, therefore, chiefly of soil origin. Near the oceans, marine organisms are dispersed into the air by the action of wind and waves. Numerous bacteria and viruses from man and animals are released into the air in saliva or mucous droplets. The microbial population is the greatest at the beginning of a rain or snowfall since the precipitation washes most of the dust to which the bacteria are attached from the atmosphere. The number is determined to some extent by the nature of the area. More dust, and consequently more bacteria, are present in the air over a city than over open country. Thus, rain or snow from an urban area ordinarily renders a higher count than that from a suburban area.

SURFACE WATER

The rivulets of run-off water from cultivated land contain large numbers of bacteria, as well as extracted organic and mineral matter. Runoff from rocky areas naturally contains fewer bacteria. Seasonal variations in rivers occurring as a result of rainfall or melting snow introduce fresh washings from the ground. A stream heavily loaded with sewage often shows the largest bacterial count when the stream is lowest and the sewage, therefore, least diluted. The flora of such a stream contains a mixture of soil and sewage forms, while a stream without sewage pollution shows a predominance of soil and water forms.

In a slow-moving stream receiving little polluted matter, a considerable reduction in bacterial population often takes place since the suspended matter to which the bacteria are attached has settled out. A similar situation prevails in lakes and impounded reservoirs. As a result of ultraviolet light, antagonisms with other microorganisms, and starvation, a further natural reduction of the microbial population takes place.

UNDERGROUND OR GROUND WATER

In some areas, ground water contains few bacteria because of the filtering action of the soil but this is not always the case. Where the subsurface is predominantly limestone, for example, the surface water may reach the underground level through caves and channels with little or no change in bacterial population, and pollution from surface sewage may move directly to the subsurface water. In general, water from shallow wells less than 100 feet deep contains more bacteria than water from deep wells.

Water from springs is comparable to water from deep wells in the number of existing bacteria. However, surface contamination is difficult to exclude, and spring water may show a flora more characteristic of shallow than of deep wells. The bacteria in special types of water, such as oilfield brines, are of considerable interest. When these waters contain sulfates and the bacterium *Desulfovibrio,* which reduces sulfates to sulfide, a special microbiological problem results. The water may become corrosive or may precipitate insoluble sulfides, a problem since the industry involved needs a water free of precipitate. Also, this water is unsuited for human consumption. The presence of excessive numbers of slime-producing organisms in water pumped into the ground to force out and recover natural gas results in rapid clogging of the porous rock formations.

Types of Bacteria in Water

The normal flora consists of representatives of the *Chlamydobacteriales,* including the iron and sulfur bacteria, certain stalked forms, the *Caulobacteria* which grow attached to some objects, bacteria involved in the cycle of nitrogen (both nitrogen fixing and the ammonia- and nitrite-oxidizing bacteria), and a variety of chromogenic rods and cocci. Soil sporeformers (*Bacillus* and *Clostridium* species) and other soil types, such as *Pseudomonas* species, are commonly found, but they probably are not true water bacteria even though they do at times find suitable growth conditions in water. They are probably adventitious invaders, temporarily present, thriving only where the water is high in impurities. The most widely studied adventitious water bacteria are those of sewage origin: the coliforms, the sewage streptococci, certain types of obligate anaerobes, and other intestinal organisms.

Certainly no one set of cultural conditions will detect all the types of bacteria present in a water sample. The type of medium, the temperature of incubation, the oxygen tension, and other cultural environmental conditions must be adjusted to the requirements of the particular organisms or group to be detected.

Sanitary Examination of Water

The normal bacterial flora of water are not of sanitary significance. However, most sources of water for municipal use are subject to undesirable contamination and, consequently, may contain pathogenic organisms of human intestinal origin. Therefore, water for domestic use must be subjected to regular bacteriological examination. Epidemiological evidence has established beyond doubt the relationship between waterborne disease and the presence of organisms of intestinal origin (See Table 17.1), and thus the value of sanitary bacteriological examinations for the evaluation of the safety of the supply is evident. Probably no single bacteriological test is more frequently performed than the examination of water.

Pathogenic bacteria in water supplies are present in such small numbers, compared with other organisms having the same source in nature and the same general requirements for growth, that the other bacteria are likely to overgrow and obscure the pathogen sought. If intestinal discharges have soiled the water, large numbers of coliforms are certain to be present, especially *Escherichia coli* which is a predominant part of the normal flora of the intestinal tract of all warm-blooded animals. These organisms are more easily recovered and identified than the pathogenic forms. No effort is made to detect pathogens in the routine examination of water but the routine test checks for an intestinal organism which, by its presence, indicates the possible presence of a pathogenic organism. Such an organism is termed an **indicator organism.** More important, its absence indicates that intestinal pathogens probably are not present. Several groups of bacteria serve as indi-

Table 17.1 The Waterborne Diseases at Which Water Treatment Is Primarily Aimed

Disease	Organism	Description
Typhoid fever	*Salmonella typhi*	Gram-negative Nonsporeforming bacillus
Paratyphoid fevers	*Salmonella paratyphi* strains	Gram-negative Nonsporeforming bacillus
Bacillary dysentery	*Shigella dysenteriae*	Gram-negative Nonsporeforming bacillus
Asiatic cholera	*Vibrio cholerae*	Gram-negative Nonsporeforming vibrio
Amoebic dysentery	*Entamoeba histolytica*	Protozoan of the genus *Amoeba*

cator organisms: the hydrogen sulfide-producing anaerobes and the intestinal streptococci. However, none has proven more reliable than the coliforms.

A water supply may contain coliforms without containing, at the same time, pathogenic bacteria. However, the accumulated epidemiological evidence indicates that water free of coliform bacteria is safe to drink. It is obvious that water contaminated with any fecal material is undesirable for human consumption. Many wells and springs containing coliforms have been used by small groups for years without ill effects. However, the contamination from the combined sewage of numbers of persons will eventually contain the excrement of one infected with intestinal disease. In sewage from large cities one pathogen usually exists for every ten million coliforms. Pollution of water almost certainly results in infection with pathogenic bacteria when a susceptible individual consumes the water.

A sanitary chemical examination of water reveals the state of the nitrogen present in the water. If contamination with organic matter has been recent, the nitrogen present is chiefly in the form of albuminoid and ammonia nitrogen. If the contamination was remote, nitrites and nitrates account for most of the nitrogen present. A higher-than-normal level of sodium chloride suggests contamination with domestic waste. However, chemical examination does not yield information from which the sanitary quality of a water supply can be evaluated with certainty. Nitrogenous material of plant origin cannot be chemically differentiated from that of animal origin. A sanitary survey of the watershed, together with a sanitary chemical examination, will supply suggestive data, but only a bacteriological examination can determine with certainty the sanitary quality of the water.

Methods for the bacteriological examination of water are recommended by a committee of the American Public Health Association and are given in the *Standard Methods for the Examination of Water and Waste Water* of the Association. The complete examination consists of three carefully specified tests, and the recommendations are periodically revised to incorporate the latest developments (See Table 17.2).

PRESUMPTIVE TEST

Only a few bacteria other than the coliforms ferment lactose with the production of gas. Graduated amounts of water are placed in tubes of lactose fermentation broth or lauryl tryptone lactose broth. Each of five tubes containing 10 ml of double-strength broth is inoculated with 10-ml portions of water. In addition, a series of five tubes of regular strength broth are inoculated with both 1.0-ml as well as 0.1-ml portions. Incubation is at 35 C for 24 hours, at which time the individual tubes are checked for the presence of gas. Gas formation is indicative of a positive presumptive test. If no gas is present in any of the tubes, incubation is continued for an additional 24 hours. The absence of gas after 48 hours terminates the examination with the report of negative (no coliforms present). However, if gas is produced within 24 or 48 hours, the presence of coliforms is indicated, and the examination is recorded as a positive presumptive test. The results of this test can be used to determine the most probable number (M.P.N.) of coliforms in the water sample by using computations tabulated in the *Public Health Reports* and reprinted in *Standard Methods*.

If only coliform bacteria gave positive results under the conditions described, additional tests would be unnecessary. Unfortunately, other organisms cause gas formation from lactose and result in a false positive test. A number of anaerobic *Clostridium* species and the aerobic *Aerobacillus* species are sporeforming rods that ferment lactose with the formation of gas. Gas formation may also result from that type of bacterial association known as **synergism.** Synergism results in the production of gas from lactose by an association of organisms, neither of which produces gas when grown separately. One member of the pair attacks lactose and produces intermediate split products, from which the second lactose-negative organism produces gas.

THE CONFIRMED TEST

Inoculum from tubes of positive presumptive tests are transferred to either a differential lactose-containing plating medium or to a selective liquid me-

Table 17.2 Synopsis of the Qualitative Microbiological Examination of Water

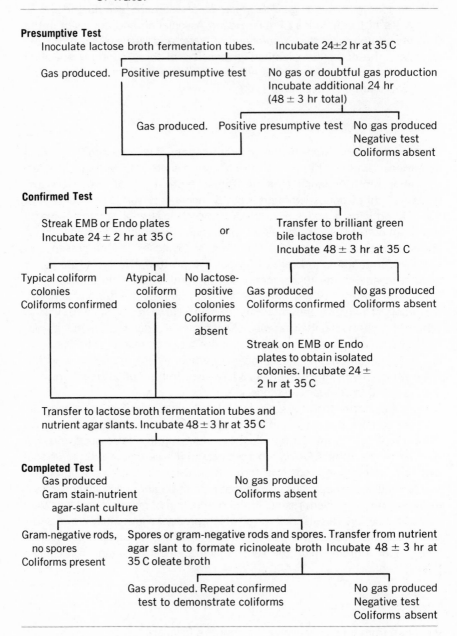

Presumptive Test

Inoculate lactose broth fermentation tubes.   Incubate 24±2 hr at 35 C

Gas produced. Positive presumptive test

No gas or doubtful gas production
Incubate additional 24 hr
(48 ± 3 hr total)

Gas produced. Positive presumptive test

No gas produced
Negative test
Coliforms absent

Confirmed Test

Streak EMB or Endo plates
Incubate 24 ± 2 hr at 35 C   or

Transfer to brilliant green
bile lactose broth
Incubate 48 ± 3 hr at 35 C

Typical coliform
colonies
Coliforms confirmed

Atypical
coliform
colonies

No lactose-
positive
colonies
Coliforms
absent

Gas produced
Coliforms confirmed

No gas produced
Coliforms absent

Streak on EMB or Endo
plates to obtain isolated
colonies. Incubate 24 ±
2 hr at 35 C

Transfer to lactose broth fermentation tubes and
nutrient agar slants. Incubate 48 ± 3 hr at 35 C

Completed Test

Gas produced
Gram stain-nutrient
agar-slant culture

No gas produced
Coliforms absent

Gram-negative rods,
no spores
Coliforms present

Spores or gram-negative rods and spores. Transfer from nutrient
agar slant to formate ricinoleate broth Incubate 48 ± 3 hr at
35 C oleate broth

Gas produced. Repeat confirmed
test to demonstrate coliforms

No gas produced
Negative test
Coliforms absent

dium. The most commonly used solid medium is probably eosin methylene blue (E.M.B.) agar. Plates are streaked so as to obtain isolated colonies. After incubation at 35 C for 24 hours, the plates are examined. Lactose-fermenting colonies produce dark purple to pink-colored colonies due to the formation of acid and the subsequent development of the colored eosin-methylene blue complex. If typical coliform colonies are observed, the test is recorded as positive. If the positive presumptive tests had resulted from a *Clostridium* species, no suspicious colonies on the aerobically incubated plate will appear. If positives resulted from an *Aerobacillus,* the colonies will be atypical but sufficiently similar to coliforms to justify further examination. The lactose-splitting member of a synergistic pair likewise may produce a colony that is atypical in appearance, but is nevertheless that of a lactose-fermenting colony. Transfers of typical or atypical colonies should be made for final identification.

Several liquid confirmatory media have been developed; the most satisfactory is brilliant green bile lactose broth. When inoculum from a positive presumptive test produces gas in this medium, the presumptive test is confirmed. The organisms responsible for false presumptive tests are inhibited by the brilliant green dye and by the bile so that gas formation is a positive indication of the presence of coliforms. E.M.B. agar is employed to secure isolated colonies from positive confirmatory tests for use in the completed test.

COMPLETED TEST

The completed test establishes that typical or suspicious colonies appearing on the plate of the confirmed test are actually of the coliform group, that is, they conform to the characteristics of a coliform. Transfers from an isolated colony are made to a lactose fermentation tube and to an agar slant. A coliform begins prompt lactose fermentation, and a gram stain from the growth on the slant will reveal a gram-negative, nonsporeforming bacillus.

PROBABLE ORIGIN OF COLIFORMS

Additional tests on pure cultures of coliforms can determine their identification and their probable origin. Since all coliforms have sanitary significance, the isolation and identification of strains has little significance in sanitation studies. However, determination of possible fecal origin or identification is accomplished by a series of biochemical tests, the **IMViC** reactions. Indol (I), methyl red (M), Voges-Proskauer (V), and citrate (C) utilization constitute the IMViC tests. All possible IMViC types have been encountered, but coliforms of known fecal origin are predominately IMViC + + − −; nonfecal types are usually IMViC − − + + (See Table 17.3).

The Public Health Service decrees that water used on interstate carriers (buses, airlines, and trains) shall have less that one coliform per 100 ml. This is determined by planting the previously mentioned multiple presumptive tests. If less than 10% of the tubes inoculated with 10 ml of water are positive, probability tables indicate that the water meets the Public Health standard. Since the ratio of coliforms to pathogens is about ten million to one, the probability is negligible that a consumer would imbibe a pathogen in such water. Proven incidents of waterborne disease have not been demonstrated in cities where the water met the coliform test. The margin of safety is greater than the figures indicate since a single intestinal pathogen seldom leads to disease.

QUANTITATIVE EXAMINATION

The total number of bacteria in water may be important but is not routinely determined except in special circumstances depending on the use of the domestic water. For example, in the food industry, the use of a high-count water for cooling processed cans results in increased spoilage. The cans leaving the retort are under pressure, and vacuum develops with cooling. If a minute amount of cooling water is sucked in through an imperfect spot in the seal as the vacuum develops, spoilage will probably appear in a few days.

The Membrane Filter Procedure

This technique is a relatively new application which has been accepted as an alternative method for determining the potability of water. The procedure has several advantages over the multiple-tube fermentation broth test: a

Table 17.3 The IMViC Reactions

	Indol	Methyl Red	Voges-Proskauer	Citrate
E. coli (fecal)	+	+	−	−
Enterobacter aerogenes (nonfecal)	−	−	+	+

greater degree of reproducibility is possible, larger volumes of water may be tested (thus enhancing sensitivity of coliform detection), and results may be obtained more rapidly. Analysis by filtration requires less equipment and is generally less cumbersome than the tube test. For example, a sample may be filtered in the field and the membrane filter then shipped to the laboratory on a preservative medium (Fig. 17.1).

However, the membrane method does have limitations. Water containing algae or other suspended matter, if filtered in volumes large enough for dependable examination, may clog the membrane. Furthermore, deposition of material on the filter may interfere with development of bacterial colonies. In water samples containing large numbers of noncoliforms, estimates of numbers of coliforms will be low in comparison to expected values.

The filter membrane is a thin layer of complex cellulose structure which has practically no absorptive capacity and functions as a physical screen in trapping bacteria. Pore size is determined primarily by the chemical composition of the membrane and the temperature of drying and humidity during the manufacturing process.

The procedure is quite simple and rapid. A test sample, the volume of which depends on suspected quality of the water, is passed through the filter. The membrane filter is placed aseptically, bacteria side up, on an absorbent pad saturated with E.M.B. broth lying in a sterile petri plate. Nutrients diffuse up through the membrane to the cells trapped on the membrane. After incubation, coliform colonies may be observed growing on the surface of the membrane (Fig. 17.2).

Water Purification

Since most waters for municipal use are subject to contamination with coliform bacteria from fecal matter and, consequently, may contain pathogenic forms as well, the water must be treated for domestic use. Since water is not the natural habitat of pathogenic bacteria, their numbers tend to decrease naturally through the lack of proper nutrients and certain environmental conditions. However, purification by sedimentation, the antagonistic effect of plankton, or the action of ultraviolet rays of sunlight cannot be depended on to yield safe water in any reasonable time.

The objective of water purification is to render water of unsafe or uncertain quality suitable for domestic use. The nature of the supply and the cost of purification under the particular circumstances determine which method should be employed. Clear water may be rendered satisfactory by suitable chemical treatment alone, but turbid water must be clarified as well as treated chemically.

The earliest method of artificial water purification was by slow sand filtra-

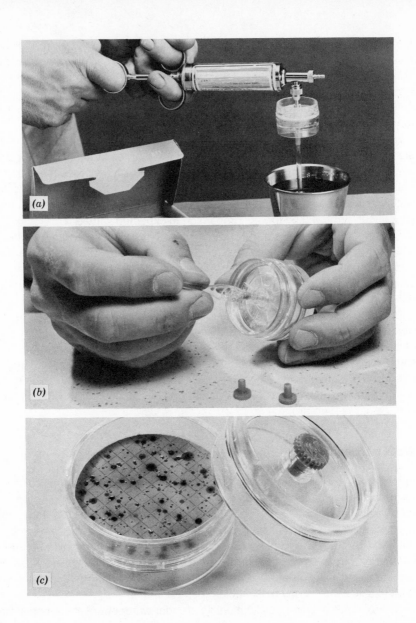

Fig. 17.1. Field test of water with membrane filters. (a) A measured amount of water is drawn through the filter. (b) A sterile ampoule of broth is applied to the back of the filter. (c) After 18 hours incubation, colonies showing typical sheen indicate polluted water. Photos courtesy Millipore Filter Corp.

Fig. 17.2. The dark colonies with metallic sheen on this plate are typical coliforms indicating pollution in the water tested here. Photo courtesy Millipore Filter Corp.

tion. Introduced into the United States about 1870, this method had been successful for some time in England. A slow sand filter consists of a bed of coarse stones layered successively with pebbles, gravel, and sand. Water, flooded on the surface to a depth of two to three feet, slowly trickles through and is collected at the bottom of the filter. The bacterial load is reduced 95 to 99%; odor and organic matter are also reduced and turbidity is lessened. Though efficient, this method has several limitations. Water which is not fairly clear will clog the filter with silt within a short time. A large area of filter bed is required since the capacity of a slow sand filter is only from 2½ to 5 million gallons per acre per day. The bed must be protected against freezing. At intervals, the sand must be washed or replaced. This relatively slow operation requires an alternate filter bed to maintain purification service while the major filter is being cleaned. The purification achieved is a combination of physical, chemical, and biological actions. The suspended particles are physically strained out. The suspended matter is biologically removed by biochemical oxidations through the action of the **Schmutzdecke,** a zoogleal mass that develops on the sand grains after a short period of use and consists of various living organisms.

Rapid sand filtration avoids most of the limitations of slow sand filtration and has almost completely superseded the older procedure. Developed in the United States to purify turbid water, the procedure essentially consists of adding flocculating chemicals, such as iron or aluminum sulfates, to the water. After holding the water for a few hours while the particles settle out, thereby removing most of the suspended matter and bacteria, the water is passed through filter beds similar in construction to slow sand filter beds, but of greater capacity. Residual suspended matter is trapped by the filter, and a clear effluent results. When the filter becomes badly clogged, the flow of water is reversed, thereby washing the filter which can then be returned to service.

Filtration is not adequate to remove all bacteria. Samples of filtered but otherwise untreated water frequently contain coliforms. Water is chemically treated with chlorine gas or with hypochlorite to kill pathogenic bacteria that might be present. Activated charcoal may be added to adsorb objectional tastes and odors. The sanitized water should contain a residue of 0.1 to 0.2 parts per million of available chlorine.

The value of water purification is emphasized by comparing the incidence rates of waterborne disease before and for a year or two after purification in a number of cities. A city with a high waterborne disease rate almost certainly has a water supply of inferior quality or has experienced a break in water treatment resulting in an epidemic. In past years, typhoid fever was a common urban disease, but today waterborne typhoid is essentially a disease of country areas where people depend on water of uncertain sanitary quality.

Sewage Disposal and Sewage Purification

Sewage is the water supply of a city after it has been used. It contains various industrial and domestic wastes and is rich in highly putrescible organic matter. Consequently, sewage requires a great deal of oxygen for microbial disintegration. A high biochemical oxygen demand, B.O.D., is needed for the oxidation of this organic material.

In some cases purification before disposal is not required, but most cities demand some degree of purification. Untreated sewage may be dumped into the sea in coastal cities not located near oyster beds or bathing beaches, but few cities are so fortunately located. Similarly, raw sewage may be allowed to flow into a river. However, the volume must not be great enough in relation to the volume of the stream to create anaerobic conditions with resulting damage to fish and wild life. Also, the operation must not affect a city using the river as a water source. With the present day problem of pollution, more rigid regulations, however, will need to be imposed.

A special problem has resulted from the great increase of petroleum wastes which find their ways into our rivers, lakes, and oceans. For example, life conditions in the deep ocean have been constant for millions of years and the introduction of pollution from spilled petroleum into this stable but fragile environment might be more than the marine life could survive. This could upset the marine food chain with severe and perhaps catastrophic implications in the deep sea. Fortunately, numerous microorganisms utilize petroleum as a source of food and convert it to less toxic products. Some ecologists doubt that these microbes will be able to deal with the situation if the level of contamination increases beyond that now existing. Methods for enhancing microbial decomposition are eagerly sought.

Under special climatic and soil conditions where toxic industrial wastes are not present, sewage may be used as irrigation water. A sandy soil, a relatively dry warm climate, and a suitable crop are necessary for the successful disposal of sewage as irrigation water. Truck crops, especially vegetables to be eaten raw, should never be irrigated with sewage.

In all sewage purification procedures, the bacteria naturally present in the soil or water transform the sewage organic matter from an unstable to a stable form. Nitrogen, for example, goes through the cycle from protein to nitrate. Much of the carbon ends as carbon dioxide, and other substances likewise approach or attain the stable end form typical of each. In the artificial purification of sewage the bacterial action occurring naturally in the cycles of the elements is harnessed in an engineering device designed to favor the development of a particular physiological type.

The simplest procedure employing the activity of anaerobic bacteria is found in the septic tank with subsurface porous tiles to distribute the effluent. These tanks are necessarily small installations used generally for single dwellings. The **Cameron tank** is a somewhat more elaborate installation operating on the same principle (Fig. 17.3). The sewage is only partially purified, the effluent still being highly putrescible and requiring further (aerobic) bacterial activity for stabilization. Preliminary anaerobic digestion facilitates final aerobic oxidation. Such a tank may be combined with an aerobic device, such as a contact bed, or the effluent may be distributed through a subsurface tile system providing low-cost purification for small installations.

The Cameron tank has been improved by combining a sedimentation chamber and a digestion chamber. The resulting structure, the **Imhoff tank,** may be either circular or rectangular (Fig. 17.4). Sewage solids settle into the lower chamber where they are digested by anaerobic bacteria. The effluent,

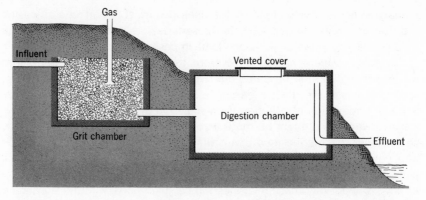

Fig. 17.3. The Cameron tank. A modified septic tank for disposing of the domestic sewage of a single family in a rural area. Effluents are of poor quality but do not become too objectionable in thinly populated regions.

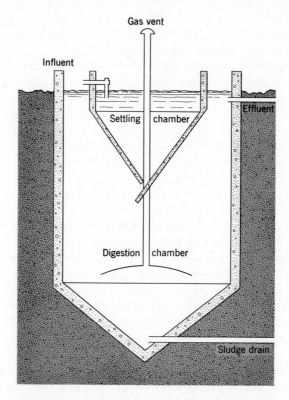

Fig. 17.4. The Imhoff tank provides for settling and sludge digestion in a single tank.

considerably reduced in suspended solids, is still putrescible and requires further treatment. The slowly digesting solids accumulate in the lower chamber, and at intervals this **sludge** is removed and dried for fertilizer.

The **contact bed** is the simplest of the several devices for employing aerobic bacteria in the purification of sewage effluent. A bed of crushed-coarse stone or similar material is treated with sewage so as to encourage a heavy growth of aerobic bacteria. The bed is filled, allowed to stand a short time, drained, and rested; then the cycle is repeated. If the sewage is distributed over a similar bed through sprinkler heads, the installation is known as a sprinkling filter. The heavy growth of aerobic organisms which accumulates on the stones traps particles, including sewage bacteria. The sewage material is then decomposed by aerobic action.

The **activated-sludge method** is a widely used purification procedure using the action of aerobic bacteria (Fig. 17.5). This method is frequently combined with a preliminary sedimentation tank to remove the large suspended particles and expose them to anaerobic digestion. Effluent from the prelimi-

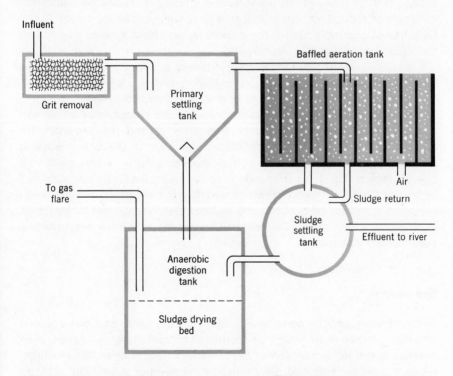

Fig. 17.5. The activated sludge system depends on pumping air through the settled sewage and returning much of the settleable solids to the tank to maintain a high inoculum of aerobic bacteria.

nary sedimentation tank is passed to an aeration tank where it is vigorously agitated with air. In the course of six to nine hours in the aeration tank, essentially all of the organic matter is changed to a stable form, the effluent having a very low B.O.D. Dissolved organic matter is utilized by bacteria to make cell bodies which settle out. The slowly digestible, flocculent sludge is settled out in a settling basin and pumped back to the aeration tank. The stable effluent is drawn off and may or may not be chlorinated before final disposal. Sludge accumulates slowly, and periodically a part must be drawn off to prevent overloading of the system. This excess sludge is added to the solids collected in the primary settling tank in an anaerobic digestion tank. The large volume of inflammable gas produced by anaerobic digestion contains corrosive gases and is customarily burned off. After several months in the anaerobic digestion tank, where microbial action compacts the sludge and modifies it to a brown humus, it is drawn off, dried, and used as fertilizer. The thermophilic conditions in the digestion tanks destroy the pathogens.

Such a plant, operating properly and efficiently, is practically odorless except in the vicinity of the influent. The effluent is usually chlorinated to eliminate or reduce pathogens that may have survived the treatment. Essentially free of oxidizable matter, the effluent is very likely of better quality than the stream into which it is discharged.

Recent engineering advances have resulted in processes of sedimentation, filtration, clarification, aerobic oxidation, and sludge treatment combined into more complex, but more or less self-contained, units.

Large volumes of industrial waste, such as those produced by a food factory processing seasonal products, frequently exceed the capacity of a purification plant. Under this condition, purification of the excess waste is carried out by the combined action of anaerobic bacteria and algae in a sewage **lagoon** (Fig. 17.6). The organic materials in the waste are stabilized by bacterial action. Photosynthetic oxygenation at the surface by algae keeps the pond from becoming putrid and restores dissolved oxygen to the liquid. In locations with a mild winter climate, lagooning may provide satisfactory year-round purification.

Summary

Because many healthy persons and animals carry organisms of infectious diseases, sewage must always be suspected of containing pathogens, even in areas where no clinical illness has been reported. Where intestinal organisms cannot be excluded, they must be removed or killed. The coliform organisms are easily detected and are removed at the same rate and with

Fig. 17.6.   A sewage lagoon requires an area of low rainfall and high sunlight. Anaerobic decomposition of the settleable solids occurs at the bottom, and the algae growing near the surface produce oxygen to satisfy the B.O.D.

the same procedures that remove or kill intestinal pathogens. Since they are ten million times as numerous in sewage as are possible pathogens, the absence of coliforms is accepted as a reliable indication of the absence of pathogens.

Sewage disposal requires microbiological action to mineralize the dissolved solids or to incorporate them into microbial cell bodies, so that they may be removed with the colloidal and sedimentable solids; the elements of organic matter must be converted to their stable, odorless forms, that is, sulfur compounds to sulfates, nitrogen compounds to nitrates, phosphorus to phosphate, and carbon to carbon dioxide. The solid materials are compacted by microbial action to produce a usable fertilizer. The liquid effluent is oxygenated so that it will not interfere with aerobic life in the water into which it is discharged.

Although sewage treatment removes undesirable materials, certain synthetic detergents and plastics are not properly mineralized. Industry often uses nonbiodegradable chemical biocides and corrosion inhibitors in waters for cooling towers and other refining processes. These undesirable contaminants appear in the effluents resulting from sewage treatment. Such polluted water is not only dangerous to water life but interferes with the recycling of municipal waters. Legal restrictions have been imposed to prevent the use of materials that are not readily degradable by microbial activity.

EIGHTEEN

Milk and Milk Products

The natural use of milk in the nourishment of a young mammal depends on direct transfer from the lactating animal to the offspring. As the milk is not exposed to the environment, little or no possibility of contamination exists. Under artificial methods of collecting and handling, milk receives bacteria from practically everything with which it comes in contact. Hence, if the milk remains at incubation temperature for any length of time after collection, a great increase in bacterial population may attack the quality of milk. If pathogenic bacteria are among those which gain entrance, the consumer's health is jeopardized. The problem of securing good milk and good dairy products through dairy sanitation involves constant vigilance and close attention to all details of operation.

Although the family with one cow faces the same bacteriological problems which assail the large dairy, the commercial dairy's problems and responsibilities result from its large-scale methods of production and processing. This chapter discusses the bacteriology of milk and milk products as a unit rather than considering the problem of large- and small-scale operations separately.

Significance of Bacteria for Dairy Products

Control of the numbers and kinds of bacteria in dairy products has four major applications. The first is in determining the market standards of fluid milk. The letter grade of milk is based on a number of factors, one of which is the number of bacteria present in the milk at the time of delivery. Obviously, a knowledge of the important sources of bacteria, of the methods for preventing their entrance in large numbers, and of procedures for interfering with their growth has much economic importance to a milk producer.

Dairy bacteriology is also concerned with prevention of spoilage of milk and milk products. Uncontrolled bacterial growth in milk and its derivatives causes deterioration ranging from slight flavor or texture losses to extensive spoilage. Here, also, a knowledge of sources of bacteria, modes of entrance, and control of activity has great value in lengthening the shelf life of the product.

Bacteria and related microorganisms are employed in the manufacture of such dairy products as fermented milks, butter, and cheese. Desirable flavor of each of these commodities is associated with the controlled activity of special microorganisms. Improvements in procedures and in the quality of the product become possible only as knowledge of the organisms concerned is developed.

Finally, dairy products, chiefly fluid milk, are instrumental in the transmission of certain infectious diseases. Investigations of many epidemics have proven the infectious agent to be milkborne. The health of the consumer can be adequately protected only by a knowledge of the sources and controls of infectious organisms.

Source of Bacteria in Milk

In dairy manufacturing operations, milk cannot be produced that is totally free of bacteria. Consequently, the dairy bacteriologist has to establish the relative importance of the various sources of milk-dwelling bacteria as a first step toward the inauguration of adequate control procedures. The bacteria of raw milk are of both endogenous and exogenous origin.

ENDOGENOUS ORIGIN

A few bacteria can be found in freshly drawn milk, even when milk herds have been thoroughly examined and found sound and healthy, and when the milk has been drawn under as nearly aseptic conditions as possible. Numerous such observations have led to the conclusion that milk as it occurs in the udder is not sterile. The normal udder contains a few bacteria which presumably have entered through the teat canal and become established in the milk-duct system. Such bacteria are of a few types and are present in relatively small numbers, commonly only a few hundred per milliliter at most.

Bacteria of exogenous origin come from a variety of sources including the barn air, the coat of the animal, the hands and clothing of the milker and milk handlers, and the utensils and equipment used in processing the milk for delivery to the consumer. During hand milking, bacteria may enter the milk more readily than with machine milking since the milk is more exposed to the dust carried in the barn air, to material from the coat of the animal, to the hands and clothing of the milker, and to flies.

Contamination from exogenous sources during hand milking can be kept at a low level by the use of a small-top milk pail, thereby decreasing surface exposure. Other important precautions include not sweeping or feeding hay shortly before milking, cleaning the coat of the animal with a damp cloth, and washing the udder. The milker should wear clean apparel and wash his hands before milking. Since bacteria are carried in droplets from the respiratory tract, the milker should avoid sneezing, coughing, even talking, over an open pail.

In all milk collection the most important sources of bacteria invasion are the items of dairy equipment. Milk on the farm is commonly strained or filtered to remove any solids which may be present in it. Of little or no value in reducing the bacteria, this operation actually may add bacteria if the strainer is not clean.

All dairy equipment should be so constructed that it can be cleaned readily and thoroughly. Irregular surfaces, seams, and joints make removal of milk residue difficult. Properly maintained milking machines, pipe lines, covered storage tanks, and similar equipment serve to reduce exposure to the environment. Such equipment should be dismantled and promptly washed with cold or tepid water, not hot water, to remove the milk, and then scrubbed in hot water containing a cleansing agent. Just before use, the equipment should be rinsed with a fast-acting chemical disinfectant such as a chlorine or quaternary ammonium compound. Pipe lines should be constructed and installed in such a way that cleaning can be done satisfactorily without dismantling.

Milk and other dairy products are now widely packaged in single-service paper containers. The accumulated evidence indicates that such containers do not pose any bacteriological problems.

Types of Bacteria Found in Raw Milk

The types of bacteria which may be present in raw milk are most conveniently considered from the standpoint of physiological groups. Although all types do not appear in every raw-milk sample, each occurs widely.

UDDER FLORA

Relatively few bacterial types are able to exist under the conditions which prevail in a healthy udder. Although rod-shaped forms have been noted, micrococci are most common. This udder flora is sometimes described as inert since it produces only minor changes in milk.

DESIRABLE FERMENTERS

The most common change occurring in raw milk is nongaseous souring resulting from the production of lactic acid which coagulates the casein. This common change, termed normal fermentation, is caused by organisms of exogenous origin, probably coming chiefly from dust from grain and other feeds. They include *Streptococcus lactis,* and related streptococci, and certain *Lactobacillus* types, chiefly *L. casei.* Natural souring may serve to check the action of other organisms, thus acting as a preservative.

UNDESIRABLE FERMENTERS

Souring by certain yeasts, members of the genus *Clostridium,* or members of the coliform group is accompanied by gas formation and in some instances by disagreeable flavors and odors. The gas may be trapped in the curd and cause extensive foaming. Coliforms, coming chiefly from animal manure or barn dust, are present in most milk by the time it leaves the milking shed. They do not have the same significance as when present in water, but their presence in appreciable numbers does indicate improper attention to dairy sanitation.

PROTEOLYTIC BACTERIA

The proteins of milk can be broken down by any proteolytic organism which finds its way into milk. Most commonly, the causal organisms are aerobic sporeformers of dust origin. *Streptococcus fecalis* var. *liquefaciens* grows over the range of 10 C to 45 C, best at about 37 C. By a renninlike enzyme it causes a rapid coagulation of milk of low acidity. Proteolysis begins shortly after coagulation and progresses with the growth of the organism (Fig. 18.1).

In raw milk, the aerobic sporeformers are of little significance because they are overgrown by the lactic acid bacteria which survive pasteurization; however, they can germinate and grow without restraint if the milk is held for any length of time within the range of about 20 C to as much as 55 C. A slight amount of protein breakdown imparts a bitter flavor to milk. Even the very slow growth of psychrophilic bacteria occurring in refrigerator storage gives an off-flavor in four to seven days.

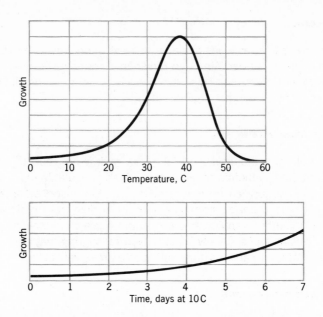

Fig. 18.1. Growth of proteolytic bacteria. Although the optimum temperature of this fecal strepto-coccus is 37 C, it will make significant growth in a week at 10 C.

BACTERIA PRODUCING UNUSUAL CHANGES

In rare instances the growth of certain bacteria may impart a color to milk. *Pseudomonas synxantha* produces an intense yellow to orange color in cream, *Pseudomonas syncyanea* growing in sour milk produces a blue color, and various species of *Serratia* produce a red color. Ropy or slimy milk may be caused by several organisms, a common one being *Alcaligenes visco-lactis.*

PATHOGENIC BACTERIA

Studies of dairy products as vehicles for the transmission of pathogenic bac-teria have so developed and improved dairy production that large outbreaks of milkborne infectious disease are now uncommon. Diseases which are spread through dairy products fall into two groups: (1) those of animal origin affecting animals only, or both man and animal; and (2) those of human or-igin.

The bovine disease organisms transmissible to man through milk are those involved in cattle mastitis, tuberculosis, brucellosis, and Q fever. Masti-tis, or inflammation of the udder, may be caused by any one of several organisms, the most common being *Streptococcus agalactiae,* which is not pathogenic for man. The closely related human pathogen, *Streptococcus pyogenes,* can also cause mastitis. Septic sore throat or scarlet fever may develop in persons who drink raw milk containing this organism. *Staphylococcus aureus* may also cause mastitis. Some strains of the organism produce an exotoxin which causes severe human gastroenteritis.

At one time, bovine tuberculosis was the most significant animal disease transmitted to humans by milk, but this is no longer true in the United States. The detection and elimination from the herd of tubercular cows and the pasteurization of milk have greatly lessened the incidence of milkborne tuberculosis in areas of the world where these control measures are practiced.

Clearly the most important animal disease transmitted by dairy products at present is brucellosis. Three species of *Brucella* cause brucellosis, or undulant fever, in man: *B. abortus,* the cause of contagious abortion, or Bang's disease, in cattle; *B. suis,* the cause of a similar condition in swine; and *B. melitensis,* the cause of diseased condition in goats. *B. abortus* is less virulent for man than either of the others but is of greatest importance in the United States because of its prevalence.

Q fever is a milkborne febrile disease of man caused by a rickettsial organism *Coxiella burnetii.* This organism is sufficiently resistant to heat to occasionally survive pasteurization.

The intestinal pathogens, easily transmitted human-disease organisms, are introduced into the milk by handlers and can infect the consumer. Diphtheria has been milkborne in a number of instances; the causal organism usually gains entrance directly from a carrier who is a milk handler or occasionally from a diphtheritic lesion on the udder or teat of the cow. Septic sore throat, scarlet fever, and gastroenteritis may follow direct infection of the milk by a milk handler.

Control of Bacteria in Milk

The contamination of milk can be greatly reduced by guarding against the recognized sources of contamination. As quickly as possible, after collection, the milk should be chilled and held at a low temperature to restrain the growth of organisms unavoidably present in it.

PASTEURIZATION

Though low initial bacterial content can be produced by attention to production sanitation, it remains uncertain that any raw milk, or milk product prepared from raw milk, is completely safe for human consumption. The occurrence of undetected disease in the animal, the possibility of infection from human sources, and the fact that milk is an excellent medium for the survival and growth of bacteria combine to make raw milk a potential hazard to the health of the consumer. Sanitary care in production must be supplemented to insure a raw milk safe for domestic use. Pasteurization effectively treats the milk, ridding it of almost all harmful microbes.

Pasteurization is a modification of the process developed by Pasteur for the treatment of beer and wine to prevent bacterial deterioration. In the dairy industry, pasteurization consists of heating at the minimum time-temperature combination adequate to destroy the most heat-resistant pathogenic organism which might be present—the tubercle bacillus. In practice, a slight increase above the minimum is added as a safety margin to compensate for unavoidable variations in operation which might affect the results. Pasteurization does not sterilize the milk, but it does destroy most of the nonspore-forming bacteria present.

Pasteurization is carried out in two ways: (1) by the low temperature–long time, or holding method, in which milk is heated in a vat for 30 minutes at 143 F (62 C); or (2) by the high temperature–short time, or flash method, in which the milk is heated at 161 F (72 C) for at least 15 seconds. The holding method is best for small lots of milk; the flash method, which permits continuous operation, is preferable for large lots. After either process, the pasteurized milk should be cooled promptly to below 50 F (10 C) and protected against recontamination in all subsequent operations.

The effectiveness of the pasteurization process can be determined by the **phosphatase test.** Raw milk always contains the enzyme phosphatase, which is slightly more heat resistant than the tubercle bacillus. If pasteurization has been properly done the enzyme will be destroyed. In the test, a small quantity of milk is added to a solution of disodium phenylphosphate. If active phosphatase is present, the compound splits, and the liberated phenol reacts with an indicator giving a blue color.

The public health value of pasteurization has become so well established that most market milk today is pasteurized. When properly done, the nutritive value of the milk is not significantly affected.

Bacteriological Examination of Dairy Products

Milk may be examined bacteriologically for the total number of bacteria and for special groups of bacteria. Recommended procedures are described in *Standard Methods for the Examination of Dairy Products* of the American Public Health Association.

The most widely used procedure for taking the bacterial count of milk is to culture an aliquot in liquefiable solid medium, count the colonies which develop, and calculate the number on the assumption that each colony comes from a single cell. The results of the test are greatly influenced by slight variations in laboratory technique. In any case, the procedure detects only those bacteria which can grow under the conditions of the test. Thus, at best, the recorded count is only an approximation of the true number of bacteria in the sample. Despite the limitations of the plate count method, the data it yields are adequate to evaluate the bacterial condition of a milk supply. Counts are frequently accomplished using the membrane filter technique (Fig. 18.2.).

Fig. 18.2. A membrane filter through which was passed 10 ml of milk diluted 1:1000. The filter was transferred to a peptone-glucose medium and incubated for 24 hours. The grid marked on the filter facilitates counting the colonies. This filter method is now widely used as a refinement of the plate count. Photo courtesy Millipore Filter Corp.

DIRECT MICROSCOPIC COUNT

The most rapid method of determining the total bacterial content of milk is by direct microscopic counting. A measured quantity of milk is spread uniformly over a marked glass slide and stained. The bacteria in a number of fields of known diameter are counted. Calculation of the total count requires only simple arithmetic.

Direct microscopic counts allow a rapid determination of the bacterial quality of milk as received at a collecting station; thus satisfactory and unsatisfactory lots are not mixed. Counts of clumps of cells rather than of individual cells give results which most nearly approximate plate counts.

DYE REDUCTION TESTS

Certain dyes which show a difference in color between the oxidized and reduced forms have been used in a rapid test of milk quality. During growth, bacteria exhaust the dissolved oxygen in a medium, thereby causing a lowered oxidation-reduction potential at a rate directly related to the number of active cells present. Thus, a short reduction time indicates many bacteria, a long time indicates few. The test does not give an accurate quantitative measure of the bacterial content but does permit the grading of raw milk.

The dyes used may be either methylene blue, which is blue in the oxidized form and colorless when reduced, or resazurin, which changes from blue through pink to colorless with reduction.

EXAMINATION FOR SPECIAL ORGANISMS OR GROUPS

Coliforms in milk can be enumerated by plating an aliquot in a differential medium which yields characteristic colonies, such as violet red-bile agar, or by inoculation of dilutions of the sample into a sufficient number of fermentation tubes of lactose broth to permit statistical calculation of the most probable number. Coliforms are almost always present in raw milk and are not objectionable in very low numbers. Large numbers suggest negligence in production. Their presence in pasteurized milk indicates recontamination after pasteurization.

WISCONSIN CURD TEST

A test for gas-forming bacteria that interfere with cheese making can be made by coagulating an aliquot of the milk with rennet, pressing out the whey, incubating the curd, and observing for evidence of the growth of undesirable types. Heat resistant or **thermoduric** bacteria, which survive pasteurization, indicate a lack of cleanliness of equipment if present in appreciable numbers. These can be detected by a comparison of results from cultures made before and after pasteurization. True thermophiles can be detected by incubating cultures at an elevated temperature.

Grading of Milk

The standards of quality for milk and cream are established by local ordinance and consequently may vary from city to city. In an effort to establish uniformity in production and processing practices, the Public Health Service of the Federal Government recommends the adoption of a standard milk ordinance which specifies in detail the requirements for both farm and dairy plant. The suggested bacteriological procedures are those of the *Standard Methods*.

Bacterial content is important in determining the grade of market milk. Grades A, B, and C for both raw and pasteurized milk are recognized in the standard ordinance, but most milk today is grade A pasteurized. The maximum numbers of bacteria permitted per milliliter at the time of delivery to the consumer for grades A, raw and pasteurized are:

	Before Pasteurization	After Pasteurization
Grade A raw	50,000	—
Grade A pasteurized	200,000	30,000

The numbers of bacteria permitted are doubled in the case of cream and are disregarded for buttermilk and sour cream.

CERTIFIED MILK

Certified milk is produced under the supervision of a medical milk commission only by dairies operated under the rigid specifications of the American Association of Medical Milk Commissions. Developed when much of the market milk was raw milk of indifferent quality, certified milk production was designed to provide safe milk for infant and invalid feeding. It was originally raw milk but today may be either raw or pasteurized. The total count may not exceed 10,000 per milliliter for raw, or 500 for pasteurized, and the coliform count must be less than 10 per milliliter.

UNGRADED MILK

Ungraded milk is a milk which, for some reason, usually a lack of sanitary inspection at the source of production, cannot be assigned a grade. The term indicates a lack of full information and nothing else.

Dairy Products

Fermented milks are produced by growing lactic acid bacteria, either in pure or mixed culture, in raw or whole milk. The acid produced curdles the milk and gives it an agreeable sourness. The characteristic flavors of different types of soured milk result from the types of organisms responsible for the souring, the composition of the milk, or the temperature at which it was incubated.

The use of soured milk goes back beyond recorded history and has been common to all peoples in all countries in which milk was available. Since variations in flavor among batches may sometimes occur with natural souring, dairy plants have adopted standardized production methods with starter cultures. Usually skim milk is heated for about 30 minutes at 180 to 190 F (82 to 88C), cooled to 70 F (21 C), and inoculated with a starter culture which has been selected on the basis of flavor, production, and activity. The milk is incubated at about 70 F (21 C) until properly acid; the curd is broken to a smooth consistency and then held at about 40 F (5 C) for several hours and bottled. The flavor of the product can be improved by adding 0.1 to 1.2% citric acid or sodium citrate to the milk to serve as a source for diacetyl formation during fermentation. The organisms most commonly used to produce commercial buttermilk are a mixture of *Streptococcus lactis* and *S. citrovorus.*

Of the many fermented (soured) milks, each is characteristic of specific peoples or areas. Bulgarian buttermilk, which is prepared with *Lactobacillus bulgaricus,* once had considerable vogue as a remedy for autointoxication. Acidophilus milk, a popular fermented milk, is prepared by souring sterilized skim milk with *L. acidophilus.* Yoghurt, extensively recommended for special diets, is prepared by fermenting a concentrated milk with a mixture of *Lactobacillus* (probably *L. bulgaricus*) and *Streptococcus thermophilus* at a temperature of from 104 to 115 F (40 to 46 C), followed by fermentation by a film-forming yeast.

Butter has a very delicate and easily affected flavor. It presents many interesting bacteriological problems in production and in preservation. Commercially produced butter is of two types, sweet cream and sour cream butter. Formerly, most butter was produced from cream which had been soured by lactic acid bacteria, either naturally or following inoculation with a starter culture; however, now sweet cream butter is more popular.

The flavor of sour cream butter is greatly influenced by the culture used in ripening the cream. Earlier, special cultures selected for their ability to produce desirable flavor compounds were strains of *Streptococcus lactis* or related types. Later, mixed cultures of *S. lactis, Leuconostoc citrovorum* (*S. citrovorus*), *L. dextranicum,* and *S. diacetilactis* were introduced. Mixed cultures are used almost exclusively at present. The cream is ripened at 70 F (21 C). The most important desirable flavor component of butter is diacetyl. Sweet cream butter can be given a desirable flavor by incorporating a distillate of starter cultures containing volatile flavor compounds, or by working the culture into the butter.

Organisms which can tolerate the conditions of salt and low moisture found in butter may produce undesirable flavors attributable to the end products of growth or may actually lead to decomposition. Bacterial deterioration is best controlled by the use of only high-quality raw product, pure water, and careful attention to sanitation of plant and equipment.

Butter is not a sterile product. Restraining of bacterial activity by low temperature storage is essential to preserve butter that must be held for extended periods between manufacture and consumption.

CHEESE

The essential ingredients for the manufacture of cheese are milk, microorganisms, salt, and, in most cases, rennet. There are about 400 varieties of cheese, representing some 20 general types, prepared by varying the conditions of manufacture and curing.

Cheeses can be divided into three main groups — soft, semihard, and hard cheese — on the basis of the moisture content of the finished product. Cottage cheese and cream cheese are unripened soft cheeses. Other varieties of cheese are ripened by bacteria, molds, or a combination of bacteria and molds.

Good quality cheese is a product of good quality milk. Cheese was once made from raw milk, but, since numerous undesirable types of bacteria may interfere with the ripening process, pasteurized milk is now used for most cheese manufacturing.

In cheese manufacture, milk is warmed to growth temperature, inoculated with lactic acid bacteria, *Streptococcus lactis,* or *S. cremoris* (Fig. 18.3.). These bacteria are allowed to grow for a short time; the milk is then coagulated with rennet, and a firm curd is allowed to form. The curd is cut, the whey allowed to separate, and the remainder is salted, shredded, and pressed into forms for ripening. The appearance and flavor of the ripened cheese are determined by the kind and manner of action of the organisms present, and the technology of manufacture (such things as the way the curd is prepared for ripening and the temperature and time of storage). Most of

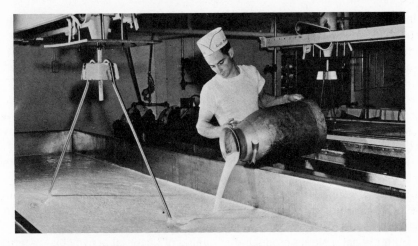

Fig. 18.3. Starter culture, one of the agents that causes milk to coagulate, is added to the vat as paddles rotate to stir it thoroughly into the milk. Stainless steel vats of this type hold 18,000 pounds of milk—enough to make 1800 pounds of cheese. Photo courtesy Kraft Foods.

the semihard and hard cheeses are virtually the same in appearance and flavor before ripening.

Practically all the hard cheeses, such as Cheddar or American cheese, are ripened under conditions which favor growth throughout the mass and not on the surface. The soft cheeses, such as Limburger and Camembert, are ripened by a growth of organisms on the surface; the enzymes concerned in the changes diffuse into the cheese. Semihard cheeses, such as Roquefort, are ripened by a combination of surface and interior growth. The latter is obtained by piercing the cheese with needles inoculated with mold spores to provide access of the oxygen needed for the obligately aerobic mold. The molds used are carefully selected strains of *Penicillium roquefortii*. This organism produces the distinctive flavor as well as the color of the "blue" cheese (Table 18.1).

Microbial deterioration of cheese results largely from molds. Paraffin coating or other wrapping which excludes oxygen prevents mold growth. Wrappings dipped in calcium propionate solution also have been used.

Summary

The microbiologist is interested in milk from the point of view of (1) grading, (2) spoilage, (3) disease, and (4) products. Pasteurization, aimed primarily at the tubercle bacillus, has eliminated the hazards involved in milk consumption. In the current production of foods that require microbes, such as cheese, modern microbiology imitates the conditions permitting the natural fermentation encountered in the historical development of the product.

Table 18.1 Microbiology of Cheese

Type of Cheese	Name	Organisms Involved in Production and Main Function
Hard	Cheddar	*Streptococcus lactis* – acid production *Lactobacillus casei* – curing
	Swiss	*Propionibacterium freundenreichii* *P. shermanii* – flavor and holes
Semihard Soft	Roquefort Limburger	*Penicillium roquefortii* – flavor and color *Brevibacterium linens* – primarily (actively proteolytic) thought to be responsible for flavor
	Camembert	*Penicillium camembertii* – primarily responsible for flavor *Geotrichum sp.* and *Brevibacterium linens* may by present during ripening Curd inoculated with *P. camemberti*

NINETEEN

Food Microbiology

Whatever is food for man or animal is also food for microorganisms. Following the slaughter of a food animal or the harvesting of a fruit or vegetable, decomposition begins. Natural enzymes of the food cause part of the decay, but most of it results from the growth of microorganisms which are naturally present or are added in the handling of the food. If the activity of the food enzymes and microorganisms is not inhibited or controlled, decomposition proceeds until the foodstuff is no longer acceptable to a discriminating consumer. The extent to which decomposition may proceed before food is inedible varies with the foodstuff and the consumer. Some products, beef especially, may be improved by a limited amount of enzymatic and microbial activity. Both nationally and individually, certain items of food are preferred which evidence a deteriorative change. However, any foodstuff may finally reach a stage at which it is unacceptable. Prevention of deteriorative changes is the primary objective of food preservation.

The activity of microorganisms often produces desirable changes in the flavor or texture of a foodstuff. Thus decomposition may be controlled and directed to yield a product of distinctive character. The tenderizing of beef, the ripening of cheese, the fermenting of wine, and production of soya sauce are examples.

Occasionally, a food may become contaminated with a pathogenic organism. Although most pathogenic bacteria can sometimes be foodborne, any of the waterborne and milkborne diseases may be spread readily through food. Only a few are frequently transmitted by food and are classified as foodborne illness. These organisms are likely to contaminate food because of their origin in nature, the way the food preparation is conducted, and their ability to survive and grow in the food.

Food Spoilage

No food, whether in its natural environment or in its preparation for use, is free of microorganisms. Yeasts, molds, and bacteria are always present, and any one or all of these microbes may take part in food deterioration. The type of organism concerned in food spoilage is determined by the nature of the specific product. Likewise, the rapidity of spoilage is affected by the characteristics of the food. Yeasts, molds, and bacteria do not grow at the same rate under similar natural conditions of exposure, and an environment which favors one will rarely, if ever, equally favor the others.

All natural foodstuffs are acid in reaction. If the food is high in acid (pH 4.5 or less), virtually all bacteria will be inhibited since few can tolerate acid (Fig. 19.1). Yeasts and molds are most often encountered in spoilage of such foods. In some instances molds are more common, and in others, yeasts; in a few cases, yeasts, molds, and acid-tolerant bacteria may all occur. Microbial spoilage under acid conditions is chiefly a breakdown of nonproteinaceous constituents of the food. The food gives off the rather distinctive odors of fermentation and its texture is softened.

Foods low in moisture are subject to mold attack since molds generally grow at lower levels of moisture than do yeasts or bacteria. Thus fats and fatty foods, which have lower free water than do the protein or carbohydrate portions of plants or animals, may support only mold growth. Bacteria sometimes grow in microscopic droplets of water within a fatty foodstuff, but growth is rarely if ever as luxuriant as in food with more abundant moisture. Short-chain organic acids, such as butyric, may occur among the breakdown products and impart rancid odors and flavors.

Bacteria are most active in foods with a reaction nearer the neutral point (pH 5.5 to 6.5) and with a moisture content of 45% or more. They act chiefly on the protein component of the food. Protein-breakdown products may serve as flavor contributors when present in low concentrations, for example,

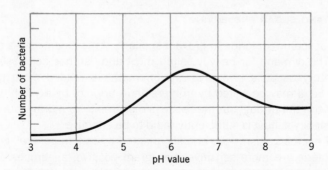

Fig. 19.1. A pH curve for a bacterial culture isolated from spoiled food. Food spoilage bacteria grow slowly at pH values below 5, and under such conditions they usually yield to yeasts and molds.

in the flavorful cheeses. However, when deterioration is well advanced, the nitrogenous- and sulfur-containing volatile end products render the food too disagreeable in odor for consumption. Changes in texture and a darkening in color characterize advanced decomposition.

Food Preservation

Fresh, succulent foods contain enzymes and carry microorganisms which cause spoilage unless their activity is restrained. Food preservation inhibits or delays the deteriorative changes attributable to these agencies without impairing the nutritive value of the food. The palatability of a foodstuff is important in preservation. Little is gained if preservation processes yield a product which is unsavory to the consumer. Food preservation, then, is designed to preserve the physical structure of the food in such a way that its nutritive value and overall quality are maintained.

Theoretically, any procedure, chemical, physical, or biological, which will interfere with microbial activity is adaptable to food preservation. In actual practice, not all are. A substance such as formaldehyde, for example, interferes with microbial activity but does not meet the minimum requirements for an acceptable food preservative.

Under any method of preservation, food must be handled carefully to keep the load of contaminating microorganisms at as low a level as possible. Practical food preservation operations are geared, at best, to moderate contamination and may fail if the contamination is excessive or includes unusually resistant organisms. The useful life of many foodstuffs would be measured in days if it were not for the intervention of man. The procedures which man has found suitable for extending storage life include: (1) salting and sugar preserving; (2) pickling and fermenting; (3) dehydration; (4) refrigeration and freezing; (5) chemical preserving; (6) canning; and (7) radiation sterilization.

SALTING AND SUGAR PRESERVING

The use of sodium chloride to prevent spoilage of meats antedates recorded history. The demand for heavily salted meat and fish has decreased in the past century because new preservation methods yield a more palatable product. Food may be salted by immersion in brine or by dry salting. Distribution of the salt throughout the tissue is most rapid from brine, but penetration during dry salting is sufficiently rapid to be effective.

Not all microorganisms are equally affected by salt. In general, gram-negative bacteria are more sensitive than gram positive; anaerobes are more sensitive than aerobes or facultative forms, and rods are more easily inhibited than cocci. Most bacteria are restrained by about 5 to 15% salt concentration, some being inhibited by as low as 3% salt. A few, however, grow in salt up to saturation. Yeasts and molds may grow at relatively high salt levels (greater than 15%) if conditions are otherwise suitable. As a food preservative, salt is generally a bacteriostatic agent, not a bactericide. Although the action of sodium chloride is not fully understood, it results in part from osmosis or dehydration. Interference with enzyme action, especially proteolytic enzymes, prevents putrefaction and the accompanying evil odors associated with it.

Chiefly meats and fish are preserved by salting. Salt-tolerant microorganisms, some of which are pigmented and some luminescent, may cause deterioration of properly salted foods. Frank spoilage of heavily salted foods is uncommon.

Salted foods are often smoked. This is done now for flavor but was originally an additional method of preservation. Smoke deposited on the surface of meat adds pyroligneous acid which has some bacteriostatic and fungistatic action. One of the useful side effects of the smoking process is to dry the outer layer of the flesh, thereby preventing the growth of putrefactive bacteria. Modern taste in the United States rejects heavily smoked and salted foods and, except for such specialty items as "country" ham and bacon, the usual preserved pork products must be refrigerated to prevent spoilage.

The most important use of sugar in the food industry is to improve flavor and maintain good appearance in fruits. In high concentration, it acts as a preservative in such products as fruit jams and jellies and sweetened condensed milk. The storage life of the heavily sweetened product is greatly extended over that of the same material without added sugar, but slow growth of yeasts (Fig. 19.2) and molds at sugar concentrations up to as much as 70% can eventually cause spoilage. Therefore, mild heat treatment is usually employed to destroy these low-resistant forms in sugar preserving. The finished product, of course, must be protected against contact with the air to prevent reinfection and to inhibit the growth of obligate aerobes such as molds.

FERMENTATION AND PICKLING

The storage life of some foodstuffs is considerably extended by creating a pH in the food which is unfavorable for the growth of most spoilage-causing bacteria. In pickling, a preformed edible acid, such as acetic acid in the form of vinegar, is added to the food; in fermentation, the food carbohydrates are acted on by microorganisms yielding lactic acid, which gives the final product a desirable flavor and texture. Minor changes in the fat and protein components of the food during fermentation also may contribute to the flavor.

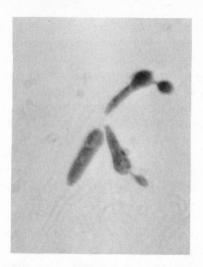

Fig. 19.2. A yeast isolated from fruit preserves where it caused spoilage of the highly sugared product.

The organisms which ferment most foods are members of the family *Lactobacilliaceae* (Fig. 19.3). These bacteria are somewhat tolerant of both salt and acid. Thus, salt added to a food material to be fermented inhibits most of the contaminating organisms, especially the proteolytic ones, and selectively favors the growth of the lactobacilli. Acid production renders the food unfavorable for the growth of all except acid-tolerant microorganisms, chiefly

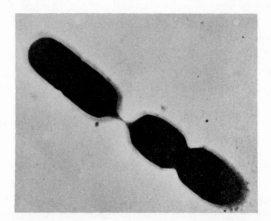

Fig. 19.3. Electron micrograph of the gram-positive, nonsporeforming, acid-producing rod, Lactobacillus acidophilus. Like all lactic acid bacteria, the lactobacilli are weakly or not at all proteolytic and are relatively tolerant of acid. Photo by Mudd, Polevitzky, and Anderson: J. Arch. Pathol., **34**, 199, 1942, courtesy A.S.M. LS-78.

yeasts and molds. Through its osmotic action, salt extracts from the food the cell fluids containing sugar and other soluble cell constituents. The sugar is acted on by the microorganisms and broken down to various products of metabolism. The brine solution is diluted by the fluid drawn from the tissue cells. Consequently, close control of the salt level during the early days of a fermentation is essential to maintain its selective action.

Vegetables which are preserved by lactic fermentation in salt brine include cabbage (as sauerkraut), cucumbers (as salt stock for pickles), okra, corn, green beans, and olives (Table 19.1). Fermented dairy products include butter and cheese.

Silage is an important fermented animal food, especially for dairy herds. Silage is produced from a variety of such agricultural plants as young corn, maize, or pea vines. The fermentable carbohydrate content of the plant material that is being made into silage is sometimes fortified by the addition of agricultural molasses, which produces greater acidity and, thus, better keeping quality.

In some products (such as catsup) in which acidity has an important role, a preformed acid, usually acetic acid in the form of vinegar, is added instead of producing acid in the food by fermentation. Citric acid is also extensively used in the food industry.

Undesirable contaminants may interfere with the fermentation and cause serious damage in instances where the proper salt concentration has not

Table 19.1 Some Natural Food Fermentations

	Sauerkraut	Pickles	Olives
Raw Material	Cabbage	Cucumbers	Green Olives
Preparation	Cabbage is shredded and packed in salt, and weights are applied. The action of the weight and salt withdraws juice from the cabbage.	Fresh cucumbers are packed in tanks and covered with brine.	Green olives are washed first in dilute lye to remove some bitterness and then in water. Washed olives are packed in oak barrels, and brine is added at once.
Salt Concentration	2.5%	10-20%	12%
Organisms involved in fermentation	Leuconostoc mesenteroides Lactobacillus brevis L. plantarum L. pentosus	Lactobacillus plantarum	Lactobacillus pentosus or L. plantarum or L. brevis

been maintained. Spoilage of the finished product by acid-tolerant organisms will occur if the food is not protected against contamination. If such organisms use much of the acid of the food in their growth, the pH becomes favorable for the growth of a variety of bacteria, and decomposition results. Encapsulated variants may produce a slime in the fermented food product and pigmented organisms may produce unsightly spots on the food.

DEHYDRATION

Microorganisms require a certain level of moisture for growth — about 40% for bacteria, probably about the same for yeasts, and substantially less for molds. The keeping quality of a foodstuff, obviously, is closely associated with its water content. Dehydration as a means of food preservation is based on reducing the water content of the food below the level necessary for the growth of microorganisms.

Dehydration is, perhaps, the oldest form of food preservation. Such foods as peas, beans, rice, and cereal grains are preserved naturally by water loss with maturity. At some prehistoric time, man began consciously to preserve fruits, vegetables, and meats by drying. In warm dry climates, foods were dried by exposure to the sun's rays. In humid regions, fire and smoke were used. Each method is still used today, both commercially and privately.

Theoretically, dehydration has much to offer as a method of food preservation. Reduction of bulk by water removal at the place of production, and restoration at the place of consumption, permits tremendous savings in transportation and warehousing. Dehydration has been successfully and extensively used for many years for such foods as fruits and fish. Considerable manufacturing use has been made of dried onion flakes, garlic, and chili peppers in preference to the fresh product. Modern procedures of **lyophilizing** (drying from the frozen state) are yielding more acceptable products, such as dried milk and potatoes, and more convenient products, such as instant coffee and tea.

PRESERVATION AT LOW TEMPERATURE

Storage at low temperature is an old and effective method of food preservation, either by frozen storage at a temperature well below freezing, or by refrigeration at a temperature near, but still above, freezing. A low temperature prevents, or retards, the growth of many spoilage microorganisms. Since enzyme action continues slowly at low temperature, those foods which are subject to enzymic deterioration must be blanched before freezing to inactivate the enzymes.

Development of the frozen food industry has been encouraged by the introduction of mechanical refrigeration into the market and home. Frozen foods are not sterile but do have a long storage life if held in the frozen condition. Once food is thawed, the cellular juices leak from the frost damaged cells and the contaminating bacteria resume activity and quickly cause spoilage unless arrested by cooking.

There are two methods for preparing frozen foods. Quick freezing is accomplished in 30 minutes or less by blowing air at −34 C over the food. In slow freezing, greater deterioration of vitamins takes place; autolysis is not immediately stopped, and the formation of larger ice crystals results in greater tissue damage.

PRESERVATION BY CHEMICALS

In preservation by a chemical additive, an antimicrobial substance not toxic for the consumer is incorporated in the food. There are few such substances. Antioxidants, which do not have antimicrobial properties, are not ordinarily regarded as chemical preservatives. Salt, sugar, vinegar, spices, alcohol, or wood smoke have preservative action but are usually added for their flavor or sometimes to mask deteriorative flavors of the food.

Prior to the development of the Pure Food and Drug Act, a number of substances were used as food preservatives which are no longer permitted. Some of these, particularly boric acid, or borax, and salicylic acid, persisted in home use long after elimination at the commercial level. Chemical preservatives are never a substitute for care in selection, and cleanliness in preparation, of the foodstuff for processing. Use of chemical food preservatives is controlled by rigid specifications and supervision by federal and state agencies. Some of the prohibitions are probably more restrictive than necessary, but their enforcement is aimed at encouraging the rejection of inferior raw materials and discouraging sloppy and unwholesome methods of production.

Sodium and calcium propionates are widely used as mold inhibitors. The shelf life of bakery products can be extended by incorporation of propionate in the mix. Surface molding of moist packaged commodities can be restrained by wrappings which have been dipped in propionate solution. Sodium benzoate is permitted in salad dressings and related products. Sorbic acid is an effective inhibitor of yeasts and molds (Table 19.2).

Some broad-spectrum antibiotics have been approved as a dip for poultry, and their use may in time extend to other foodstuffs which require cooking before use. The shelf life of poultry under refrigeration can be extended several days by antibiotic dipping. Close control of the operation is essential. If drug-fast variants of spoilage significance develop in the dipping vat, the procedure will, of course, cease to be effective. Concern for undue prolifera-

Table 19.2 Chemical Preservatives Currently in Use in the United States

Chemical	Examples of Foods Where Used
1. Sulfur dioxide	Dried fruits, fruit juice
2. Ethylene oxide	Fruits, spices
3. Smoke	Meat, fish
4. Propionic acid and propionate salts	Bread, cake, pie crust, figs, berries, food wrappers
5. Benzoic acid and benzoate salts	Beverages, relishes, jams, ice for fish storage
6. Sorbic acid	Syrups, cheeses, margarine, salads, candy
7. Sodium hypochlorite	Washing fruits and vegetables
8. Broad-spectrum antibiotics (Chlorotetracycline, etc.)	Poultry, fish

tion of bacteria resistant to those antibiotics has prompted many to question this method of preservation.

Gaseous ethylene oxide is very effective in the sterilization of foods, such as spices, which cannot be treated satisfactorily by other methods. The food material is sealed in a chamber; a vacuum is created and released either with pure ethylene oxide or with ethylene oxide mixed with carbon dioxide to reduce the explosion hazard. Sulfur dioxide treatment of dried fruits eliminates insect infestation as well as mold growth.

CANNING

In the canning process, food is preserved by heat sterilization in a hermetically sealed container. Heating destroys the organisms present in the container; hermetic sealing prevents the entrance of others. The combination of time and temperature required to destroy the most heat-resistant spoilage microorganism for a particular food is termed the **process** for that food. Processes vary, depending on the chemical and physical natures of the foodstuff and the size of the container. Acid foods, such as fruits and tomatoes, require only mild heating to destroy yeasts, molds, and acid-tolerant bacteria. Low-acid foods, such as corn and peas, require a process severe enough to destroy heat-resistant spores. The sterilization process requires more heat and/or time than is necessary to cook the food to edibility. The nutritive value of a food is not affected by processing, although the flavor and texture may be. Just as the water purification process is aimed at *E. coli* and the pasteurization of milk is aimed at *Mycobacterium bovis,* so the canning processes are aimed at the destruction of the spores of *Clostridium botulinum.* Adverse publicity accrues to the canning industry from an outbreak of botulism and this accounts for the overcooking commonly found in canned foods.

Of the three main types of spoilage due to underprocessing, the most frequently observed is the **flat sour,** in which acid without accompanying gas is produced by *Bacillus stearothermophilus.* Since gas is not formed, the container is flat, hence the descriptive name. This spoilage is most commonly encountered in low-acid foods, such as peas and corn (Table 19.3).

The second most common cause of spoilage is *Clostridium thermosaccharolyticum.* Since this organism produces large amounts of gas as well as acid from the food sugar, the container becomes swollen (**hard swell**) and frequently bursts. This spoilage occurs in semiacid foods, such as greens and asparagus, as well as in low-acid foods.

Both types of spoilage are caused by a thermophilic organism, and neither presents any hazard to health. The large amount of lactic acid which is formed as the major metabolic end product renders the food too sour to eat. Microscopic examination of the spoiled food shows large numbers of bacteria, but cultures will not yield growth unless made early. This condition of high acidity which quickly kills out the bacteria is known as **autosterilization.** Occasionally, contamination with heat-resistant spores of mesophilic soil anaerobes may cause the third type of spoilage which is characterized by a swollen container and putrefactive decomposition of the food.

In all cases of underprocessing the causal organism is present in pure culture, and the food is uniform in appearance and odor. However, when spoilage is caused by container leakage, the food is quite varied in appearance and odor and a variety of organisms are present, including nonspore-forming types such as cocci.

RADIATION STERILIZATION

Ultraviolet radiation is used to control surface contamination of some foods and to sterilize sugar under special conditions. In recent years other types of radiation have been used experimentally. Perfected methods of radiation sterilization would avoid the difficulties of handling food at low temperature and might have some advantages over thermal processing.

Sterilization can be effected by exposure to a beam of electrons, beta ra-

Table 19.3 Microbial Spoilage of Canned Foods

Organism	pH Range	Type of Change
Bacillus stearothermophilus	5.3 and above	Flat sour
Bacillus thermoacidurans	4.2 and above	Flat sour
Clostridium thermosaccharolyticum	4.8 and above	Acid and gas (hard swell)
C. nigrificans	5.3 and above	Putrefaction, blackening
C. sporogenes	Above 4.5	Putrefaction, blackening
Yeasts	3.7 and below	Yeasty odor

diation, or by exposure to electromagnetic radiation in the form of gamma rays. Beta rays do not penetrate deeply and hence have little importance for food treatment. Gamma rays, which are more penetrating, have a limited practical use. Unfortunately, some foods develop disagreeable flavors when treated sufficiently to inactivate the food enzymes and kill contaminating microorganisms. Effective radiation doses are now known, that is, those ranges necessary for enzyme denaturation, sterilization, pasteurization, and reduction of total bacterial population. However, much more research is necessary before radiation sterilization can be widely used.

Illness Resulting from Food

Bacteriologists use the term **food poisoning** to describe illness arising from the consumption of (1) food infected with certain types of living bacteria, or of (2) food in which certain bacteria have grown and produced toxic substances (Table 19.4). Illness following the consumption of poisonous plants, fungi, fish and shellfish, or food accidentally or maliciously contaminated with toxic chemicals are not included.

Although food may serve as a vehicle for the transfer of almost any pathogenic microorganism, food poisoning is limited to illnesses caused by a few microorganisms. The chief offenders, in decreasing order of frequency of occurrence in the Western World, are: *Staphylococcus, Salmonella,* and *Clostridium botulinum.* Before the true relationship of bacteria to food poisoning was recognized, illness resulting from unwholesome food was attributed to basic nitrogenous products of protein breakdown known as **ptomaines.** It is now accepted that food illness is not due to ptomaines and that the term "ptomaine poisoning" is a misnomer. Ptomaines are not toxic when taken by mouth, even in high concentration. In low concentration, they serve as flavor contributors in such foods as ripened cheeses.

Bacteriological examination of foods suspected of having caused food poisoning sometimes shows large numbers of one or another organism of uncertain etiological relationship to the illness. These are *Pseudomonas, Proteus,* alpha-hemolytic streptococci, molds, coliforms, or sporeforming rods. The ingestion of large amounts of microbial protein causes a minor digestive upset in some persons and especially in young children. Consequently, baby food is handled with extreme care to exclude even the nonpathogenic bacteria and to minimize any microbial growth. But, except in infants, no one of the organisms cited is significant as a cause of food poisoning. In the entire United States only a few cases of botulism occur each year but the high mortality associated with the disease has marked it for special concern. Most instances of food poisoning in this country involve the staphylococci or the salmonellae.

Table 19.4 General Outline for Laboratory Diagnosis of Food Poisoning

Food Sample

→ A

→ B

→ C. (Special procedures for identification of botulinum toxin)

A

Direct microscopic examination: estimation of contamination load, types of organisms

B

Enrichment broth →

Plating on differential and/or selective media →

Incubation 20-24 hr at 37 C →

Isolated organisms:

Biochemical tests
Phage typing (*Staph.*)
Serotaxonomy (*Salmonella*)

Animal injection or feeding (used primarily in diagnosis of botulism), e.g., Dolman Kitten Test for *Staph.* enterotoxin

C. (Special procedures for identification of botulinum toxin)

Animal feeding or injection

1. Test animal unprotected

2. a. Control animal heated food sample

   b. Control animal protected with specific antitoxin →

Note condition of animals:
Development of flaccid paralysis
Death

→ Enrichment cultures for *Clostridium botulinum*

→ Proceed with animal tests

Food is contaminated with staphylococci by food handlers with boils, cuts, or sores on the hands or fingers, or by handlers who sneeze or cough frequently. A typical staphylococcus food poisoning incident involved a food handler with a low grade throat infection with *Staphylococcus aureus* (she was recovering from a cold) who spent a warm afternoon separating left-over meat from the bones and fat, grinding and mixing it into a meat salad, and preparing sandwiches. During this time a few staphylococci contaminated the preparation; this would have been of no consequence at all except for the fact that before the task was completed the food had been incubating at a warm temperature for many hours which resulted in a massive growth of the staphylococci in the meat salad. The food-poisoning strain of *Staphylococcus* excreted a powerful exotoxin which attacks the enteric (digestive) tract and is therefore called an **enterotoxin.** Within one to five hours after the sandwiches were consumed symptoms of severe nausea, vomiting, abdominal cramps and diarrhea appeared. Except for the aged and the debilitated, persons almost never die of "staph" food poisoning. Recovery begins after a few hours and is usually complete in a day. The toxin is produced at temperatures between 15 and 50 C and is fairly resistant to heating. Not all strains of staphylococci produce toxin; it is limited to coagulase-positive strains and usually to those that liquefy gelatin. The staphylococci that are the normal residents of the healthy skin are usually gelatinase and coagulase negative so regular inspection of food handlers is required to restrict only those with colds or skin lesions from contact with foods susceptible to staph toxin production.

Protection against *Salmonella* infection of food is not so simple. *Salmonella* carriers can be found among healthy food animals, chiefly swine and poultry, and *Salmonella* infection of eggs is common. Thus, naturally infected food, which cannot be differentiated from noninfected food, may be encountered. *Salmonella* may be introduced into food through poor personal sanitary habits of a carrier who is a food handler, and by rodents and vermin. *Salmonella typhimurium,* the causative agent of mouse typhoid, is a common offender. This and other species of *Salmonella* that are associated with animals are much less virulent for humans than are the salmonellae of human typhoid and paratyphoid fever. They cause distress lasting a day or two but only if ingested in massive doses. This means that they must grow in the food if they are to be present in sufficient numbers to produce disease. A typical *Salmonella* food infection is as follows. Chicken which has these bacteria on the inner surface of the body cavity due to rupture of the intestine at slaughter is prepared on the kitchen work table and then cooked thus destroying the bacteria on the meat. When the cooked chicken is repacked into another container it is contaminated by a few salmonellae surviving on the table top. This is of no consequence from the health viewpoint except that the warm food serves as an ideal growth medium so that if several hours elapse before the cooked chicken is eaten it will then contain millions of salmonellae. When the organisms reach the intestine of the consumer they continue to make some growth but the endotoxin released from the digestion of the bacterial cells irritates the lymphoid tissue of the intestine and a low grade fever develops which is accompanied by violent diarrhea. This usually occurs six to 12 hours after eating and recovery is complete in two to four days. The illness is really a food infection but it is generally called a food poisoning because of the endotoxin. It is a common disorder but, because of the short duration of the illness, and because the death rate is almost zero, most cases are never reported in the public health statistics on disease. Other foods that may be involved include those that are not eaten promptly after cooking, especially if there is opportunity for contamination by animal strains of salmonellae, followed by an unrefrigerated incubation period to permit the necessary growth. Complete reliance cannot be placed on cooking as a sterilizing measure, but thorough cooking will serve to reduce the contamination even if it does not eliminate it.

Foods, which are "fingered and fussed over," such as croquettes, filler for cream-filled pastries, and salads that are prepared well in advance of the meal and not stored properly either before or after cooking are responsible for most cases and outbreaks of *Staphylococcus* and of *Salmonella* food illness. Leftover foods are potential sources of food poisoning even when handled with great care and should not be used, especially for mass feeding, if they have remained unrefrigerated for even a few hours. In institutional food preparations the factor that is often neglected is that food stored in large bulk requires many hours to cool to the center even if placed in the refrigerator promptly. In such cases the bacteria can produce massive growth in the warm part before the food finally cools to below the minimum temperature for growth or toxin production.

Botulism is limited to foods that provide an anaerobic environment with a neutral pH, that have access to contamination with soil or with the bottom sediments of lakes, and that have not been heated adequately to destroy all of the very resistant spores of *Clostridium botulinum*. A typical case of botulism involved home-canned string beans. Because garden soil is imbedded in the crease in the side of the bean pod it is almost impossible to wash out completely this source of spores of *C. botulinum*. The uncontrolled heating in the home canning failed to kill all the spores. The can was packed closely with solid material so the heat penetrated only slowly to the center of the can. The cans were then stored for months. This permitted the germination of some of the surviving spores and they grew in the liquid, which in the case of beans is of neutral pH and rich in proteinaceous food. On opening the can the housewife noticed a slightly unusual appearance of the food, but she tasted it and found it acceptable. She therefore heated the beans and served them to the family. After about 24 hours she developed symptoms of headache and double vision followed by progressive paralysis of the central nervous system and death during the third day. The family ate only the heated food and showed no symptoms at all. The powerful neurotoxin is very heat labile, being destroyed by boiling for 5 minutes. Since it is not formed in acid foods, most fruits and juices, many acid vegetables, and pickles will never develop toxin regardless of how highly contaminated the food may be. Meats and fish products have been implicated, and the earliest recorded case involved a sausage. Commercial food processing in the United States has been adequate so that in the last 50 years only a few cases of botulism have occurred that have been traced to that source.

The mold, *Aspergillus flavus,* produces a toxin, called **aflatoxin,** which has a cumulative effect producing tumors on the liver, especially in young children. In certain populations in Africa where the people subsist largely on tubers or groundnuts and the moist climate is conducive to mold growth, there have been disastrous epidemics that resulted from consumption of this moldy food. The mold, *Claviceps purpurea,* also called ergot, grows on rye and other grains and produces a drug that stimulates the contraction of smooth muscle. In large doses it paralyses the sympathetic motor nerve endings and, in countries where rye is used as a major food, ergot poisoning is common. In this country, deaths are reported in cattle that are fed moldy hay, but there has been no clinical evidence of human disorders from mold toxins. This is probably due to the care used in the commercial preparation of food and to the rejection of materials showing incipient spoilage.

Summary

Microbiology in the food industry is involved in food spoilage by bacteria, molds, and yeasts. It is also concerned with food and utensils as vehicles for carrying pathogens from the sick to the well. The illnesses which result from eating foods that have been made toxic by microbial growth are referred to as food poisoning. In food processing where microbes modify flavor and textures, the microbiologist must inhibit the undesirable organisms and promote the ones that are useful to his purpose. In most of these processes, the organisms or sequences of organisms are well known, and many processes are carried out by selected strains cultured in the microbiology laboratory.

TWENTY

Industrial Applications of Microbiology

The development of microbiology into an important scientific discipline is a consequence of investigations initially concerned with infectious diseases. The studies of pathogenic bacteria led to a better understanding of the activities of microorganisms in general and, specifically, of the ones concerned in such historic processes as the leavening of bread, the retting of flax, and the manufacture of wine and beer. Modern microbiology has improved these processes and developed many others in which the activity of microorganisms is economically significant. This aspect will be the subject matter of this chapter. Obviously, many of the microbiologically trained employees in industry work at the same problems that confront persons in hospitals, universities or public research institutions. Others work on fundamental biological problems, the understanding of which may be essential for the industrial goals. Many microbiologists are employed by industry to put microbes to profitable use.

Several factors determine whether a microbiological industrial process is practicable and profitable. Organisms may be required which will produce a desired end product by their metabolic activity. Sometimes this is a fairly easy task, but other times the search may be long and tedious. Even if an organism is found, the yield of end product may be too low to permit its profitable use. In such cases, "training" the organism, or mutating it by exposure to a mutagen such as ultraviolet light, may succeed in developing a high-yielding strain. The medium must be studied to determine optimum conditions of aeration, concentration of medium components, pH, and temperature. Successful fermentation technology is the proper manipulation of the regulating mechanism of the cell to yield the maximum of the desired product. The "survival of the fittest" doctrine which governs "wild type" situations does not apply because the culture is protected from competition with contaminants. Therefore, high-yield mutants are employed under growth conditions modified for optimum yield, without concern for the survival qualities of the inoculated strain.

Industrial microbiology can be divided into several categories:

1. The production of microbial cells or cell material.
2. The production of fermentation products.
3. The modification of materials to make more valuable products.
4. The use of microorganisms to eliminate undesirable materials.
5. The deteriorative effects of microorganisms as they affect industry.

THE PRODUCTION OF MICROORGANISMS

Yeasts are produced in large quantities for use in leavening bread. A pure culture of the selected yeast strain is transferred from a test tube to a flask and, after good growth is obtained, this inoculum is "built up" by transfer to larger volumes of broth until sufficient inoculum is obtained for a large fermentation tank. Great care must be taken to exclude contamination at all stages in the process. The yield of cells is favored by cultivation under aerobic conditions. The crop of cells is harvested by centrifuging and filtering and is either sold as a moist cake, dried and sold as dry yeast cake, or processed to yield an effective complex of enzymes.

Yeasts may also be cultivated for consumption as food. Yeast cells are rich in most of the nutrients that are required for animal nutrition, and special strains are selected for their high vitamin content. Enormous quantities of yeasts, produced in the brewing of beer, are reclaimed and processed into animal food. In the manufacture of paper, wood chips are cooked in sulfite liquor to extract the soluble constituents of the wood; *Torula utilis,* a yeast with a satisfactory flavor, has been grown in such waste sulfite liquor and harvested for food use. The flavors of some of these products could restrict them from the human food market but the high-quality proteins and the abundant vitamin content of microbial cells make them especially valuable as animal food supplements. Reports indicate that success has been attained in the economic production of microbial cells in a medium where waste petroleum products are supplied as the source of carbon. The culture of algae for food, employed commercially in Japan, has received extensive study. This work has been stimulated by space research; algae could be especially valuable in interplanetary voyages since they require no energy other than sunlight and they use up the respired CO_2 of the spacemen, yielding oxygen as well as edible cells in return.

Rhizobium strains or species which live in the roots of legumes fix nitrogen but are specific for that particular plant and are usually present in the soil only in low numbers. Furthermore, these organisms in their natural environment are often inefficient fixers of nitrogen. Hence, the inoculation of seed of various legumes with selected, highly efficient strains of *Rhizobium* increases yield and is commonly used wherever a legume is grown as a main crop. Industrial laboratories devoted to the production of cultures of *Rhizobium* are an important adjunct to agriculture.

Laboratories for the production of starter cultures for the manufacture of dairy products are of great industrial importance. Such cultures are used in the manufacture of buttermilk and the special types of soured milk (such as acidophilus and yoghurt), in the ripening of cheese, and in the ripening of cream for processing into sour cream butter. The organisms used are strains of *Lactobacillus* and *Streptococcus* that are selected to contribute the desired characteristics of flavor, odor, and texture to the final product.

One important phase of the biological industry is the mass cultivation of organisms to be used in vaccines. The bacteria may be grown in the usual way, as in the preparation of the vaccine for prophylaxis against pertussis or the vaccine against typhoid and paratyphoid fevers. Viruses and rickettsiae for processing into antiviral vaccines can be grown in fertile eggs (Fig. 20.1), in tissue culture, or in animals. Toxins and toxoids are prepared for prophylaxis and for diagnostic reagents. All such products must be tested for effectiveness and for safety. The tests usually involve the use of great numbers of laboratory animals that must be challenged with a virulent strain following vaccination to make sure that the product induces a useful level of immunity. Conditions for manufacture are carefully controlled. The expanded use of biologicals in veterinary medicine has opened a huge market in that field. Recently biological insecticides have been produced by mass cultivation of microorganisms. Sprays of *Bacillus thuringiensis* are very effective against the alfalfa caterpillar, the tobacco hornworm, and the worm variously known as the corn earworm, the cotton bollworm, and the tomato worm. The bacterial insecticide is harmless to the plant and to other animals, including humans. Because of this selectivity, such biological insecticides would be welcome replacements for the nonselective chemicals such as D.D.T.

THE PRODUCTION OF FERMENTATION PRODUCTS

The fermentations industry is the branch of industrial microbiology that produces useful products through the action of microorganisms. These processes may involve carbohydrate breakdown to produce the solvents, alcohol, acetone, and butanol; or the acids, lactic, citric, gluconic, and acetic; or the sugar, sorbose. Similar processes yielding the vitamins, riboflavin and cyanocobalamin (B_{12}); the amino acids, lysine and glutamic acid, as well as

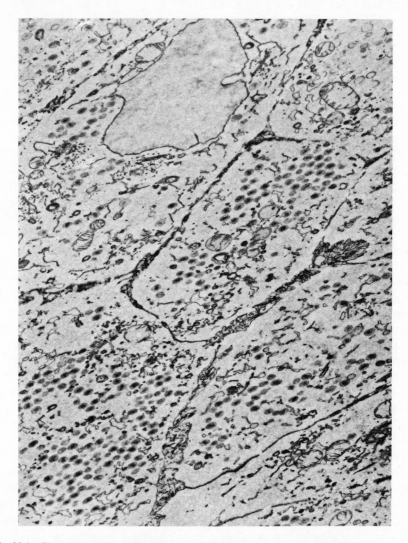

Fig. 20.1. The cells from the chick embryo shown in this electron micrograph are heavily infected with the virus employed for making smallpox vaccine. Photo by N. McDuffie.

those yielding the antibiotics, require specialized media for the formation of these nitrogen-containing substances. Many other products can be produced by fermentation, and this list will be expanded to include new products as new processes are developed. Some products will be removed from the list as the chemists work out methods for synthesis that are cheaper than production with microbes. But the versatility of the microbe, especially in the production of complex molecules will probably result in an expanding list of products produced by fermentation.

Alcohol fermentation, wherein the yeast *Saccharomyces cerevisiae* converts sugar almost quantitatively to ethyl alcohol and carbon dioxide, is still the largest volume process, although in the United States about 75% of the industrial alcohol (this excludes beverage alcohol) is produced from petroleum by chemical synthesis. Fermentation alcohol production may utilize a cheap sugar source, such as molasses, or it may employ grain in which the starch is first converted to sugar by amylase from sprouted barley (malt). Some alcohol is produced through fermentation of the paper mill wastes (the sulfite liquors), which contain sugars extracted from wood pulp. In northern Europe cellulose waste, such as sawdust, is utilized following conversion to sugar by boiling with mineral acids.

In all these processes, a mash is made by adding such yeast nutrients as ammonia and phosphate. The pH is adjusted to about four to inhibit most bacteria and, since the fermentation is an anaerobic process, molds fail to develop. Furthermore, the large inoculum of a yeast strain, selected for rapid development in the mash, outgrows any contamination, and a successful fermentation is complete in two days without costly sterilization of the mash. Cooling coils in the tanks remove the heat produced by the fermentation and maintain the temperature at about 25 C. The alcohol is recovered by distillation and residual "slops" may be concentrated for feed or fertilizer.

The butanol-acetone fermentation was developed during World War I, based on research done by Chaim Weizmann, a scientist-politician who became the first president of Israel. The sporeforming anaerobic *Clostridium acetobutylicum* is inoculated into a mash prepared by cooking 10% ground corn in water. No malt is necessary since *Clostridium acetobutylicum* produces its own starch-digesting enzyme. The pH at the beginning of the fermentation is about six and at the end it is about four. Calcium carbonate is added, if necessary, to neutralize excess acid. The temperature is maintained at 37 C, and after 48 hours the solvents are separated by fractional distillation. From 100 pounds of corn, the yield will be about 16 pounds of butyl alcohol, seven pounds of acetone, and two pounds of ethyl alcohol. When molasses is used as a substitute for corn, other strains of clostridia are employed.

Although many lactic acid bacteria (streptococci and lactobacilli) and molds of the genus *Rhizopus* yield lactic acid almost quantitatively from sugar, most fermentation lactic acid is produced anaerobically using strains of *Lactobacillus delbrueckii*. The sugar mashes are adjusted to a pH of five and the temperature is maintained at 45 C. The organism requires a rich medium containing vitamins and amino acids. The acid is neutralized with calcium carbonate and when the fermentation is complete, the liquor is concentrated by evaporation to precipitate the crude calcium lactate. Edible and pharmaceutical grades of lactic acid are purified from this crude product.

Aerobic fermentations usually involve incomplete oxidations. A number of these processes involve *Acetobacter suboxydans,* which carries out restricted oxidations to give useful products. Among the industrial processes now in use, this *Acetobacter* converts glycerol to dihydroxyacetone, a substance widely used in suntan lotions. It will also convert sorbitol to sorbose, a sugar subjected to further chemical manipulations to produce vitamin C (ascorbic acid). *L. delbrueckii* has been used to ferment glucose to gluconic acid and to 5-ketogluconic acid, although the gluconic acid is produced more efficiently by molds of the genera *Penicillium* and *Aspergillus.*

The partial oxidation of ethyl alcohol to acetic acid by the highly aerobic gram-negative rod, *Acetobacter aceti,* is the basis of vinegar production. The commercial vinegar fermentation is usually carried out in a generator where a 10% alcohol solution is sprayed over and percolates through a 10-foot layer of wood shavings. A countercurrent of air is pumped up through this container to supply the oxygen to the *Acetobacter,* which grows as a gelatinous zoogeal mass attached to the wood. The raw material is usually hard cider or wine, but diluted alcohol, enriched with peptone or other organic material to provide a nitrogen source, may be used and stimulates the growth of the organism.

Citric acid was once obtained from cull citrus fruits. Now most of it is produced by *Aspergillus niger* growing in shallow uncovered pans on a 15% sugar solution, adjusted to about pH two and enriched with ammonium and mineral salts. Because of the simple medium, the high sugar content, the low pH, and the large inoculum, contamination is not a problem. The mold forms a pad on the surface; after almost quantitative conversion of the sugar to citric acid, the medium may be replaced and the mold pad may ferment a new batch of sugar. The citrate is recovered from the broth by precipitating it as a calcium salt. The fermentation product is believed to result from a defective respiration involving a failure to convert citrate efficiently to the next compound in the citric acid cycle. Therefore, citrate accumulates in the cells and is excreted into the medium. The mold used is a strain selected for this defect, or the medium may be made deficient in minerals so as to cause the failure in further metabolism of the respiratory intermediate. This illustrates our earlier statement that fermentation technology requires an understanding of the regulation of metabolism in the cell.

The production of vitamins, amino acids, and antibiotics involves moderate pH and rich nitrogenous media wherein contamination cannot be tolerated. The problems of maintaining sterility in this industry have taxed the ingenuity of fermentation engineers (Fig. 20.2). These products are usually produced in small amounts in the broth, and methods of recovery may involve countercurrent extraction with solvents, adsorption on colloidal substances such as charcoal, as well as precipitation as insoluble salts.

Fig. 20.2. Filtration. When the yield of fermentation products is at its peak in the huge fermentor, the contents of the fermentor are drawn off and passed through rotary vacuum filters, where the mycelium is filtered off. This is the first step in the recovery of the fermentation product from the culture liquor. Photo courtesy Pfizer, Inc.

Vitamin B_{12}, a chemical containing cobalt, is produced in a cobalt-enriched organic medium by a wide variety of microbes. The processes actually employed by the industries are often secret, but it is known that strains of *Propionibacterium freudenreichii, Bacillus megaterium,* and *Streptomyces olivaceous* produce high yields of B_{12} under industrial procedures. Riboflavin for the pharmaceutical industry is produced mainly by chemical synthesis, but that used for animal feed is produced by fermentation. The mold, *Ashbya gossypii,* or the yeast-like *Eremothecium ashbyii* is inoculated into a sterile medium containing approximately 1% of a crude sugar, 1% corn steep liquor, 0.5% of some peptone derived from an animal source, together with an antifoaming agent. After aeration for four days the broth, together with the cells and their products, are dried for incorporation into animal feed; such concentrates may contain up to 2.5% riboflavin.

Lysine is an amino acid which is often deficient in the foods of people living on a restricted diet. A mutant of *Escherichia coli* grown under suitable conditions will release into the medium large quantities of the distinctive bacterial cell wall constituent, diaminopimelic acid. Removal of a carboxyl group from diaminopimelic acid yields lysine; this can be done by an enzyme produced by *Enterobacter aerogenes.* Glutamic acid is produced by many organisms in a glucose-ammonia medium that is rich in solubilized protein, but a successful commercial process employs the gram-positive nonsporeforming short rod, *Brevibacterium luteum.*

Currently, the antibiotic, chloromycetin, is produced by chemical synthesis; all others are produced by fermentation. Successful commercial production of antibiotics (Chapter 9) depends on the selection of high-yielding strains, the utilization of optimum conditions of growth (including nutrition, temperature, pH, and degree of aeration), the utilization of efficient methods of extraction, and careful testing for efficacy and safety. The search for strains producing new products or higher yields of accepted products is never-ending, and this involves not only the global search for new wild-type strains but also the application of mutagenic procedures to produce strains that never existed before. The medium usually employs corn steep liquor plus carbohydrate, peptone, buffer, antifoam, and mineral enrichment. Many of the details are trade secrets.

MICROBIOLOGICAL MODIFICATION OF MATERIALS TO MAKE MORE VALUABLE PRODUCTS

Microorganisms were used empirically to modify or transform material for centuries before the existence of microorganisms was recognized. Bread dough has been leavened since before recorded history. Bits of dough from one fermenting mixture were preserved and used to inoculate the next lot of dough. Today the manufacture of bread is no longer an art but an industry in which various steps are controlled on the basis of experimentally derived information.

Fibers of flax and hemp are separated for weaving into linen in the process known as **retting.** The fibers are held together in bundles in the plant stem by pectin, a cementing material. When the stems are submerged in water, the bundles are broken down by the enzyme pectinase, which is released by *Clostridium* species growing in the soaking vats. The individual fibers can then be recovered and processed.

Hides for tanning are prepared by removal of hair, blood, and as much other extraneous material as possible. They are then soaked in tanks where bacteria digest out materials which must be removed to yield a product which can be treated with the tanning chemicals to give a durable leather.

The manufacture of alcoholic beverages comprises one of the largest industrial applications of microbiological activity. The nondistilled alcoholic beverages produced in largest volume are wine and beer. Wine can be made from many fruit and berry juices, but most of it is made from grape juice.

Fermented apple juice gives hard cider, and fermentation of a solution of honey yields the historic beverage, **mead.**

The organism responsible for the wine fermentation is *Saccharomyces ellipsoides* (Fig. 20.3). The juice is treated with enough sulfur dioxide to inhibit undesirable organisms but not interfere with the growth of the yeast. Starter cultures can be added to the vat, or fermentation can be carried out by the yeasts naturally present on the fruit. Under the latter condition, the wines are more variable. If all the fruit sugar is used up, a dry wine is formed. If fermentation is stopped while some unfermented sugar remains, the product is a sweet wine.

Beer is prepared by the fermentation of grain carbohydrates with selected strains of *Saccharomyces cerevisiae.* The starch of grain to be fermented by yeasts is first hydrolyzed enzymatically. In the germination of barley, an abundance of amylase is formed. The barley is dried and ground to give malt. Malt is steeped with ground corn and rice to give wort, which is then fermented to give beer. Sake is a beer made from rice. The starch of rice is hydrolyzed to a fermentable sugar by fungal enzymes.

In recent years, steroid hormones such as cortisone, hydrocortisone, and

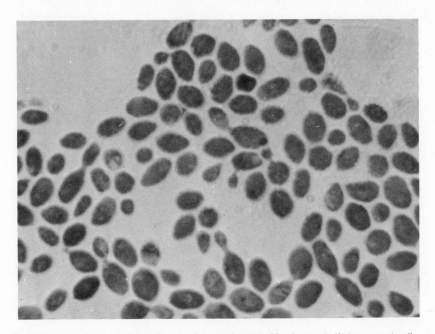

Fig. 20.3. A wine yeast. Although French wines are fermented by the yeasts that occur naturally on the grapes, American wines are often fermented by carefully selected strains cultured in the laboratories of the wineries.

several other related hormones from the adrenal gland have occupied a prominent place in the therapy of a number of organic diseases. The naturally-occurring hormones are difficult to recover from animal tissues and thus are too expensive for large-scale production. Synthesis likewise is difficult and expensive. A number of abundant, naturally-occurring steroids can be converted to the hormone structure in part by microbial action. The industrial development of this transformation of the steroid molecule has made cortisone and related hormones available at a price the average patient can afford.

USE OF MICROORGANISMS TO ELIMINATE UNDESIRABLE MATERIALS

Industrial wastes containing substances toxic to the bacteria concerned in sewage purification cannot be disposed of in the same system as the domestic wastes. Such wastes, however, may be purified under conditions favoring the growth of types of organisms which destroy the toxic waste material. Phenolic wastes are disposed of successfully by an activated sludge type of treatment. The organisms concerned are probably species of *Pseudomonas.* Dilute formaldehyde can be rendered harmless by bacterial action. Detergent wastes, introduced into sewage systems both by housewives and industry, pose a threat when present in any appreciable amount. These compounds are also susceptible to bacterial decomposition. Such chemicals as D.D.T. are only slowly degraded by microbial action and tend to accumulate in the environment. Although when the chemical reaches a sufficient concentration in the soil or water microbial populations will flourish and decompose it, some of these chemicals exert harmful effects on other living systems when present in concentrations too low to serve as a satisfactory microbial substrate level. Patents have been issued on the disposal of garbage by fermentation. The compost which results is a valuable soil conditioner. At the present time, investigations are in their infancy for improved microbial methods for solid waste disposal and the recycling of man's waste products in interplanetary space vehicles.

DETERIORATIVE EFFECTS OF MICROORGANISMS

Microorganisms are responsible for much economic loss from the deterioration of textiles, leather, paper, painted surfaces, rubber, concrete, and metals. The problems are often resolved by the incorporation of microbial inhibitors, but there are numerous situations where this is impractical. One of the most important organisms concerned in deteriorative processes is *Desulfovibrio desulfuricans.* This obligately anaerobic organism reduces sulfate to sulfide. The reduction process is extremely corrosive to all buried or submerged metal.

Summary

The problems of industrial microbiology are extremely varied. Some industrial microbiologists deal entirely with medical problems, for example, in the production of vaccines and other biologicals. Others are occupied in the production of useful chemical substances, and the microbes are used as an adjunct to the chemical industry. Their work is being extended to improving techniques for carrying out such historic processes as the retting of flax, tanning of hides, and brewing of beer. The development of new approaches in microbiology has revealed new methods for controlling deterioration. As mankind becomes more crowded on this planet, scientific modifications of the microbiological conversion of industrial wastes help relieve some of the problems of congestion.

TWENTY-ONE

Microbiology of the Soil

The soil is the mother of all life—microbes, plants, animals, and men. The hypothesis on the origin of life that has the greatest scientific support suggests that life originated in a "soup" of organic matter which collected on the surface of the earth during the early stages of the formation of this planet. Formed as a result of chemical reactions catalysed by the mineral elements, the organic matter was acted upon by the high temperatures, radiations, electrical discharges, and lack of oxygen in the atmosphere. Laboratory experiments imitating those conditions show very clearly that such synthesis does occur. As the earth cooled, complex aggregates were formed that maintained themselves and "grew," using the chemically synthesized organic matter as building material. Gradually, these forms became more distinct from the organic soup and can be regarded as the primitive forms of life. At this level they lacked the ability to synthesize their own substance; they could only assimilate it from the environment, and, in this respect, they were autocatalytic nucleoproteins similar in some respects to the modern viruses (Fig. 21.1).

Eventually, certain necessary substances in the organic soup were depleted by being tied up in the bodies of these primitive microorganisms. Growth ceased until in some organisms an enzyme developed which could catalyze the formation of the necessary substance from smaller organic molecules in the soup. The acquisition of additional biosynthetic reactions and the necessary accompanying source of energy eventually yielded forms related to our heterotrophic anaerobes. Further evolution led to microbial photosynthesis by the photoorganotrophs; this was followed by the advent of

Fig. 21.1. The influenza virus is simple in structure and in a suitable environment (the mucous membranes of man) behaves as an autocatalytic nucleoprotein. Photo courtesy Virus Laboratory, University of California, Berkeley.

photolithotrophic bacteria, and finally by green-plant photosynthesis. Green-plant photosynthesis produced oxygen, and, with the appearance of oxygen in the atmosphere, aerobic life was possible (Table 21.1).

The aerobes rapidly decomposed the nonliving organic matter in the organic soup, thus creating conditions favorable for autotrophic bacteria. In the other direction, with the development of higher animals and plants, strains of parasitic microbes developed. Further organic evolution is of direct concern to the microbiologist only insofar as higher plant and animal cells provided environments in which completely dependent viruses could again exist.

Soil Ecology

Myriads of microorganisms persist and thrive in the soil and the ocean sediments. The soil is composed of a mineral structure covered with a smear of organic matter. The spaces between the particles contain water and air. If the soil is very wet, the air is restricted and it is anaerobic; if it is dry, it is aerobic. Every soil particle presents a special environment for the growth of microorganisms. One side of a particle may be aerobic, the other anaerobic. One particle may have high sugar content; another may be suitable only for autotrophic growth. Where organic matter is decomposing, heat is produced and a thermophilic environment is available. Not far away, the temperature may be suitable for mesophiles. One soil particle may be acid, another alka-

Table 21.1   A Recent Theory on the Origin of Life

Probable Sequence of Events

1. The earth condenses and cools to the liquid and solid state.
2. Organic matter is synthesized by physicochemical reactions.
3. Organic soup condenses to aggregates which accumulate more of the organic material to themselves.
4. These aggregates develop membranes and acquire a chemical environment inside of the "cell" different from that outside. Catalytic reactions proceed to form complex molecules not present in the soup.
5. Anaerobic heterotrophic metabolism develops.
6. Photosynthetic pigments permit use of light for energy by photoorganotrophs.
7. When organic matter is used up, photolithotrophic bacteria (purple and green sulfur bacteria) develop.
8. The algae acquire additional enzymes and as a result gain independence of both organic matter and reduced sulfur. They give off oxygen (green-plant photosynthesis).
9. Aerobic life appears.
10. A few microbial types (autotrophs) acquire all enzymes necessary for chemolithotrophy (energy from oxidizing inorganic substances; carbon from CO_2).
11. Higher plants and animals provide an environment of organic compounds. Microbes invade these and lose synthetic abilities which are superfluous in parasitism. Some obligate parasites and viruses lose all or almost all synthetic ability, thus approximating the first forms.

line. Thus, the soil is not analogous to a large test tube of broth but rather to millions of tiny test tubes each with distinctive growth conditions. Thousands of diverse types of microbes can and do persist and flourish in the soil (Fig. 21.2). It is estimated that less than 10% of the population is recovered by plating procedures.

Such conditions are not static. When the dead body of a worm begins to decay, aerobic species rapidly use up the oxygen in the soil, and other organisms which are anaerobic continue the putrefaction process. If the area is well insulated and enough heat is produced, the anaerobic mesophiles may yield to thermophilic species. As the products of decomposition diffuse, they are in turn attacked by still other organisms. When the thermophiles complete their life cycle, their dead cell structures and the residue of food unavailable to them are attacked by other organisms. As the processes slow down, oxygen diffuses into the area faster than it is consumed, and aerobic microorganisms finally convert all the carbon to CO_2. In the absence of organic matter, autotrophic bacteria oxidize the ammonia or sulfur for energy, and use CO_2 to build their structure. So cycles of growth and death continue, each type or organism growing explosively when conditions are favorable and yielding to a better-adapted growth type as the environment is changed.

The Soil Organisms

The outer layers of the earth contain the soil microbes. An acre furrow slice (the upper six inches) of a rich upland soil contains about 400 pounds of bacteria, 4000 pounds of fungi, and an equal amount of protozoa and actinomycetes. The bacteria exist as microcolonies attached to the soil particles. These can be observed by direct microscopic examination as aggregates of 10 to 1000 organisms. In plate counts of soil, the colonies are broken up in the dilution bottle, and the number of living organisms can be counted. Up to 100 million bacteria may be found per gram of rich soil, but even on the best medium these make up only one-tenth the number that can be seen microscopically. Counts of molds and actinomycetes are less significant because the numbers depend on fragmentation of the mycelium. Wet swampy soil contains mostly anaerobes; under such conditions organic

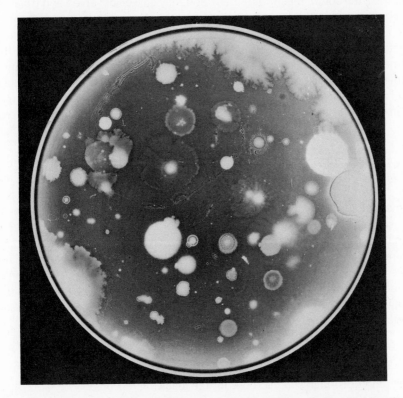

Fig. 21.2. A plate count of soil shows numerous and diverse organisms. On this plate are only the aerobic organisms that will grow on nutrient agar at 37 C. For every organism that gave rise to a colony on this plate there are in the soil at least 10 that failed to grow under these conditions.

matter accumulates. Organic accumulations in such wet areas are called peat; the slowly decomposing residual organic matter of upland soils is called **humus.** If the soil is well aerated, aerobic bacteria are more abundant, and filaments of molds and actinomycetes bridge the gaps between the soil particles. Especially high bacterial counts are found on particles of animal and plant material in the soil; plant roots excrete food material for microbes, and higher numbers of organisms are found near roots of living plants.

ENRICHMENT CULTURES

Although the soil contains a vast array of organisms, certain individual processes can be studied, and the organisms can be isolated by the **enrichment-culture** or "baiting" procedures. For example, adding relatively high concentrations of ammonia to the soil favors the growth of the ammonia-oxidizing bacteria; the soil will teem with them, and they can be observed and isolated. Such studies can be carried out with any other material on which microbes are active. The soil may be enriched with organisms that destroy plastics, bacterial capsules, mold mycelium, vitamins, or any other material by the same procedure. Or an enrichment culture may be developed by inoculating some soil into a medium containing the material on which microbes are active. After several days, a transfer is made into a new flask of the same medium. Repetition will yield a subculture predominantly of one type of microorganism, which can be isolated into pure culture by plating or dilution techniques.

Soil microorganisms may be controlled by (1) limitation of desirable growth conditions, (2) bacteriophages that invade the cells, and (3) protozoa and slime molds that eat the cells. In addition to decreased food supply, the same factors which affect growth and death in the test tube are operative in the soil. Soils that become acid as a result of deposits of pine needles contain many less microbes than neutral soils, such as those enriched by yearly accretions of oak leaves. Very dry or very wet, very cold or very hot soils have lowered counts. Although data on the virus population of the soil is scanty, we do know that soil bacteria and actinomycetes are susceptible to bacteriophage attack and phage can be isolated from soil. The total weight of the soil protozoa probably exceeds the total weight of the soil bacteria.

OUR INTEREST IN SOIL ORGANISMS

We are interested in soil microbes for the following reasons:

1. A few soil microorganisms are pathogenic to man. These include the bacteria causing botulism, tetanus, and gas gangrene. They also include cer-

tain actinomycetes and molds that cause skin diseases and the deep mycoses. Many plant pathogens reside in the soil. Some of the plant pathogens die off in the soil as a result of the activities of other microbes, especially when the plant they parasitize is not grown in that field for several years.

2. The production of microbial capsules improves the soil tilth. The physical condition of the soil, especially the formation of a desirable crumbly soil for a seed bed, is promoted by microbial growth.

3. The production of acids by soil organisms dissolves minerals necessary for plant nutrition. The mineral particles which contain insoluble forms of the important plant foods, potassium and phosphorus, are dissolved into the soil solution from corners of soil particles where microorganisms produce a local acid reaction.

4. The cycles of growth and death of microorganisms tie up plant nutrient materials in the bodies of microbes, conserving them against leaching from the soil.

5. Trace nutrients are released in forms that can be used for the growth of plants in the cycles of growth and death. For example, some plants can absorb nitrogen only in the form of nitrates. Microbes split ammonia from decomposing proteins, and autotrophic bacteria oxidize the ammonia to nitrate.

6. Almost 90% of the carbon dioxide in the air (the raw material of photosynthesis) is released from organic matter through the decomposing activities of soil microorganisms.

7. Soil organisms are responsible for the decomposition of noxious chemicals, such as pesticides and atmospheric pollutants deposited by precipitation.

8. The microbes also function in other cycles of the elements. There is reason for concern that pollution of the environment by man-made toxic chemicals could inhibit some of these processes and upset the natural balance.

The Carbon Cycle

The building of organic matter from CO_2 is primarily the function of green plants and algae. A few bacteria are photosynthetic, and the chemolithotropic bacteria use CO_2 as their sole source of carbon. However, the amount of organic matter built by these two groups of bacteria is negligible. The major chemical activity of the other microbes is the release of CO_2. The amount of CO_2 produced by the respiration of animals is negligible compared to that released by respiration and fermentation of microbes. Without this return of CO_2 to the atmosphere, the carbon cycle would be interrupted and photosynthesis would cease since all carbon would be tied up as organic matter. The carbon cycle is integrated with the oxygen cycle since photosynthesis by green plants uses CO_2 and produces oxygen. Plant, microbial, and animal respiration reverse this process.

The Sulfur Cycle

In living material sulfur is found in the proteins. After death, putrefactive bacteria split the sulfur from the protein in the form of evil-smelling hydrogen sulfide. Hydrogen sulfide is also produced in marine environments and in the soil by the anaerobes, *Desulfovibrio,* and *Desulfotomaculum,* which use sulfate in their anaerobic metabolism as the hydrogen acceptor thus:

$$H_2SO_4 + 8H \rightarrow H_2S + 4H_2O$$

Hydrogen sulfide is oxidized to elemental sulfur (S) and to sulfate by photolithotrophic and chemolithotrophic sulfur bacteria. Sulfate is assimilated by green plants and by bacteria and built into the protein molecules.

Nitrogen Transformation

The protoplasm of living cells is primarily protein. Some microorganisms and all plants are able to manufacture their amino acids from ammonia. All animals and many microorganisms need preformed amino acids as part of their food and they secure them from the organisms having synthetic ability. Most microbial proteins are nucleoproteins, composed of protein combined with one or the other of the nucleic acids, RNA and DNA. Both of these acids contain nitrogen, which enters the molecule as ammonia or as an amino acid. When amino acids or nucleic acids are decomposed by microorganisms, ammonia is released. Ammonia plays a central role in the nitrogen conversions in nature (Fig. 21.3).

AMMONIFICATION

Most of the nitrogen in the biological world makes the circuit from ammonia to amino acids to protein and back again to amino acids and ammonia. While in protein form, it may be digested and rebuilt many times from plant to herbivorous animal to carnivorous animal proteins. A small amount is excreted by animals in the form of urea or uric acid from which microorganisms split the nitrogen as ammonia. The proteins from dead animals and plants are decomposed by microorganisms and much is used to build the protein of microbial cells in the soil. When proteinaceous material is present in excess in the soil, the microbes respire the carbon portion from some of the protein to secure energy and the ammonia is released. **Ammonification,** the process of splitting ammonia from proteinaceous material, is carried out by a wide variety of microorganisms.

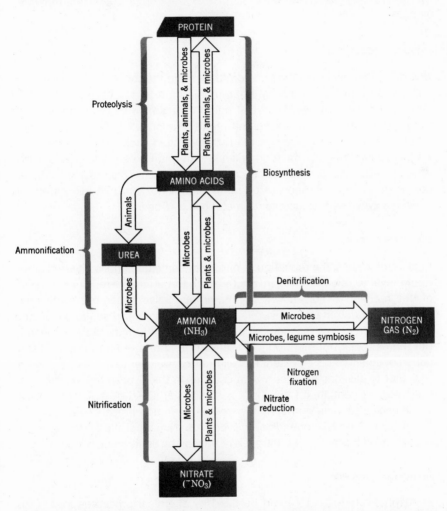

Fig. 21.3. As nitrogen metabolism is more clearly defined, it is evident that we are dealing with a series of cycles, or rather excursions, from ammonia and the amino acids. Plants may take in nitrates through their roots; this must be reduced to ammonia before building it into amino acids and proteins. Many of the processes are carried out only by microbes; plants and animals are more restricted.

NITRIFICATION

When ammonia accumulates in the soil and carbohydrate material is in short supply, autotrophic microorganisms thrive and secure their energy by oxidizing ammonia to nitrite and oxidizing the nitrite to nitrate. This process of **nitrification** supplies nitrates for those plants which cannot assimilate ammonia through the plant roots. The chemolithotrophic *Nitrosomonas* and *Nitrobacter* are associated with this process, but other organisms may participate.

NITRATE REDUCTION

Any plant that secures nitrogen from the soil in the form of nitrate must be capable of reducing the nitrate to ammonia in order to utilize it in building amino acids. A wide variety of soil microorganisms also have the enzyme, nitrate reductase, which reduces nitrate to ammonia. Under anaerobic conditions, when nitrate is used as the final hydrogen acceptor, much more nitrate is reduced by the microorganisms than can be built into amino acids; as a consequence, ammonia is excreted into the soil. This process of anaerobic respiration is carried out by many organisms that can use nitrate instead of oxygen as the acceptor of metabolic hydrogen.

DENITRIFICATION

In anaerobic soils, the reduction of nitrate by some bacteria is diverted at the nitrite level to release nitrogen gas (N_2) into the atmosphere. The chemistry of this reaction is not known, but many facultative anaerobes participate. The bacteria dominant in this release are of the genera *Pseudomonas* and *Achromobacter*. This reaction is detrimental to soil fertility since nitrogen is utilized as a plant nutrient only in the form of nitrate or ammonia. Denitrification occurs to a lesser extent in well-drained, well-aerated soils. Soil chemists refer to gaseous nitrogen which escapes to the atmosphere as "free" nitrogen to differentiate it from the "fixed" nitrogen which remains in the soil. Some geologists believe that all the free nitrogen now present in air (80% of the atmosphere or 12 pounds over every square inch of the earth's surface) was released from the fixed nitrogen of the soil by denitrifying bacteria.

NITROGEN FIXATION

Atmospheric nitrogen (N_2) can be fixed by chemical processes into forms such as ammonia for use as plant fertilizer. A small amount is also fixed by lightning during electrical storms. But unless some other mechanism had been available for replenishing the fixed nitrogen of the soil, life would have persisted on this earth only with great difficulty until such time as chemists learned the secret of nitrogen fixation. Biological nitrogen fixation restores to the soil a large part of the nitrogen lost by denitrification; since modern agricultural practice encourages nitrogen fixation, soils actually become richer in nitrogen in spite of the removal of quantities of the element in agricultural products.

A. **The Symbiotic System.** The most important fixation of nitrogen in agricultural areas involves the use of **legume plants,** such as alfalfa, clovers, peas, beans, peanuts, and lentils. These plants cannot fix nitrogen by themselves; only when gram-negative bacteria of the genus, *Rhizobium,* invade their roots and form **nodules** is nitrogen fixed by the combined efforts of plant and bacteria. This is called symbiosis since both partners benefit; the bacterium secures carbohydrate, and the plant becomes independent of soil nitrogen.

When the bacterium invades a root hair of the plant, it induces the plant to produce a tumor-like swelling (the nodule) in which the bacteria multiply. The vascular system of the plant connects the nodule with the food-bearing transport system of the root and stem. Although most soils contain rhizobia which would form nodules on legume plants, the modern farmer does not depend on these (wild type) strains. Instead he buys cultures of bacteria, selected for their efficiency in fixing large amounts of nitrogen when in symbiosis with legumes. The legume inoculation, consisting of the bacteria mixed with moist humus, is spread over the seed to be available for invading the rootlets as soon as the seeds germinate.

Not all legume plants are susceptible to invasion and nodule formation by the same rhizobia. The genus, *Rhizobium,* has been divided into species on the basis of the legume species which are invaded; for example, one species associates with alfalfa, another with the clovers, and another with soybeans. The legume-*Rhizobium* symbiosis appears to have evolved from an earlier parasitic association. Early ancestors of the bacterium may have produced a disease involving tumors of the plant roots; from this evolved a parasitic association not harmful to the plant, and, finally, modifications in the plant and bacterium led to the symbiosis.

B. **Nonsymbiotic Fixation.** Several types of free-living microorganisms in the soil can use gaseous nitrogen as their only nitrogen source. Where they grow vigorously, they make a significant contribution to the fixed nitrogen of the soil. On the surface of the soil and in surface waters the blue-green algae and the photosynthetic bacteria make a small contribution to the total fixed nitrogen. These organisms use both CO_2 and N_2 from the air as their principal source of carbon and nitrogen when light is available for energy. They live essentially on air and water.

Under anaerobic conditions certain species of the genus, *Clostridium,* (gram-positive, anaerobic, sporeforming rods) fix significant amounts of nitrogen and restore at least part of that lost by denitrification. Under aerobic conditions, organisms of the genus, *Azotobacter,* (gram-negative aerobic organisms with an unusually large cell of two to five micrometers), fix nitrogen vigorously when carbohydrate is abundant. Several other bacterial species have been found to fix nitrogen under laboratory conditions, but whether they make any contribution to the nitrogen economy of the soil has not been assessed.

The Soil Microbes

The molds exist as a network of mycelium and of fruiting bodies. Plate counts overemphasize those types that produce many spores or whose mycelium fragments easily in the dilution bottle. The association of fungus mycelium with plant roots where they form a special structure, the **mycorrhiza,** appears to be an association beneficial to both the plant and the mold. The mold invades the root of the plant but also grows out into the soil serving as auxiliary root hairs for absorbing nutrients. In some forests, the molds connect the decomposing leaves on the soil surface with the plant roots, thus producing a sort of closed system wherein the nutrients of the fallen leaves are returned to the plant. In the absence of the fungi, those trees that depend on mycorrhiza, do not thrive. Some molds are able to trap, kill, and decompose nematodes and protozoa and thus restrict those organisms in the soil. The soil actinomycetes are able to attack many of the resistant organic compounds from plant tissues and function in the slow decomposition of humus. The present view is that their production of antibiotics has no great influence on the biological balance in the soil. Protozoa, on the other hand, are important scavengers of bacteria and algae and are a significant factor in ecology. Insects and soil nematodes may feed on molds. A very unusual bacterium, *Bdellovibrio* (**bdellus,** the Latin term meaning leech), was discovered in 1962 that parasitizes certain other bacteria. The small organism is highly motile due to a single, polar flagellum. The curved rod adheres to the cell wall before penetration and replication (Fig. 21.4). Progeny are released by subsequent lysis of the host cell. A number of *Bdellovibrio* have been isolated from their normal habitat, soil and water.

Algae are found on the surface of all soils but they usually play a significant role only where the soils have been denuded by erosion or by volcanic activity or by excessive drought. Under such situations, they serve as the pioneering organisms and produce organic matter to make the soil more hospitable for other forms. As a group, algae are better adapted to a liquid environment so their often massive growth in flooded rice paddies make an important contribution to soil fertility. Under such conditions, the nitrogen-fixing ability of some of the blue-green algae is especially significant.

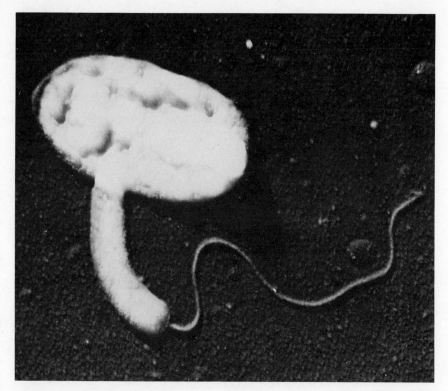

Fig. 21.4. **The tiny, curved rod of** Bdellovibrio **is attacking a larger bacterial host cell in this electron micrograph.** Photo courtesy H. Stolp and M. P. Starr: Antonie van Leeuwvenhoek, **29,** 217, 1963.

Animal viruses are inactivated in the soil. When heavily inoculated, a few may be recovered after a month or more. Some viruses that produce plant diseases may persist for years and a rotation of crops is necessary to keep down epidemics. Greenhouse soils are sterilized by chemicals or steamed to control plant viruses. Bacteriophages are a normal part of the soil biota and may be responsible for the degeneration of the population of the root nodule bacteria in soils cropped for long periods with the same legume plant.

Summary

Soil microbiology deals with the organisms occupying the thousands of different environments which exist on and around the soil particles. The microbial population of the soil is a dynamic system involving myriads of sequences of growth and death. With the exceptions of the algae and the autotrophic bacteria, soil microbes are involved primarily with the destruction of organic matter. This releases the elements tied up in the biological material so that they may participate further in living systems. Although the soil serves as a reservoir for plant pathogens, the overall result of the activities of soil microbes is beneficial to the growth of plants whose root systems draw nutrients from the soil. For students of the environment, the soil presents endless varieties of associations of microorganisms. These serve admirably as model systems for ecological studies.

Glossary

Aerobe An organism that utilizes molecular oxygen.

Agglutination Aggregation of particulate antigenic particles (agglutinogen) (cells, flagella, etc.) by antibody molecules (agglutinin).

Allele An alternative form of a gene.

Allergy A pathological state resulting from reactivity to an antigen or hapten. The term used to describe both immediate and delayed reactions to exposure to the antigen. Allergens elicit skin-sensitizing antibodies.

Amino acid A subunit of protein structure.

Anaerobe An organism that grows in the absence of molecular oxygen.

Anamnestic reaction The "remembrance" reaction; the apparent accelerated response of an animal to an antigen upon secondary or subsequent exposure to the antigen.

Angstrom A unit of measurement equal to 0.1 nanometer; that is, 10^{-10} meters.

Antibody Gamma globulins produced by the animal body in response to antigenic stimulation and that will react specifically with those antigens in some demonstrable way.

Antigen Any macromolecule (usually a foreign protein or polysaccharide) that can induce the animal body to form gamma globulins which will react specifically with that antigen in some demonstrable way.

Antitoxin Antibodies produced against exotoxins.

Apoenzyme   Protein moiety of an enzyme.

Ascospore   Sexual spore produced inside a spore case, the ascus, by certain fungi.

Autoimmune disease   A pathological condition resulting from immunization of an individual against his own antigen(s).

Autolysis   Dissolution of the cells by self-produced lytic enzymes.

Autotroph   An organism that uses carbon dioxide as its principle carbon source.

Auxotroph   A nutritional mutant which requires one or more nutrients in addition to those required by the wild type.

Axial filament   A bundle of fibers attached to the ends of the spirochete cell, which affords it motility.

Bactericidal   Agent that kills bacteria.

Bacteriostatic   An agent that inhibits bacterial growth by its presence.

Barophile   An organism capable of life under high barometric pressure; as at the bottom of the sea.

Binary fission   Cell division where mother cell splits into two daughter cells.

Biosynthesis   Manufacturing of chemical compounds from simpler building blocks; anabolic activity.

Buffer   A chemical substance (for example, phosphates, amino acids, and proteins) that resists a drastic change in the pH when small amounts of acid or alkali are added.

Burst size   The average number of viral progeny per cell released from a population of infected cells.

Capsid   The protein coat or head of certain viruses.

Capsomere   Protein subunits that make up the capsid.

Capsule   A detectable layer, usually polysaccharide in nature, external to the cell wall. Several functions are associated with this structure, for example, resistance to dryness, factor of virulence, and impedence to phage infection.

Carrier   An individual that is infected with a disease-producing organism and disseminates it, yet shows no physical symptoms of the disease.

Chemosynthesis   Process in which a chemical compound is used to obtain energy for growth.

Chemotherapeutic agent   A chemical agent used systemically to treat a pre-existing infection in the host.

Chemotherapeutic index Ratio of the minimum curative dose to the maximum tolerated dose.

Chemotherapy Treatment of a preexisting infection with a chemical agent applied systemically.

Chlamydiae Obligate, parasitic procaryons once believed to be large viruses, as the psittacosis group.

Chromatophore Cytoplasmic organelle where photosynthetic pigments and enzymes of photosynthetic bacteria are localized.

Cilium Multistranded, short, blunted organelle of locomotion possessed by certain eucaryons.

Coenocytic A hypha without septa and thus a multinucleated mass of cytoplasm.

Coenzyme The nonprotein part of the enzyme which is easily separated from the enzyme protein.

Coliform A gram-negative, nonsporeforming bacillus that ferments lactose with the production of acid and gas.

Commensalism Phenomenon whereby an organism benefits by living in association with another organism without doing harm to the latter; for example, nutrients.

Complement A heat-labile protein fraction of mammalian serum, consisting of nine or more components, capable of mediating lysis of certain cells (for example, red blood corpuscles and certain gram-negative bacteria) after such cells have reacted with their specific antibody.

Conidium An exogenous, asexual spore produced on the distal tip of a fertile hyphe, the conidiophore, by certain fungi and ascomycetes.

Conjugation A means of genetic transfer between certain bacteria; genetic material of donor cell is transferred via a pilus or conjugation tube to the recipient cell.

Contagious Communicable, or spread from one individual to another; for example, a contagious disease.

Cyst A resting dormant cell produced in a few bacterial groups by surrounding the vegetative cell with a thick envelope, called an exine.

Cytoplasmic membrane A thin, delicate structure internal to the cell wall; it is composed of phospholipoprotein and functions primarily as the determinant of transport of molecules in and out of the cell.

Decay Decomposition of organic products under aerobic conditions.

Desensitization The act of rendering a hypersensitive individual nonsensitive to the allergen by treatment with that allergen in minute doses through an alternate portal of entry.

Desiccation The removal of water.

Diatom Unicellular, nonflagellated alga possessing a cell wall impregnated with silica.

Dimorphism Filamentous growth under one environmental condition and unicellular growth under another.

Diploid A cell that carries two chromosomes of each linkage group.

Eclipse period Phase during viral replication within a host cell in which there are no detectable, infectious viral particles.

Endemic A disease situation wherein there are always a few cases within a given population.

Endospore A dormant or resting cell, produced primarily by the bacterial genera *Bacillus* and *Clostridium* which shows extreme resistance to adverse conditions.

Enterotoxin A poison secreted by certain bacteria that causes gastrointestinal disorder or food poisoning.

Enzyme An organic catalyst or reactor molecule of biological origin. **Constitutive** — always present in the cell; substrate independent. **Induced** — present in the cell only when needed; substrate dependent. **Extracellular** — activity is outside the cell; digestive or hydrolytic in its activity. **Intracellular** — activity is inside the cell; usually involved in the biosynthesis of macromolecules and energy-yielding reactions.

Epidemic An outbreak of a disease affecting many persons in a given population.

Episome An extra piece of genetic material capable of existing free or as a part of the normal chromosome; for example, the F factor.

Etiological agent The cause of a disease.

Eucaryon An organism that displays a more highly evolved cellular differentiation and has a true nucleus.

Facultative Having characteristics that permit alternate responses under different conditions; for example, facultative anaerobe and facultative psychrophile.

Fermentation Generally, the anaerobic decomposition of carbohydrates.

Flagellum Appendage or organelle of locomotion. In eucaryons, a multi-stranded structure; in procaryons, a single fibril which is below the resolving power of the microscope. In the latter cells; **polar**

flagellation—the flagella are located at one end of the cell; **lophotrichous**—both ends of the cell; **peritrichous**—all over the cell's surface.

Fomites Inanimate objects, such as eating utensils, linens, toys, and handkerchiefs, that are contaminated with disease-producing organisms and thus serve as secondary vehicles in the transmission of disease to a susceptible host (singular, fomes).

Gamma globulin Proteins of a given electrophoretic mobility, composed of diverse molecular weights (150,000-900,000) which include the immunoglobulins.

Generation time The time required for a mother cell to divide into two daughter cells.

Genotype Sum total of the genes of an organism; that is, its genetic code.

Halophiles "Salt loving"; an osmophile that prefers to grow in the presence of high salt concentrations.

Haploid A cell that carries only one chromosome of each linkage group.

Hapten A chemical substance that can induce antibody formation if coupled with a high molecular weight carrier but cannot do so alone. Haptens can, however, react specifically with such antibodies even in the absence of the carrier; the determinant group of an antigen.

Hemolysis Dissolution of red blood corpuscles.

Heterocaryon A coenocytic organism that contains nuclei derived from both parental strains.

Heterothallic Conditions in which either the male or female sex structures, not both, is formed in a single mycelium.

Heterotroph An organism that uses predominantly organic carbon for energy and protoplasmic building blocks.

Heterozygous The presence of two different alleles for the same characteristics in a diploid cell.

Homothallic Condition in which both male and female sex structures can develop on the same mycelium.

Homozygous Identical gene pairs of a particular characteristic within a diploid cell.

Hypersensitivity An acquired reactivity to antigens resulting in harm to the tissues of the animal. Such reactions may be immediate or delayed.

Hypha A single filament of a mold or actinomycete mycelium (plural, hyphae).

Icosahedral A twenty-sided structure; the shape of many viruses.

Immunity   Relative term inferring the ability to resist or to overcome a harmful antigen such as an infecting bacterium or virus. The word has come to mean any state in which the animal has produced an immunoglobulin to a specific antigen. **Active:** immunity produced by an animal in response to exposure with an antigen. **Passive:** immunity produced by the administration of preformed antibodies formed in another animal's body (for example, in the maternal circulation). **Natural:** immunity that occurs without intentional exposure of the individual to an antigen. **Acquired:** immunity that occurs following an apparent or subclinical case of disease or an intentional injection of an antigen.

Immunology   The study of those chemical modifications in the blood and other tissues which are involved in immunity to disease.

Interferon   Viral-induced protein that interferes with viral replication at the cellular level.

L-form   Pleomorphic bacterial form that is devoid of certain cell wall materials due to the presence of a chemical inhibitor or a genetic block.

Lag phase   Early period of growth curve where cells do not demonstrate a constant growth rate.

Latent period   The time from virus infection of the cell to release of mature, viral progeny.

Logarithmic growth   Cells demonstrate a constant growth rate; cells divide exponentially.

Lyophilization   A process whereby a material is dried in the frozen state under vacuum; used to preserve microbial cultures.

Lysogeny   A condition in which the bacterial cell contains a prophage.

Mesophile   An organism whose optimum temperature for growth is within the moderate temperature range (that is, 22–44 C).

Mesosome   Invagination of cell membrane into the cytoplasm of many bacteria that appears to function in the synthesis of the septum.

Microaerophile   An organism that prefers a reduced oxygen tension and sometimes an increased carbon dioxide tension.

Micrometer   A unit of measurement equalling 1/25,400 inch or 0.001 mm; this term replaces the older micron (μ).

Mitosis   Asexual process occurring in eucaryotic cells whereby each chromosome is duplicated and distributed to each daughter cell.

Morbidity   Cases of a disease for unit population.

Murein   Unusual protein responsible for rigidity of cell walls of procaryons.

Mutagen   Agent that increases the mutation rate.

Mutation A chemical change in the genetic code of the cell.

Mycelium A mass of hyphae of a mold or actinomycete colony.

Mycoplasma Genus of bacteria that are devoid of cell wall material due to the cell's genetic inability for its synthesis; natural protoplasts.

Mycosis A disease caused by molds; may be either a dermatomycosis or a systemic mycosis.

Nanometer A unit of measurement equal to 10^{-9} meter; older term, millimicron.

Necrosis Death of the cell.

Negri bodies Aggregate of rabies viruses appearing as inclusion bodies within the infected animal cell; they are of diagnostic significance.

Nomenclature The act of calling by name.

Nucleoid A primitive nuclear apparatus, the DNA molecule, of procaryotic organisms.

Nucleotide Subunit of nucleic acids.

Obligate Strict or restricted; for example, obligate aerobe, obligate thermophile.

Osmophile An organism that prefers to grow at a high osmotic pressure, that is, high concentrations of sugar or salts.

Oxidation A chemical reaction involving the loss of electrons or hydrogen atoms or addition of oxygen.

Pandemic A world-wide epidemic.

Parasite An organism that lives on living tissue.

Pathogen An organism that has the potential to cause disease.

Phenotype Morphological appearance and physiological behavior of a genotype as influenced by environment.

Photodynamic sensitization: Damage to biological material by exposure to visible light in the presence of certain dyes.

Photoreactivation The process of exposing to visible light to repair DNA damaged by ultraviolet light.

Photosynthesis Process where light is used for energy for growth.

Phycocyanin Blue carotenoid pigment possessed by blue-green algae.

Pilus Appendage possessed by certain bacteria that is used for adhesion to surfaces; also, specialized pili serve as the conjugation tube in transfer of genetic material.

Pinocytosis   Process of engulfing a liquid into the cell.

Plasmodesm   A protoplasmic connection between cells.

Plasmolysis   Shrinking of the cell's cytoplasm due to the loss of water by osmosis.

Poly-β-hydroxybutyric acid   A lipidlike storage granule produced by certain bacteria.

Polysome   A chain of ribosomes attached to an RNA messenger molecule.

Precipitation   Aggregation of soluble antigen particles by antibody molecules (precipitin).

Procaryon   Microorganism displaying a less differentiated cell organization with a primitive nuclear apparatus.

Prophage   Temperate phage's nucleic acid that is integrated into the bacterial chromosome; the cell is lysogenized.

Prophyloxis   Agent that prevents diseases; for example, a chemoprophylactic agent.

Protoplast   That part of a cell that includes the cell membrane and its contents.

Prototroph   An organism exhibiting wild-type nutrition.

Prosthetic group   The nonprotein constituent of an enzyme.

Psychrophile   An organism whose optimum temperature for growth is around 10 C; that is, cold loving.

Putrefaction   Decomposition of protein under anaerobic conditions.

Receptor site   The external cell structure or chemical moiety that serves for the attachment of viruses.

Reduction   A chemical reaction involving the gain of electrons or hydrogen atoms or loss of oxygen.

Respiration   Oxidation of a chemical compound by an organism with the release of energy. **Aerobic** — the final electron acceptor is molecular oxygen. **Anaerobic** — the electron acceptor is an inorganic substance ($SO_4^=$, $NO_3^=$) other than molecular oxygen.

Ribosomes   Cytoplasmic structures of cells involved with protein synthesis.

Rickettsiae   Obligate parasitic protists that are classified between viruses and bacteria; many are disease producers, which are transmitted to man by arthropod vectors.

Saprophyte   An organism that obtains its nutrients from dead, organic matter.

Septum   A transverse cross wall between cells.

Serotype Separation or identification of an organism on the basis of its antigenic makeup.

Sexduction Transfer via conjugation of donor's genetic material by an F episome.

Spheroplast Osmotically-sensitive bacterial cell devoid of part of its rigid cell wall.

Sporangium Spore case that contains asexual sporangiospores, and is borne on the distal tip of a fertile hypha, sporangiophore.

Synchronized culture Population of cells all dividing at the same time.

Synergism The action of two organisms to produce an end product not formed by either one alone.

Substrate A chemical compound that undergoes a chemical change due to enzyme activity.

Taxonomy The study of the principles of classification.

Teichoic acid A phosphate polymer of either glycerol or ribotol that is believed to confer the net negative charge of bacterial cells; found principally in gram-positive bacteria.

Temperate phage A bacterial virus capable of either infecting and lysing its host cell or establishing a symbiotic relationship with the infected cell, that is, lysogeny.

Thermophile An organism whose optimum temperature for growth is above 45 C; that is, heat loving.

Toxin A biological poison. **Endotoxin** — an intracellular, lipopolysaccharide toxin released only upon lysis of the cell. **Exotoxin** — a toxic protein produced and secreted by metabolizing cells.

Transduction A means of genetic transfer between certain bacteria wherein a small amount of genetic material of donor cell is transferred to recipient cell by a temperate phage.

Transformation One means of genetic transfer between bacteria where the freed DNA of the disrupted donor cell is taken into the recipient and part of it is incorporated into the genome.

Tuberculin An extract of the tubercle bacillus that is used to skin-test an individual's sensitivity to the organism.

Virion An extracellular, infectious viral particle.

Virulent The ability to overcome the defense mechanisms of the host and elucidate pathological conditions.

Zoonosis A disease of animals that is transmittible to man.

Index

Growth: definition of, 55; exponential, 101; factors affecting, 102-115; measurement of, 55-60; nutritional requirements, 103-105; synchronous, 100
Growth curve, bacterial, 100-102, 104
Growth factors, 47-48, 105
Growth phases: death, 102; lag, 101-102; logarithmic, 101-102; maximum stationary, 102
Guarnieri bodies, 209
Gymnoplasts, 72

Halophiles, 106
Hanging-drop method, 43
"H" antigens, 134-135; phase variation, 135-136
Haploid, 131
Haptens, 233
Hay fever, 244
Heat, destruction by, 162
Hemagglutination, 276
Hemolysin, 228, 238, 240, 261
Hemolysis: alpha 261, 266; beta, 261-262; gamma, 261
Hemophilus influenzae, 248, 267-268, 280
Hemp, 346
Hepatitis, infectious, 210, 212, 252, 285
Hepatitis, viral, 162, 249
Herpes simplex, 209-210
Herpesvirus, 212, 250
Heterocaryon, 171-172
Heterothallic, 171
Heterotrophs, 104-105, 111, 116, 118, 350
Heterozygous, 132
Hexachlorophene, 153
High frequency recombining cells, 142
Hippocrates, 221
Histamine, 245
Histoplasma capsulatum, 250, 281, 283-284
Histoplasmosis, 250, 281, 283
History, microbiological, 25
Hives, 244
Hog cholera, 212, 229
Holdfast, 188
Homothallic, 171
Homozygous, 132
Hoof and mouth disease, 29, 198

Hormones, 347-348
Host-parasite relationships, 223-224
Humus, 354
Hyaluronic acid, 261
Hyaluronidase, 228, 261
Hydrogen bacteria, 121
Hydrogen-ion concentration, 109-110. See also pH
Hydrogen peroxide, 165
Hydrogen sulfide, 195
Hydrolases, 88
Hydrostatic pressure, 115
Hymenoptera, 215
Hypersensitive reactions: and tissue damage, 244; delayed, 244; immediate, 244
Hypersensitivity, state of, 228, 258
Hypertonic solution, 106
Hypha, 168-169
Hyphomicrobiales, 126
Hyphomicrobiates, 195
Hypotonic solution, 106

Imhoff tank, 305, 307
Immobilization test, 241
Immunity: acquired, 229-231; active, 229-231, 243; artificial, 229-231; 243-244; duration, 243-244; natural, 243-244; passive, 229-231, 243-244; phenomenon, 232; placental transfer, 244, types of, 243
Immunology, 1, 29-31, 232
Impetigo, 252, 259, 264
IMViC reactions, 299-300
Inducible enzyme, 143
Industrial microbiology: categories, 13, 340; production of microorganisms, 340-341
Industrial wastes, 348
Infection, frequency of, 276
Inflammatory response, 229
Influenza, 210, 212, 245, 250, 267, 279-280
Influenza virus, 201, 276, 351
Infusions, culture media, 48
Insect control: by bacteria, 341; by viruses, 215
Interferon, 210
International unit, 159
Involution forms, 102

Serum, blood, 7
Serum sickness, 244
Sewage: activated sludge, 307-308; B. O.
D., 304; Cameron tank, 305-306; contact beds, 307; definition, 304; disposal,
304-305; Imhoff tank, 305-307; irrigation water, 305; lagoon, 308; microbial
activity in, 305, 307-308; nonbiodegradable chemicals, 309; purification,
189, 305-308; septic tank, 305
Sexduction, 142
Sheep red blood cells, 240
Shigella, 248, 269
Shigella dysenteriae, 227, 296
Shigellosis, 248, 251, 269
Shingles, 278
Silage, 329
Silicon dioxide, 187
Silver nitrate, 256
Sinusitis, 262, 280
Slime producers, 167, 194, 294
Smallpox: 30, 277-278; etiological agent,
212; immunity to, 210; in tissue culture,
342; transmission, 225, 249-250; vaccine, 30, 230, 245, 278
Smallpox virus, electron microscopy, 226
Smith, Erwin F., 29
Smuts, 167, 176
Soil: and legumes, 16; carbon cycle, 355-
356; ecology, 351-352; enrichment
cultures, 354; fertility, 16-17; 177, 181;
humus, 354; microbes activity in, 16-
17, 177, 181, 353-355, 360-361; peat,
354; sulfur cycle, 356
South African tick bite, 290
Soya sauce, 15
Spallanzani, 26
Specimens, types of, 273-274
Sphaerotilus, 195
Spheroplasts, 72, 106
Spirillum, 65, 193
Spirillum rubrum, 64
Spirillum volutans, 75
Spirochaetaceae, 193, 194
Spirochetes, 5, 193, 271
Spontaneous generation, 26-27
Sporangia, 173-174
Sporangiophore, 173
Sporangiospores, 171, 174
Spores, bacterial, *see* Endospore

Spores, mold: aplanospores, 171; ascospores, 175; asexual, 170-172, 174-175,
177; classification of, 170; conidia, 171,
177; distribution, 168; endogenous,
171; exogenous, 171, 174-175; sporangiospores, 171; zoospores; 171
Sporozoa, 190-191
Stains: acid fast, 45-46; capsule, 44; differential and diagnostic, 44; endospore,
44-45; flagella, 44; for electron microscopy, 42; gram, 46-47; lipid, 44; negative, 44; polyphosphate, 45; simple, 43
Staphylococcus: 259-260; coagulase,
260; common colds, 280; diagnosis of,
260; diseases caused by, 225, 259;
enterotoxin, 259, 275; Fleming's strain,
159; food poisoning, 252, 334, 336-
337; leucocidin, 227-228; nutrition, 105;
phagocytosis, 230-231; pneumonia,
267
Staphylococcus aureus: food poisoning,
252, 334, 336-337; lobar pneumonia,
265; mastitis, 315; mutation rate,
133; nomenclature, 127; phenol coefficient, 152; puerperal fever, 264;
resistance to drugs, 133
Staphylokinase, 228
Sterilization: 51-53, 145; autoclave, 51-
53; chemical, 53; cold, 53; definition of,
51; dry heat, 51; filtration, 53; incineration, 53; modes of action of, 53; moist
heat, 51
Storage inclusions, 74-75
Strain, 124
Streptococcal sore throat, 248, 261-263
Streptococcus: 260-265; anaerobic,
264; buttermilk, 341; common cold,
280; intestinal, 296; leucocidin, 228;
milk, 313-314, pneumonia, 267;
sewage, 295
Streptococcus agalactiae, 127, 315
Streptococcus citrovorus, 320
Streptococcus cremoris, 321
Streptococcus diacetilactis, 321
Streptococcus equi, 127
Streptococcus fecalis, 313
Streptococcus lactis, 313, 320, 321
Streptococcus pyogenes: diseases caused
by, 250; electron scanning micrographs
of, 263-265; hemolysins, 261-262; in-

Watson, J.D., 90-91
Weil-Felix reaction, 289, 291
Weizmann, Chaim, 343
Wet mount, 43
Whooping cough: 254; *Bordetella pertussis*, 254; complications of, 267; diagnosis, 7; incidence, 248; toxin, 227; transmission, 249-250; vaccine, 245
Widal test, 238
Wine, 15, 346-347
Wisconsin curd test, 318-319
Wood-Werkman reaction, 111

X-rays, 165

Yaws, 194, 252, 271-272
Yeasts: 179-180; *Ascomycetes*, 179; as food, 340; baker's, 15; brewer's, 15, 179; budding, 179-180; characteristics, 5, 179; distribution, 179; importance, 179; in industry, 340; *Saccharomyces cervisiae*, 179, 343, 347
Yellow fever, 212, 225, 252, 288
Yogurt, 320, 341

Zoonoses, 352
Zoospores, 171